Lecture Notes in Artificial Intelligence   12186

Subseries of Lecture Notes in Computer Science

More information about this series at http://www.springer.com/series/1244

Don Harris · Wen-Chin Li (Eds.)

Engineering Psychology and Cognitive Ergonomics

Mental Workload, Human Physiology, and Human Energy

17th International Conference, EPCE 2020
Held as Part of the 22nd HCI International Conference, HCII 2020
Copenhagen, Denmark, July 19–24, 2020
Proceedings, Part I

 Springer

Editors
Don Harris
Coventry University
Coventry, UK

Wen-Chin Li
Cranfield University
Cranfield, UK

ISSN 0302-9743   ISSN 1611-3349   (electronic)
Lecture Notes in Artificial Intelligence
ISBN 978-3-030-49043-0   ISBN 978-3-030-49044-7   (eBook)
https://doi.org/10.1007/978-3-030-49044-7

LNCS Sublibrary: SL7 – Artificial Intelligence

This Springer imprint is published by the registered company Springer Nature Switzerland AG
The registered company address is: Gewerbestrasse 11, 6330 Cham, Switzerland

Foreword

The 22nd International Conference on Human-Computer Interaction, HCI International 2020 (HCII 2020), was planned to be held at the AC Bella Sky Hotel and Bella Center, Copenhagen, Denmark, during July 19–24, 2020. Due to the COVID-19 coronavirus pandemic and the resolution of the Danish government not to allow events larger than 500 people to be hosted until September 1, 2020, HCII 2020 had to be held virtually. It incorporated the 21 thematic areas and affiliated conferences listed on the following page.

A total of 6,326 individuals from academia, research institutes, industry, and governmental agencies from 97 countries submitted contributions, and 1,439 papers and 238 posters were included in the conference proceedings. These contributions address the latest research and development efforts and highlight the human aspects of design and use of computing systems. The contributions thoroughly cover the entire field of human-computer interaction, addressing major advances in knowledge and effective use of computers in a variety of application areas. The volumes constituting the full set of the conference proceedings are listed in the following pages.

The HCI International (HCII) conference also offers the option of "late-breaking work" which applies both for papers and posters and the corresponding volume(s) of the proceedings will be published just after the conference. Full papers will be included in the "HCII 2020 - Late Breaking Papers" volume of the proceedings to be published in the Springer LNCS series, while poster extended abstracts will be included as short papers in the "HCII 2020 - Late Breaking Posters" volume to be published in the Springer CCIS series.

I would like to thank the program board chairs and the members of the program boards of all thematic areas and affiliated conferences for their contribution to the highest scientific quality and the overall success of the HCI International 2020 conference.

This conference would not have been possible without the continuous and unwavering support and advice of the founder, Conference General Chair Emeritus and Conference Scientific Advisor Prof. Gavriel Salvendy. For his outstanding efforts, I would like to express my appreciation to the communications chair and editor of HCI International News, Dr. Abbas Moallem.

July 2020   Constantine Stephanidis

HCI International 2020 Thematic Areas
and Affiliated Conferences

Thematic areas:

- HCI 2020: Human-Computer Interaction
- HIMI 2020: Human Interface and the Management of Information

Affiliated conferences:

- EPCE: 17th International Conference on Engineering Psychology and Cognitive Ergonomics
- UAHCI: 14th International Conference on Universal Access in Human-Computer Interaction
- VAMR: 12th International Conference on Virtual, Augmented and Mixed Reality
- CCD: 12th International Conference on Cross-Cultural Design
- SCSM: 12th International Conference on Social Computing and Social Media
- AC: 14th International Conference on Augmented Cognition
- DHM: 11th International Conference on Digital Human Modeling and Applications in Health, Safety, Ergonomics and Risk Management
- DUXU: 9th International Conference on Design, User Experience and Usability
- DAPI: 8th International Conference on Distributed, Ambient and Pervasive Interactions
- HCIBGO: 7th International Conference on HCI in Business, Government and Organizations
- LCT: 7th International Conference on Learning and Collaboration Technologies
- ITAP: 6th International Conference on Human Aspects of IT for the Aged Population
- HCI-CPT: Second International Conference on HCI for Cybersecurity, Privacy and Trust
- HCI-Games: Second International Conference on HCI in Games
- MobiTAS: Second International Conference on HCI in Mobility, Transport and Automotive Systems
- AIS: Second International Conference on Adaptive Instructional Systems
- C&C: 8th International Conference on Culture and Computing
- MOBILE: First International Conference on Design, Operation and Evaluation of Mobile Communications
- AI-HCI: First International Conference on Artificial Intelligence in HCI

Conference Proceedings Volumes Full List

1. LNCS 12181, Human-Computer Interaction: Design and User Experience (Part I), edited by Masaaki Kurosu
2. LNCS 12182, Human-Computer Interaction: Multimodal and Natural Interaction (Part II), edited by Masaaki Kurosu
3. LNCS 12183, Human-Computer Interaction: Human Values and Quality of Life (Part III), edited by Masaaki Kurosu
4. LNCS 12184, Human Interface and the Management of Information: Designing Information (Part I), edited by Sakae Yamamoto and Hirohiko Mori
5. LNCS 12185, Human Interface and the Management of Information: Interacting with Information (Part II), edited by Sakae Yamamoto and Hirohiko Mori
6. LNAI 12186, Engineering Psychology and Cognitive Ergonomics: Mental Workload, Human Physiology, and Human Energy (Part I), edited by Don Harris and Wen-Chin Li
7. LNAI 12187, Engineering Psychology and Cognitive Ergonomics: Cognition and Design (Part II), edited by Don Harris and Wen-Chin Li
8. LNCS 12188, Universal Access in Human-Computer Interaction: Design Approaches and Supporting Technologies (Part I), edited by Margherita Antona and Constantine Stephanidis
9. LNCS 12189, Universal Access in Human-Computer Interaction: Applications and Practice (Part II), edited by Margherita Antona and Constantine Stephanidis
10. LNCS 12190, Virtual, Augmented and Mixed Reality: Design and Interaction (Part I), edited by Jessie Y. C. Chen and Gino Fragomeni
11. LNCS 12191, Virtual, Augmented and Mixed Reality: Industrial and Everyday Life Applications (Part II), edited by Jessie Y. C. Chen and Gino Fragomeni
12. LNCS 12192, Cross-Cultural Design: User Experience of Products, Services, and Intelligent Environments (Part I), edited by P. L. Patrick Rau
13. LNCS 12193, Cross-Cultural Design: Applications in Health, Learning, Communication, and Creativity (Part II), edited by P. L. Patrick Rau
14. LNCS 12194, Social Computing and Social Media: Design, Ethics, User Behavior, and Social Network Analysis (Part I), edited by Gabriele Meiselwitz
15. LNCS 12195, Social Computing and Social Media: Participation, User Experience, Consumer Experience, and Applications of Social Computing (Part II), edited by Gabriele Meiselwitz
16. LNAI 12196, Augmented Cognition: Theoretical and Technological Approaches (Part I), edited by Dylan D. Schmorrow and Cali M. Fidopiastis
17. LNAI 12197, Augmented Cognition: Human Cognition and Behaviour (Part II), edited by Dylan D. Schmorrow and Cali M. Fidopiastis

38. CCIS 1224, HCI International 2020 Posters - Part I, edited by Constantine Stephanidis and Margherita Antona
39. CCIS 1225, HCI International 2020 Posters - Part II, edited by Constantine Stephanidis and Margherita Antona
40. CCIS 1226, HCI International 2020 Posters - Part III, edited by Constantine Stephanidis and Margherita Antona

http://2020.hci.international/proceedings

17th International Conference on Engineering Psychology and Cognitive Ergonomics (EPCE 2020)

Program Board Chairs: **Don Harris, Coventry University, UK, and Wen-Chin Li, Cranfield University, UK**

- Shan Fu, China
- Crystal Ioannou, UAE
- Peter Kearney, Ireland
- Peng Liu, China
- Heikki Mansikka, Finland
- Lothar Meyer, Sweden
- Ling Rothrock, USA
- Axel Schulte, Germany
- Lei Wang, China
- Jingyu Zhang, China

The full list with the Program Board Chairs and the members of the Program Boards of all thematic areas and affiliated conferences is available online at:

http://www.hci.international/board-members-2020.php

HCI International 2021

The 23rd International Conference on Human-Computer Interaction, HCI International 2021 (HCII 2021), will be held jointly with the affiliated conferences in Washington DC, USA, at the Washington Hilton Hotel, July 24–29, 2021. It will cover a broad spectrum of themes related to Human-Computer Interaction (HCI), including theoretical issues, methods, tools, processes, and case studies in HCI design, as well as novel interaction techniques, interfaces, and applications. The proceedings will be published by Springer. More information will be available on the conference website: http://2021.hci.international/.

General Chair
Prof. Constantine Stephanidis
University of Crete and ICS-FORTH
Heraklion, Crete, Greece
Email: general_chair@hcii2021.org

http://2021.hci.international/

Contents – Part I

Human Physiology, Human Energy and Cognition

Contents – Part II

Human Factors in Human Autonomy Teaming and Intelligent Systems

Cognitive Psychology in Aviation and Automotive

Mental Workload and Performance

Towards a Multimodal Measure for Physiological Behaviours to Estimate Cognitive Load

Muneeb Imtiaz Ahmad[1(✉)], David A. Robb[1(✉)], Ingo Keller[1(✉)],
and Katrin Lohan[1,2(✉)]

[1] Edinburgh Center for Robotics, Computer Science Department,
Heriot-Watt University, Edinburgh, UK
{m.ahmad,d.a.robb,i.keller,k.lohan}@hw.ac.uk
[2] NTB University of Applied Sciences in Technology, Buchs, Switzerland
katrin.lohan@ntb.ch

Abstract. We present an experiment investigating the relationships between different physiological measures, including Mean Pupil Diameter Change, Blinking-Rate, Heart-Rate, and Heart-Rate Variability to inform the development of a measure to estimate Cognitive Load. Our experiment involved participants performing a task to spot correct or incorrect words and sentences which successfully induced Cognitive Load. Our results show that participants' task performance predicts their subjective rating of Cognitive Load and that there was a decrease in participants' performance with an increase in Cognitive Load. Furthermore, Mean Pupil Diameter Change was able to predict Blinking-Rate, and Heart-Rate was able to predict Heart-Rate Variability. This prediction is evidence that collecting data on physiological behaviours synchronously and analysing the trends can be an effective way of estimating Cognitive Load, and will help the future development of an online measure of Cognitive Load useful for responsive user interfaces.

Keywords: Cognitive load · Mental load · Human-computer interaction · Pupillometry

1 Introduction

Cognitive Load (CL) refers to the load placed on the user's working memory also viewed as short-term memory during a task [27]. CL has been categorised in three types: 1) Intrinsic Load, 2) Extraneous Load, and 3) Germane Load [26]. Intrinsic load is an effort imposed on the user's working memory due to the complexity of the task and its association with the user, whereas extraneous load is caused by the style of presentation of the material. Germane load refers to the ability of the learner to understand the material fully. Sweller et al. [27] also defined CL as a construct that can be measured in three dimensions: 1) Mental Load (ML), 2) Mental Effort and 3) Performance. In essence, CL and ML are

© Springer Nature Switzerland AG 2020
D. Harris and W.-C. Li (Eds.): HCII 2020, LNAI 12186, pp. 3–13, 2020.
https://doi.org/10.1007/978-3-030-49044-7_1

linked with one another in regard to the working memory. Consequently, in this paper, we use them interchangeably.

We understand that intrinsic load and extraneous load are particularly crucial factors in interactions with a multimodal interface. Both of these loads are strongly related to the ML or mental effort that could impact user task performance [26]. For example, an interface showing a specific kind of visualisation of data in a specific manner can result in enhancing both intrinsic and extraneous load of a particular user and hence would result in increasing CL. Therefore, such a circumstance would also support the idea of creating an online metric of CL or ML and later adapting the interface in real-time according to the CL or ML being experienced in order to decrease the level of ML in the user. However, to the best of our knowledge, it remains a challenge to estimate CL or ML in real-time in a non-intrusive manner.

Existing methods for measuring CL or ML are based on using subjective-questionnaires [12], performance-based metrics such as response time, error-rate or accuracy [21], speech-based [5,16,17,30], physiological behavior-based methods based on measuring human organs [21]: 1) brain, through measuring electroencephalogram (EEG) or electrocardiogram (ECG), 2) heart, through measuring Heart-Rate (HR), or Heart-Rate Variability (HRV), 3) skin, through measuring Galvanic Skin Conductance (GSR), and 4) eyes, through measuring eye movements, Pupil Diameter (PD), or Blinking-Rate (BR) to measure CL or ML in humans during an experimental setting. With the advancements in the field of machine learning and artificial intelligence, CL can be estimated in real-time. A few studies have used machine learning to estimate CL [11,20,29,31]. However, three of the aforementioned papers [20,29,31] used data from brain-specific (EEG) sensors. We understand that EEG sensors can tend to be intrusive and are impractical to utilise in real-world settings.

We, on the contrary, are working to create a model that, through observing the changing patterns of various physiological behaviours in real-time, can then be used to estimate CL. Our proposed method varies from the past research as we want to understand the relationship between the measures and later intend to use sensor fusion, and then regression, to estimate CL [4]. This paper presents results from one of our first experiments that involved collecting data from a range of physiological sensors synchronously in a non-intrusive manner during a task. The task involved identifying a correct or incorrect English words or sentences and categorising them as such. The focus of the paper is three fold: firstly, to show and validate that our task was able to induce CL in the participants; secondly, to direct our research forward towards our proposal [4] to create an estimate of CL in real-time; and lastly the data collected in this experiment is used to understand the relationships between different physiological behaviours through performing correlation- and regression-based analysis. This paper summarises these relationships and predictions and discusses our future work on creating datasets that can be employed to estimate CL.

2   Research Method

Our research explores the following aspects. Firstly, we focus on evaluating the contribution of task to CL through subjective ratings and task performance. Secondly, we investigate the relationship between the different physiological behaviours between themselves. Leading to the following Hypotheses (H) derived based on the literature on the measurement of CL [21]:

H1a: Participants' subjective ratings of CL will predict their overall task performances (H1a1) and time spent to complete the tasks (H1a2).

H1b: Lower participants' ratings of CL will predict better task performance or vice versa.

H1c: Lower participants' ratings of CL will predict lower time to complete the tasks or vice versa.

H2: Participants' changes in one physiological behaviour will predict a change in another behaviour.

H2a: Changes in the overall mean value of HR will predict overall mean HRV and vice versa. Similarly, BR will predict overall mean PD and vice versa.

2.1   Stimuli

To prepare the stimuli for the Cognitive Load task, we looked into the list of words from the British National Corpus [6]. We selected nouns of length 10 characters and sorted these words based on their frequency in the British National Corpus. Afterwards, we chose words with frequency ranging from 1013 to 1026. We developed a simple script in Python to perform that process. We selected the first 20 as stimuli.

To create incorrect word stimuli, we rearranged the words in two different ways for 20 words each:

1. as one word with the middle letters switched
2. as one word with scrambled letters.

Additionally, we added 20 arbitrary mnemonic words.

To prepare the sentence stimuli, we looked into the movie review dataset [23] and selected sentences consisting of 10 words each. During the selection process, all sentences that contained anything other than "normal" words (e.g., apostrophes, quotes, and numbers) were removed as were sentences containing very short words such as "a" or "I". After cleaning the dataset, we were left with 54 sentences. We selected the first 20 as stimuli.

To create the incorrect sentence stimuli, we rearranged the words by moving the middle words to the end, hence, rendering the sentences nonsensical.

The task, we set participants was to categorize as either correct or incorrect the word or sentence stimuli presented to them. All the participants were presented with stimuli from the following categories consisting of 20 words or 20 sentences each. In total, they were presented with 120 items of words and sentences.

2.2   Participants and Procedure

We conducted our study with 41 participants (21 males & 20 females with a mean age of 23.3). The study had 23 native and 18 non-native English language speakers and also had 2 participants with reading difficulties. 15 out of 41 participants wore glasses. Our participants were required to wear Tobii Pro Glasses 2 Eye Tracker (Tobii Eye Tracker). Therefore, we asked if they wear glasses. We were not able to capture eye-tracking data for one of the participants. Therefore, we are reporting the analysis of 40 participants for the eye-based data in the paper.

The study was conducted in the following steps:

1. participant completes a consent form,
2. participant completes a) a general questionnaire to report information on age, number of languages they speak and whether they have reading difficulties and b) an International Physical Activity Questionnaire (IPAQ) [9] to control for any bias in the HR and HRV measurements,
3. participant wears Tobii Eye Tracker and performs a simple calibration to get the values for their PD,
4. participant puts the CorSense HRV device on their finger (ring finger of left hand),
5. participant plays spot the real or made-up word and sentence game task,
6. during the game task ML is measured from Tobii Eye Tracker (BR, PD), Heart Rate Monitor (HR, HRV), and Webcam facing the participant, and
7. participant completes the NASA Task Load Index (NASA-TLX) questionnaire [12].

Fig. 1. Experimental set-up with a participant doing the task. HRV sensor and eye-tracking glasses are circled red. (Color figure online)

2.3   Setup and Materials

The setup as shown in Fig. 1 involved a participant performing the word game task on a computer screen while wearing Tobii Eye Tracker along with CorSense HRV device. We used PsychoPy [2], an open-source application, to program our experiment. To collect data on changes in eye and heart behaviour, we used Tobii Eye Tracker (see Fig. 2) [3] and a CorSense HRV device (see Fig. 4) [1] respectively. We also used an external webcam to collect data on the BR (see Fig. 3).

The NASA Task Load Index questionnaire [12] was used to collect subjective ratings of the amount of CL generated by the task. In addition, we used the IPAQ [7] to get relevant data on health-related physical activity. We collected this data on physical activity because the literature suggests that participant's physical activity index can create an experimental bias [10], bringing heart rate results into question if it is not taken into account. It is also significant to know that the results of the IPAQ showed that all participants were not involved in highly physical activity or training before performing the task.

To analyse our Tobii Eye Tracker data, we created the Tobii API software that eases the access to the data and allows to run the same analysis over all participants. The software can be found on *GitHub*[1]. The scripts for the generation of the stimuli and analysis of the here presented data can be found on *GitHub*[2].

Fig. 2. Tobii Eye Tracking Glasses

Fig. 3. Webcam

Fig. 4. CorSense HRV

2.4   Measurements

Our experiment used the Tobii Eye Tracker to get data on the PD. We recorded the whole session, including an elementary pre-calibration presenting a changing full-screen display of white, black and grey to get a first estimation for minimum and maximum PDs. We tracked the PD during this calibration, during the explanation of the task, and during the task itself. For the analysis, we extracted the data for both eyes and cleaned it by removing invalid samples; samples that got flagged as invalid by the Tobii software or contained negative values. Afterwards,

---

[1] https://github.com/BrutusTT/tobii_api.
[2] https://github.com/BrutusTT/ml_study.

we manually annotated the start and end of the task by finding the corresponding frames from the front-view camera stream. Segmentation of the pupil data was done by converting the frame IDs to time stamps and using these to determine the start and the end of the task segment in the PD readings as provided by the glasses. To account for different pupil sizes, we used the normalized PD using the guidelines described in this paper [18]. In our paper, we report the Mean Pupil Diameter Change (MPDC) as the ratio between the overall mean PD and the mean PD while performing the task. A similar method has also been used in the past [22].

For calculating the Blinking-Rate, we also used the Tobii Eye Tracker to record the eyes stream of the full session. We reused the aforementioned manual annotation to get the task segment by finding the correct frames in the front-view stream and calculating the corresponding frame IDs for the eye stream. To detect the total number of blinks, we applied the following mechanism. Firstly, we converted each frame into greyscale and applied a Gaussian blur to it. Secondly, we applied a binary threshold to the frame and used the blurred frame to find contours in it. The convex hull was calculated for all contours. Lastly, we computed the ratio between the squared circumference and the area of the convex hull to remove all non-spherical hulls. We used a threshold of 150 to 1200 as a limit for the area and values between 10 to 17 for the ratio to exclude non-pupil hulls. Mathematically, the ratio value should be $4\pi \approx 12.57$, but due to noise in the data, we had to widen the ratio range.

To collect data on HR and HRV, we used the EliteHRV CorSense device [15]. The device comes with an API and works in the following manner. The HRV is calculated through receiving the R-R intervals directly from the device. R-R intervals refers to the small changes (milliseconds) in the intervals between successive heartbeats. After performing necessary data cleaning, a Root Mean Square of Successive Differences (RMSSD) calculation is applied to the R-R intervals. Later, a natural log(ln) is applied to RMSSD [14].

3   Results

3.1   Hypothesis H1a1

To test our hypothesis (**H1a1**[3]), a simple linear regression was calculated to predict performance based on the NASA-TLX for CL. A significant regression model was found ($F(1,39)= 10.162$, $p < .003$), with an $R^2 = .207$, adjusted $R^2 = .186$, $\beta = -.456$. We also conducted simple linear regression to predict sentence-based- and word-based- task performance based on the NASA-TLX of CL. A significant regression model was found for sentence-based task performance ($F(1,39)= 11.747$, $p < .001$), with an $R^2 = .231$, adjusted $R^2 = .212$, $\beta = -.481$. Additionally, a partially significant regression model was found for word-based task performance ($F(1,39)= 3.406$, $p < .07$), with an $R^2 = .08$, adjusted $R^2 = .057$, $\beta = -.283$.

---

[3] H1a1: Participants' subjective ratings of CL will predict their overall task performances (H1a1).

3.2   Hypothesis H1b

A *Pearson* correlation between the NASA-TLX rating of CL and participants' overall-task performance, their word-based task performance and their sentence-based task performance was performed in order to test our hypothesis that lower participants' subjective CL predict higher task performance (**H1b[4]**). We found that there was a negative correlation between subjective CL and overall task performance $r(41) = -.456$, $p = .003$, word-based task performance $r(41) = -.283$, $p = .07$ and sentence-based task performance $r(41) = -.481$, $p = .001$. This suggests that the lower rating of participant's subjective CL did predict better task performance. This suggests that our hypotheses H1a1 and H1b can be accepted.

3.3   Hypothesis H1a2

To test whether participants' rating of CL can predict the time spent on the task (**H1a2[5]**) another simple regression was conducted to predict the total mean time taken to complete the task based on the CL. We did not find a significant regression model (F(1,39)= .011, $p < .917$), with an $R^2 = .000$, adjusted $R^2 = -.025$, $\beta = .017$ between the CL and total time to complete the task. Similarly, simple regression was computed for time spent on word-based and sentence-based tasks based on the participant's rating of CL. Once again, we did not find a significant model both for word-based task time (F(1,39)= 1.141, $p < .292$), with an $R^2 = .028$, adjusted $R^2 = .004$, $\beta = -.169$ and for the sentence-based task time (F(1,39)= .335, $p < .566$), with an $R^2 = .009$, adjusted $R^2 = -.017$, $\beta = .092$. This suggest that our hypotheses H1a2 and H1c were rejected.

3.4   Hypotheses H2 and H2a

To test that participants' changes in one physiological behaviour will result in predicting another behaviour (**H2[6], H2a[7]**), we conducted a simple regression to predict HR based on HRV. We found a significant regression model between HR and HRV (F(1,39)= 51.852, $p < .000$), with an $R^2 = .571$, adjusted $R^2 = .560$. We also conducted another simple regression to predict BR based on the MPDC. A *Pearson* correlation between the HR and HRV was performed in order to see the nature of the relationship between HR and HRV. We found that there was a negative correlation between HR and HRV $r(40) = -.743$, $p = .000$. This suggests that an increase in participants' HR would result in a decrease in their

---

[4] H1b: Lower participants' ratings of CL will predict better task performance or vice versa.

[5] H1a2: Participants' subjective ratings of CL will predict time spent to complete the tasks.

[6] H2: Participants' changes in one physiological behaviour will predict a change in another behaviour.

[7] H2b: Changes in the overall mean value of HR will predict overall mean HRV and vice versa. Similarly, BR will predict overall mean PD and vice versa.

HRV during the task. We also found a significant regression model between BR and MPDC ($F(1,39)$= 7.446, $p < .01$), with an $R^2 = .164$, adjusted $R^2 = .142$. A *Pearson* correlation between the MPDC and BR was performed. We found that there was a negative correlation between MPDC and BR $r(40) = -.405$, $p = .01$. This suggests that an increase in participants' MPDC would result in a decrease in their amount of blinking during the task. In summary, our results suggest that changes in one physiological behaviour based on either eye or heart predicts another behaviour based on the same body part. This suggests that our Hypothesis H2 and H2a was accepted.

4 Discussion

We see that participants who reported a higher amount of CL showed a decrease in their task performance. This finding shows that our findings are in line with the prior findings suggesting that an increase in CL can adversely affect users' task performance during a task [28]. More importantly, it showed that the task created in our study was able to induce a high amount of CL in the participants accordingly to their subjective ratings. Our findings based on H1a2 and H1c highlight that measures based on time spent to complete the task cannot always be considered as the predictors of CL. Our findings, hence, show that we need to be careful when using these factors as predictors of CL. These factors can be task dependent in many circumstances and hence should not be deemed reliable.

We found that change in HR was able to predict HRV, and MPDC was able to predict BR. These findings are also in line with the prior research on these behaviours in terms of CL as it has been shown that situations demanding higher mental workload increases PD [24,25] and a decrease in the BR [13]. Similarly, reduction in HRV is attributed to higher CL [19] and an increase in HR with an increase in task difficulty is attributed to high CL [8]. In conclusion, we can infer from these findings that our task induced a high amount of CL and can be used to induce high CL in future studies. On the contrary, we did not find relationships between measures from different body parts; this may be specific to our experimental task. However, we intend to conduct more experiments in diverse settings to observe these relationships in the future. In line with our long-term goal on the creation of an online measure for CL, our results are promising as we were able to find a strong relationship between different physiological behaviours when collected together simultaneously. Our findings suggest that we can use different physiological behaviours in combination during a task to find trends between them and these can be used to create a model to measure CL. Building on these findings, we want to use these behaviours simultaneously, applying a regression-based model to estimate CL or ML.

5 Conclusion and Future Work

In this paper, we presented an experiment to understand the relationships between different physiological measures including MPDC, BR, HR and HRV.

Our experimental task based on identifying correct and incorrect words or sentences with different difficulties was successful in inducing CL as our results showed that there was a significant decrease in participants' performance with an increase in CL. Additionally, there was negative correlation between HR & HRV and MPDC & BR. In summary, we found that under high CL, there is a relationship between different physiological behaviours based on their type of organs. This directs us to further analyse these trends in real-time to create a model to estimate CL in real-time.

The future work will analyse the differences in the item based data for our task between HR, HRV, MPDC, BR individually. In addition, we will create our dataset based on the items-based data analysis; mainly through separating the words and sentences. We will later use the dataset to train a classification model to predict CL in real-time. It is important to note that the positive trends reported in this paper clearly directs us to use machine learning algorithms to achieve our long term goal.

Acknowledgment. The authors would like to acknowledge the support of the ORCA Hub EPSRC (EP/R026173/1, 2017-2021) and consortium partners.

References

1. Corsense elite HRV device. https://elitehrv.com/corsense
2. Psychopy. https://www.psychopy.org
3. Tobii eye-tracking glasses. https://www.tobiipro.com/product-listing/tobii-pro-glasses-2/
4. Ahmad, M., Keller, I., Lohan, K.S.: Integrated real-time, non-intrusive measurements for mental load. In: CHI 2019 Workshop: Everyday Automation Experience: Non-Expert Users Encountering Ubiquitous Automated Systems (2019)
5. Berthold, A., Jameson, A.: Interpreting symptoms of cognitive load in speech input. In: Kay, J. (ed.) UM99 User Modeling. CICMS, vol. 407, pp. 235–244. Springer, Vienna (1999). https://doi.org/10.1007/978-3-7091-2490-1_23
6. Consortium, B., et al.: The British national corpus, version 3 (BNC XML edition) (2007). Distributed by Oxford University Computing Services on behalf of the BNC Consortium. http://www.natcorp.ox.ac.uk. Accessed 25 May 2012
7. Craig, C.L., et al.: International physical activity questionnaire: 12-country reliability and validity. Med. Sci. Sports Exerc. **35**(8), 1381–1395 (2003)
8. Cranford, K.N., Tiettmeyer, J.M., Chuprinko, B.C., Jordan, S., Grove, N.P.: Measuring load on working memory: the use of heart rate as a means of measuring chemistry students' cognitive load. J. Chem. Educ. **91**(5), 641–647 (2014)
9. Fogelholm, M., et al.: International physical activity questionnaire: validity against fitness. Med. Sci. Sports Exerc. **38**(4), 753–760 (2006)
10. Gregoire, J., Tuck, S., Hughson, R.L., Yamamoto, Y.: Heart rate variability at rest and exercise: influence of age, gender, and physical training. Can. J. Appl. Physiol. **21**(6), 455–470 (1996)
11. Haapalainen, E., Kim, S., Forlizzi, J.F., Dey, A.K.: Psycho-physiological measures for assessing cognitive load. In: Proceedings of the 12th ACM International Conference on Ubiquitous Computing, pp. 301–310. ACM (2010)

12. Hart, S.G.: NASA-task load index (NASA-TLX); 20 years later. In: Proceedings of the Human Factors and Ergonomics Society Annual Meeting, vol. 50, pp. 904–908. Sage Publications, Los Angeles (2006)
13. Holland, M.K., Tarlow, G.: Blinking and mental load. Psychol. Rep. **31**(1), 119–127 (1972)
14. Hrv, E.: How do you calculate the HRV score? Webpage, July 2018. https://help.elitehrv.com/article/54-how-do-you-calculate-the-hrv-score
15. Hrv, E.: Corsense heart rate variability. Webpage, January 2019. https://elitehrv.com/corsense
16. Jameson, A., Kiefer, J., Müller, C., Großmann-Hutter, B., Wittig, F., Rummer, R.: Assessment of a user's time pressure and cognitive load on the basis of features of speech. In: Crocker, M., Siekmann, J. (eds.) Resource-Adaptive Cognitive Processes, pp. 171–204. Springer, Heidelberg (2010). https://doi.org/10.1007/978-3-540-89408-7_9
17. Khawaja, M.A., Ruiz, N., Chen, F.: Think before you talk: an empirical study of relationship between speech pauses and cognitive load. In: Proceedings of the 20th Australasian Conference on Computer-Human Interaction: Designing for Habitus and Habitat, pp. 335–338. ACM (2008)
18. Kret, M.E., Sjak-Shie, E.E.: Preprocessing pupil size data: guidelines and code. Behav. Res. Methods **51**(3), 1336–1342 (2018). https://doi.org/10.3758/s13428-018-1075-y
19. Mukherjee, S., Yadav, R., Yung, I., Zajdel, D.P., Oken, B.S.: Sensitivity to mental effort and test-retest reliability of heart rate variability measures in healthy seniors. Clin. Neurophysiol. **122**(10), 2059–2066 (2011)
20. Noel, J.B., Bauer Jr., K.W., Lanning, J.W.: Improving pilot mental workload classification through feature exploitation and combination: a feasibility study. Comput. Oper. Res. **32**(10), 2713–2730 (2005)
21. Paas, F., Tuovinen, J.E., Tabbers, H., Van Gerven, P.W.: Cognitive load measurement as a means to advance cognitive load theory. Educ. Psychol. **38**(1), 63–71 (2003)
22. Palinko, O., Kun, A.L., Shyrokov, A., Heeman, P.: Estimating cognitive load using remote eye tracking in a driving simulator. In: Proceedings of the 2010 Symposium on Eye-Tracking Research & Applications, pp. 141–144. ACM (2010)
23. Pang, B., Lee, L.: A sentimental education: Sentiment analysis using subjectivity summarization based on minimum cuts. In: Proceedings of the ACL (2004)
24. Reilly, J., Kelly, A., Kim, S.H., Jett, S., Zuckerman, B.: The human task-evoked pupillary response function is linear: Implications for baseline response scaling in pupillometry. Behav. Res. Methods **51**(2), 865–878 (2018). https://doi.org/10.3758/s13428-018-1134-4
25. Sabyruly, Y., Broz, F., Keller, I., Lohan, K.S.: Gaze and attention during an HRI storytelling task. In: 2015 AAAI Fall Symposium Series (2015)
26. Sweller, J.: Cognitive load theory, learning difficulty, and instructional design. Learn. Instr. **4**(4), 295–312 (1994)
27. Sweller, J., Van Merrienboer, J.J., Paas, F.G.: Cognitive architecture and instructional design. Educ. Psychol. Rev. **10**(3), 251–296 (1998)
28. Van Gog, T., Kester, L., Paas, F.: Effects of concurrent monitoring on cognitive load and performance as a function of task complexity. Appl. Cogn. Psychol. **25**(4), 584–587 (2011)
29. Wilson, G.F., Russell, C.A.: Real-time assessment of mental workload using psychophysiological measures and artificial neural networks. Hum. Factors **45**(4), 635–644 (2003)

30. Yin, B., Chen, F., Ruiz, N., Ambikairajah, E.: Speech-based cognitive load monitoring system. In: 2008 IEEE International Conference on Acoustics, Speech and Signal Processing, pp. 2041–2044. IEEE (2008)
31. Zhang, J., Yin, Z., Wang, R.: Recognition of mental workload levels under complex human-machine collaboration by using physiological features and adaptive support vector machines. IEEE Trans. Hum.-Mach. Syst. **45**(2), 200–214 (2015)

Mental Workload and Technostress at Work. Which Perspectives and Theoretical Frameworks Can Help Us Understand Both Phenomena Together?

José Manuel Castillo[(⊠)], Edith Galy[(⊠)], and Pierre Thérouanne[(⊠)]

Laboratoire d'Anthropologie et de Psychologie Cliniques,
Cognitives et Sociales (LAPCOS), Université Côte d'Azur,
25 avenue François Mitterrand, 06357 Nice, France
Jose-manuel@npxlab.io,
{Edith.galy,Pierre.therouanne}@univ-cotedazur.fr

Abstract. This paper is a literature review attempting to summarize the different theories and models that can deal with the phenomena of technostress and mental workload generated by the use of digital technologies in the work context. Theories including the task-technology fit model, the person-technology fit or the socio-technical perspectives such as the process of appropriation of technology, will foster a new point of view to understand the relationship between mental workload and technostress. A discussion will be proposed, comprehending a series of reflections on the above-mentioned theories, as well as an attempt to answer the question proposed in the title of this paper.

Keywords: Mental workload · Technostress · Digital technologies · Human factors · Socio technical systems

1 Introduction

The introduction and the use of technology in the work context is one of the most widely studied topics among many different research fields. Indeed, researchers are increasingly turning to the study of the relationship between digital interfaces or tools and the assessment of concepts such as Mental Workload [6, 34] and Technostress [2, 47].

Two reasons can be identified as to why the study of technology is relevant today. Firstly, technology is viewed as a tool used by individuals in carrying out their tasks within or without the work context [25]. Secondly, the current digital era is based on an infrastructure integrating information and communication technologies. This infrastructure is changing the manner in which people work and interact with others, but also the way organizations manage work and finally, entire business models [10, 11].

The structure of this literature review is as follows: in the following part, the methodology deployed for the literature review will be detailed; in the third and the fourth part, the two variables of our main study, the mental workload and technostress

© Springer Nature Switzerland AG 2020
D. Harris and W.-C. Li (Eds.): HCII 2020, LNAI 12186, pp. 14–30, 2020.
https://doi.org/10.1007/978-3-030-49044-7_2

will be outlined. In the fifth part, both phenomena will be described, from the perspective of the HCI, and in the sixth section technostress and mental workload will be addressed, from two perspectives that follow the logic of the theory of person-environment fit: Task technology (TTF) fit and Person technology fit (PTF). In the seventh part, the concept of technology appropriation in the professional environment and its possible conceptualization as a result of a situation of mental overload at work or technostress will be presented. Finally, a discussion will be proposed, comprehending a series of reflections on the above-mentioned theories, as well as an attempt to answer the question proposed in the title of this paper.

2 Methodology

Three search engines were chosen to find the bibliographical material: Google Scholar, Science Direct and Cairn.info. The process for the literature review was carried out in two steps. In the first step, the psychological variables to be studied were established: the mental workload (or cognitive load) and technostress. In this first part, the questions that helped direct this bibliographic study were: "What is currently known regarding mental workload and technostress?" and "How are both variables (mental workload and technostress) related?". In an attempt to answer this question, the terms "Mental Workload (+) (and)(with) technostress stress", "Mental workload" and "technostress" were typed in the keyword section in this order and vice versa. The following literature reviews were consulted (see Tables 1 and 2).

Table 1. Summary of literature reviews of mental workload that were consulted

Authors	Focus
O'Donnell and Eggemeier [40]	Analysis of different measures of mental workload and the development of standards from the assessment of mental workload
Mitchell [38]	Overview of mental workload concept and measure instruments
Cain [7]	Literature review on the existent knowledge about mental workload
Young, Brookhuis, Wickens and Hancock [64]	Mental workload applied to ergonomics (human factors)
Orru and Longo [43]	Literature review on the Cognitive Load Theory (CLT)

Table 2. Summary of literature reviews of technostress that were consulted

Authors	Focus
Riedl [45]	Focus on the physiological assessment of technostress
Fischer and Riedl [17]	Emphasis on the data-collection and methods of different studies on the technostress process
Mahapatra and Pillai [35]	Literature review and classification of literature in three topics: general studies, technology used and technostress in different types of professions

In the second step, the research focused on finding a relationship between both psychological variables and the introduction of new technologies to the work context. The following keywords were resorted to: "Digital technologies in the work context" "Information and communication technologies in the work context", "digital technologies + mental workload" – "digital technologies + (and) Technostress" and "mental workload and technologies".

As is well known, the concept of "use of technology" is one of the main topics in the field of HCI studies. For this reason, the following couple-terms were also searched: "HCI and Mental workload", "HCI and technostress" and "HCI" as the main topic and in the keywords section, "mental workload" or "technostress".

Finally, to encounter theories related to the use of technology, the keywords such as the following were entered: "Use of technology in socio-technical system" "use of technology" "technology used" and "technology in use". As a result, three different theories were found: Task-Technology Fit, Person technology fit and Appropriation of technologies at work.

3 Mental Workload

The very first impression drawn from the extensive review of the literature was the considerable difficulty in finding a consensus in the definition of "mental workload". Cain [7] reports that there is no commonly accepted formal definition of this concept. This author points out that it can be characterized as a mental construction that reflects the "mental tension" resulting from the execution of a task under specific environmental and operational conditions, associated with the operator's ability to respond to these demands.

Spérandio [50] stipulates that "mental workload" appeared in ergonomic literature in the mid-1950s. It is important to focus on this literature because part of this concept, or rather the variables included in the conceptual assessment of it, such as alertness, attention and memory, had already been studied during the beginnings of experimental psychology. It is a concept very close to that of fatigue, stress, anxiety and performance.

The concept of mental workload is that which englobes different sub-concepts (such as mental strain, mental effort, cognitive demands) that are necessary to understand this complex and multidimensional notion. Moreover, mental workload has always been associated with professional positions in critical working fields, such as the nuclear, transport and aviation sectors, and has become an important issue as modern technology imposes increasing cognitive demands [64] and is now occupying a greater place in the fields of human computer interaction and computer science [33, 34].

Referring to mental workload automatically raises the idea of human capacity (working memory or attention) and demands, the latter usually from the environment that impacts this capacity [22, 61]. The Multiple resource [61] theory can better describe this idea. It considers two situations, the first one being when the demand for the task does not exceed the cognitive resources of the operator. This situation is understood as a positive one; the worker will have residual resources that can be used to solve other situations. The second situation occurs when the demand exceeds the cognitive resources of the individual, which results in degraded performance.

Another perspective that can describe the previous idea is the theory of cognitive load developed by Sweller [51, 52]. In the context of learning, this theoretical model is aimed at supporting instructors in the development of novel instructional designs aligned with the limitations of the human cognitive architecture [43] and points out to three types of cognitive load: the intrinsic load or the level of difficulty in completing the task, the extraneous load or how the task is presented in the learning environment, and the germane load or the integration of information to long-term memory from the formulation of mental schemes.

As mentioned by Orru and Longo [43], the CLT goal is to reduce the extraneous load that represents an ineffective amount of load in the instructional design. The notion of capacity is related to the mental energy that will remain for the student to create more schemas. In other words, the energy available to invest the information from working memory into long term memory.

Galy et al. [21, 23] have shown that CLT can be transposed to the working context. The authors have developed the "Individual - Workload – Activity" model [20, 22], fostering an integrative approach to understanding the cognitive load of employees. This model also takes into consideration the three components that are critical in the evaluation of the mental workload borne by workers. The first component corresponds to the individual characteristics (for example, age and/or experience), the second one concerns the activity (complexity and context of execution), and the third component represents the mental workload divided into three dimensions (intrinsic, extraneous and germane load).

Another well-known fact is that the experimental research has used three types of measurements to assess mental workload, each of them accompanied by a set of measurement techniques. The types are behavioral (mental load reflected on the operator's performance), subjective (the awareness of the operator regarding mental workload) and physiological (the homeostatic changes caused by the levels of effort).

Mental workload should be measured subjectively, as it is a psychological concept [7]. It seems that the frequent use of subjective procedures by researchers lies in their practical advantages (for example, ease of implementation and their non-intrusion) and research confirms the ability of these procedures to provide sensitive measures of operator mental workload [40].

Moreover, performance measures seem to be good indicators of cognitive efficiency. Indeed, they intervene further downstream of all cognitive processes [22]. The assessment of the behavioral cost of mental workload is divided into two main streams: the primary and the secondary task measures.

Finally, since the mental workload is strongly related to the concepts of tension and effort, physiological measures are based on the premise that both variables generate internal changes, such as the increase in the level of activation or vigilance [22]. Within the repertoire of physiological measures, the measurement of the heart rate, muscle tension or the eye movement can be found. However, these measurements should be complemented by other types of assessments in order to be more precise.

4 Technostress

The study of the phenomenon of technostress has gained ground after the incremental use of the ICT within the work context [2] and was preceded by many other researches considering stress and job disconformity among the Information System professionals [35] and some attitudes facing the introduction of technology within the work context [60]. To this day, technostress has become a serious problem for IT users and other professionals due to its potential negative effect on users' mental capacity [59]. This concept has raised interest among the scholars, as it also falls under the notion that everything, including technology, has a dark side [48].

Kupersmith [32] was one of the pioneers in assessing technostress in the early 90 s. He studied and examined the way librarians reacted to the integration of technology into their workstations. As many scholars have quoted, the concept of technostress was introduced by Brod [5], who described it as a "modern disease" caused by the inability of the individual to adapt to a type of technology in a healthy manner, and that can manifest itself in the event of the difficulty in accepting computer technologies and in the event of excessive identification with these technologies. According to Nimrod [39] the former definition was complemented by Rosen and Weil [46], who established that technostress can also have a negative impact on individuals' beliefs, attitudes, and psychological and physiological characteristics. Considering previous definitions the concept of technostress can be considered as the amount of stress (constant or high) that a person feels and manifests when he or she uses a specific type of technology, or when he or she is in direct or indirect contact with it. The following table presents the main topics that have been addressed by a great number of researches on technostress (Table 3):

Table 3. Main topics that appear in the research on technostress

Topics	Description
Technostress creators [44]	Techno-overload: Increased workload, higher working speed or change in working habits caused by new technologies
	Techno-invasion: Technology invades personal lives. Individuals spend less time with family or on vacation in order to devote more of their available time to learning new technologies
	Techno-complexity: Inability to learn or manage the complexity of new technologies
	Technological insecurity: Job insecurity linked to technology (represented as the fear of being replaced by more qualified people, reflected in the constant need to update technical skills)
	Technological uncertainty: The inability to control the constant changes in characteristics of technology, such as hardware and software
Antecedents of technostress	Individual characteristics (personal traits) [26] (Age, gender education experience) [36] and Organizational factors [59]

(continued)

Table 3. (*continued*)

Topics	Description
Consequences of technostress	The negative effects of technostress on health were listed by Salanova, Llorens, Cifre and Nogareda [47] who outline that the potential consequences of technostress are manifested by psychosomatic complaints, sleep disorders, headaches, muscle pain, gastrointestinal disorders The negative effects of technostress on organizational outcomes were listed by Tarafdar, Tu, Ragu-Nathan [53] who outline that technostress is negative, correlated with other organizational outcomes such as job satisfaction, organizational commitment and continuance commitment. Furthermore, it has been shown that technostress has a negative impact on productivity [54, 59] and satisfaction with ICT use [19]

As stated in the introduction, this paper aims to focus on the research drawn from mental workload. Table 4 presents some works on technostress including the concept or related variables of mental workload.

Table 4. Research on technostress that have included aspects also related to the concept of mental workload

Authors	Description of the related concept with MWL
Ragu-Nathan et al. [44]	The techno-overload, identified as a higher working speed or change in working habits caused by new technologies and techno-complexity, identified as the inability to learn or manage the complexity of new technologies
Isaac, Kalika, Campoy [28]	Information overload and time pressure
Ayyagari et al. [2]	Complexity inherent to the utilization of ICT
Ayyagari [3]	Information overload and technology characteristics/functionalities
Sellberg and Susi [49]	Distributed cognition into socio technical systems

As indicated in the literature review, the concept of technostress being strongly related to that of mental load, has already been studied by various scholars. Technostress is one more form of stress, the only difference being that it is strictly engendered by technology, in a direct (use) or indirect (support) way. As demonstrated in the first part, a considerable number of jobs are impacted by technostress and mental workload. It can be added that no sources have yet been found regarding the complete unification of both concepts; Only sources including them partially united.

In the following section, some of the currents that can help understand both phenomena in context will be presented.

5 The HCI Perspective

The Human Computer Interaction perspective helps comprehend technostress and the mental workload in the context of the use of digital technologies. HCI is a high-performance area of research that has quickly emerged as a priority area for research and development in computer science and applied social and behavioral sciences [4]. This concept was inherent in the emergence of the computer, or more generally of the machine [30].

In this field of study, the concept of interaction refers to the notions of "input", "output" and "feedback loop" between the user and the machine [16]. The importance regarding this field is to create a greater degree of fluid interaction between the machine and the human, in other words, to develop technologies focusing on the needs of the user and on the machines, making these easier to control.

This paper will focus on two important features that can impact mental workload and technostress: the usefulness and the usability of technologies. HCI is defined as the study and practice of usability [4]. The usability is the degree to which computer systems enable the user to achieve specified goals effectively and efficiently, promoting feelings of satisfaction in a given context of use [29]. Another inherent aspect concerning HCI is the concept of the usefulness of the system. According to Davis, the aspect of usefulness is defined as the degree to which a person believes that using a system would enhance his or her performance [14].

Mental workload and HCI are both related concepts. According to Hollender et al. [27] there is a certain compatibility between the CLT and the theories and concepts of HCI. Indeed, both perspectives have the same theories of cognition as a basis and focus on the reduction of irrelevant cognitive load (extraneous load). Table 5 presents a summary of the studies evaluating the variables of mental workload and technostress in HCI:

Table 5. Studies in the HCI field that include both variables

Author	Property of HCI	Technology	Mental workload or technostress (also related concepts)
Tracy and Albers [56]	Usability	Website interfaces	Mental workload
Longo et al. [34] Longo [33]	Usability	Website interfaces	Mental workload
Yan, Guo, Lee and Vogel [62]	Usability	Computer mediated communication (telecommunication in health care)	Technostress
Ayyagari et al. [2]	Usefulness	ICT	Technostress and some concepts related to MWL
Sellberg and Susi [49]	Usability and synchronization of information	ICT	Technostress and mental workload

From the perspective of the HCI, the strain in the interaction between the user and a given type of technology results in varying levels of mental load and technostress. The elements of the HCI perspective (usability or usefulness) constitute an essential part in other models on the study of socio-technical systems, such as the task - technology fit model and the appropriation of technological instruments.

6 Task-Technology Fit and Person-Technology Fit

The Person-environment fit (P-E fit) model commenced as a theoretical model of stress. This framework has been studied for decades, having as main contributions the works of French, Rodgers and Cobb [18], who studied work stress and imbalances occurring as a result of a poor adjustment to theirs jobs, groups or organizations. It is crucial to mention this theoretical model of stress as two further models, (an extension of the P-E Fit focusing on the use of technologies at work and its impact on performance of users), will be developed within the field of IS.

Task-technology fit model (TTF) [24, 25] is a theoretical model that focuses on the prediction of performance, encompassing the analysis of matching aspects of a determined technology with the characteristics of a determined task expected to be performed by this type of technology. More specifically, TTF is the correspondence between task requirements, individual abilities and the functionality and features of technology [24, 25].

At its beginning, it took into consideration the aspects of the task and the technology. The concept of "individuals" was then introduced. The notion of "individuals" refers to the users who may use technologies to assist them in the performance of their tasks. The "use" is conditioned by the users' characteristics, such as training, computer literacy and motivation. "Technologies" are viewed as tools used by individuals performing their "tasks", actions that the individuals carry out when turning inputs into outputs [24, 25]. Table 6 shows the eight factors that determine task-technology fit in the context of IS systems that have been identified by Goodhue [25]:

Table 6. Factors of TTF according to Goodhue

Factor	Sub-factor	Definition
Quality	Currency	Data used is current enough to meet user's needs
	Right data	Maintaining the necessary fields or elements of data
	Right level of detail	Maintaining the data at the right level(s) of detail(s)
Locatability	Locatability	Facility in determining what data is available and where
	Meaning	Facility in determining what a data element on a report or file means
Authorization	Authorization	Facility in obtaining an authorization to access data necessary to do the job

(*continued*)

Table 6. (*continued*)

Factor	Sub-factor	Definition
Compatibility	Compatibility	Data from different sources can be consolidated or compared without inconsistencies
Ease of use and training	Ease of use of hardware and software	Ease in doing what the user wants to do, using the system hardware and software for submitting, accessing, analyzing given data
	Training	User can get the kind of quality computer-related training when he or she needs it
Production timeliness	Timeliness	Information systems meet pre-defined production turn around schedules
Systems reliability	Systems reliability	Dependability and consistency of access and up time of systems
Relationship with users	IS understanding of business	IS understand user's business mission and its relation to corporate objectives
	IS interest and dedication	To support customer business needs
	Responsiveness	Turnaround time for a request submitted for IS service
	Consulting	Availability and quality of technical and business planning assistance for systems
	IS performance	IS keeps its agreements

The factors that determine the adjustment of the task with the technology are of high importance, since, intuitively, a poor level in any of them can impact performance and the mental load level of the workers. By examining the TTF, the preponderant role that the notions of the HCI occupy, stands out.

The TTF model was extended to "the technology to performance chain model" (TPC) [25], a comprehensive one that integrates user attitudes as predictors of utilization and the characteristics of TTF as a predictor of performance. In this extended model, a part is derived from theories about beliefs and affects (attitudes) and behaviors focused on studying the reason people use a determined technology, and if there is a connection between task requirements and technology functionality. The TPC model gives a highly more accurate picture of the manner in which technologies, user tasks, and utilization relate to changes in performance.

The application of TTF has been carried out in different fields. Table 7 presents a few studies chosen because of their relationship with mental workload and technostress:

Table 7. Compilation of research applying the TTF theory

Author	Technology	Task	Variable
Dang et al. [13]	Search systems of social media	Search information in English and in non-English data collection	Mental workload
Ayyagari [3]	Use of the ICTs at the office	General tasks	Technostress and information overload
Tarafdar and Wenninger [55]	ICT (Emails)	Communication and checking tasks	Information overload and distraction (aspects related to MWL)
Aljukhadar, Senecal and Nantel [1]	Websites and online platforms	Search information on random websites	None
McCarthy Aronson Mazous [37]	Knowledge management systems	General tasks applied to knowledge management systems	None
D'ambra, Wilson and Akter [12]	E-books	Academic tasks	None

The perspective of TTF cannot by itself complete all the aspects that are intrinsic to the use of a certain technology. That is the reason it was extended to the TCP model, adding an extra antecedent, which is the "individual" one, and also, giving a stronger weight to the variable of utilization (and precursors of utilization) based on different theories of attitudes and behaviors.

The perspective of the person technology fit is based on the same logic of the person-environment fit, with the further precision that this extension technology comprehends the role of the environment [57]. Unlike the TTF, the PTF is a recent theory that it is still in development, or at least there is not as vast a research as for the TTF.

Person technology fit is defined as the amount of "fit" or match perceived by an individual towards technology. According to Tomer [57], the theoretical antecedents of this construction are the Technology Acceptance Model by Davis [14] as well as the works concerning technostress by Ayyagari et al. [2].

The latter has introduced the said framework as an extension of P-E fit, arguing that the fit or misfit will be dependent on the abilities (person component) and the demands (technology component). In the previous conceptualization, it can be found that some aspects are related to mental burden. In other words, some given characteristics of ICTs (i.e. usefulness, complexity or phase of change) related to mental workload will generate more cognitive demands from the user, resulting in a misfit that will be reflected through "stressors" (work overload, role ambiguity, work-home conflict, etc.).

This perspective is increasingly being applied in academic research. Below (Table 8), some examples of current studies using the P-T fit:

Table 8. Studies applying the PT model

Author	Purpose of research	Variable
Kim and Gatling [31]	Evaluate whether the organization citizen behavior (OCB) and employee engagement are both outcomes of the level of person-technology fit and other types of fit	Technostress
Tarafdar and Wenninger [55]	Evaluate whether the overload in an extended version of PT: person – person technology fit and whether this overload can be related to TTF	Mental workload
Tomer [57, 58]	Evaluate the technology allocation according to IT professional needs	Technostress (stressor: Techno uncertainty)
Yong, Shiau and Hou [63]	Evaluate the motivations and characteristics of students towards the services provided by the cloud computing classroom, and if this fit contributes to the outcome of student performances	None

7   Appropriation of Technologies

Appropriation can be considered a situation of "subversion", because users "adapt" to (transform) and "adopt" (choose) the technology in a way that is never fully contemplated by the designers [15]. Carroll et al. [8, 9] developed the appropriation cycle model. In their model, the authors distinguish "technology-as-designed" from "technology-in-use", as well as the concepts of "non-appropriation" from "disappropriation". Some notes about their definitions follow:

- Technology-as-designed refers to technological artefacts designed for a target audience.
- Appropriation is not a static concept or a single activity, but a process that is maintained over time. In this model, ownership refers to the process where users test and evaluate the technology, selecting and adapting its functionalities and attributes based on their needs.
- Non-appropriation refers to the moment when users "ignore" or show themselves to be disinterested in the tool, artifact or in one or more of its functions.
- Disappropriation means the "choice" of users not to use or to refuse the technology implemented. According to the authors, it can appear at any phase of the appropriation cycle.
- Technology-in-use refers to the way in which groups and users use the technology designed in accordance with their perceptions and experiences. Technology-in-use is a result of the process of appropriation by which the artifact has been evaluated, adapted and integrated into the routine of the user.

Orlikowski [41, 42] proposes a theoretical model which seeks to examine the interaction between technology, individuals and organizations by integrating the deterministic and social perspective. This author also introduces the concept of "technology-in-practice", a substantial concept in the fields of sociology and the sciences of technology and information. This concept refers to the situation in which a worker uses a specific technology, the specific resources for his use (means), the standards (conditions and rules) and the interpretative schemes (direction of attitude towards use, either positive or negative) involved.

The concept of appropriation of technology is interesting for the fields of HCI computer science and ergonomics. A selection of works shows that appropriation (among other variables) is conditioned by technostress and is the result of it [53]. However, to date and within this review, no bibliographic material demonstrating that the mental workload is closely linked to appropriation has been encountered.

8 Discussion and Conclusions

Nowadays, to understand how digital technologies are changing the way employees do their jobs, it is essential to adopt a systematic and representative vision (employee perception). Different concepts can provide a relevant framework for analysis, such as the concept of appropriation, person-technology fit and task-technology fit.

This section seeks to answer the question brought up in this article, keeping in mind that the present study is a bibliographic review and its main premise is to determine at what level, or in what manner the mental workload generated by the use of technology can be a mediating or risk effect of technostress.

The relationship between the two concepts is factual for the following two reasons. The first being that the definition of both mental workload and of technostress comprise common elements. Second, within the experimental evaluation of technostress, aspects related to the mental workload have been used (for example, information overload) as independent variables.

Mental workload is clearly distinct from technostress, yet the two concepts are strongly linked in scientific literature. Both can be measured using the same methods (via physiological, subjective and behavioral techniques). Intuitively, technostress causes can affect the mental workload of employees in the use of digital technologies within the workplace.

In this paper, HCI studies were considered as a preliminary perspective in understanding the MWL and Technostress. As is known, the use of technology is a type of behavior [25] and in the case of the HCI perspective, this behavior is considered as an interaction between the user and the machine, and the achievement of this interaction will be conditioned by the functionalities or characteristics of the technology-in-practice [41]. Within the working context, technology represents a resource for the operators to carry out their various tasks. Considering the theory of Sweller [51, 52] brought up in the works of Galy [23], the aspects of the machine (such as usability) can be considered as the elements of the extraneous cognitive load, therefore hypothesizing that any difficulty in the interaction between the machine and the user will demand extra mental resources and energy that could otherwise be used for the main task.

As second and third perspectives, the theoretical models of Task-Technology fit (as its extension) and Person – Technology fit were discussed. For the former, the loss of performance in the task can be caused by a mismatch between the functionalities of the technology and the characteristics of the task to be accomplished. Functionalities of the technology such as usability, utility and usefulness, are comprised in this perspective.

One primary connection between mental workload and the TTF perspective is the notion of performance. It can be hypothesized that poor functionalities or aspects of technology could trigger a high level of misfit between this technology and the pre-scribed task, equally resulting in poor performance.

The model Person-Technology fit stipulates that the misfit between individual characteristics and technological features can lead to a higher risk of suffering from technostress [4]. It is well known that the technology chain model also considers the individual's characteristics (motivation or computer literacy). However, it can be reflected upon that the P-T Fit model gives more weight to individual characteristics, introducing personal needs and personal values. The mental workload could be gen-erated by a misfit between the individual components and the technological functionalities.

As previously discussed, the appropriation of technologies is not a cause of the variables of technostress or mental workload. However, as observed in the scientific research, both variables can condition the misappropriation or non-use of technology in the context of work. It is crucial to outline is that the appropriation of technology fosters a holistic view of the use of technology, focusing on how people use the technology, why they use it and which institutional factors (norms, values or requirements) within context will influence this use. The following graph attempts to summarize the ideas mentioned above (Fig. 1).

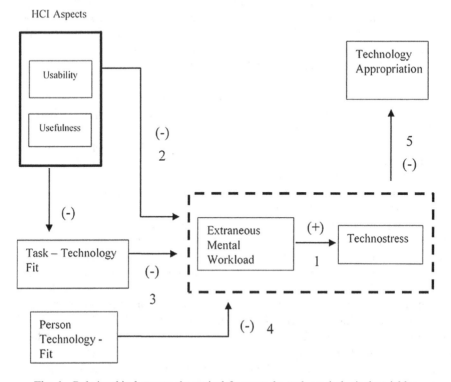

Fig. 1. Relationship between theoretical frameworks and psychological variables

From the graph, we can establish the following hypotheses:

1: The use of digital technology impacts (increases or decreases) the mental workload of workers, and a mental overload generated by the use of technology in the work context is a trigger for technostress.
2: Poor characteristics (such as usability or usefulness) of the "technology in use" will increase the extraneous mental load, generating a mental overload for the worker, which will be a trigger in the development of technostress.
3: A misfit between the task and the "technology in use" will increase the mental load of the workers, provoking technostress.
4: A misfit between the individual characteristics and the technological functionalities will foster a mental overload, that will be a trigger for technostress.
5: A mental Overload and technostress will negatively impact the process of appropriation of technology in use.

To conclude, the theoretical framework that can provide a holistic perspective of the technostress genesis and the mental workload is that of task-technology fit, and this for several reasons. Firstly, because it integrates the concepts of interaction between the person and the machine, giving an important value to the functionalities of the technology and to the support behind the given technology. Secondly, because in its most extended model, it also takes into account the individual variable (training, motivation, etc.). Thirdly, owing to the Technology chain model, it gives an important weight to the variable of use and to the precursors of the use of technology.

As is well known, technology in itself cannot represent a risk in developing technostress within the working context. It is the "technology in use" or the indirect interaction between operator and technology that represents a risk of technostress. The aim of this work was to bring to light different perspectives in understanding mental workload and technostress. It is important to consider the evaluation of both aspects, since the presence of one or the other not only compromises the performance and efficacy of workers, but also their health.

References

1. Aljukhadar, M., et al.: Is more always better? Investigating the task-technology fit theory in an online user context. Inf. Manag. **51**(4), 391–397 (2014). https://doi.org/10.1016/j.im.2013.10.003
2. Ayyagari, R., et al.: Technostress: technological antecedents and implications. MIS Q. **35**(4), 831 (2011). https://doi.org/10.2307/41409963
3. Ayyagari, R.: Impact of information overload and task-technology fit on technostress. In: Proceedings of the Southern Association for Information Systems Conference, Atlanta, GA, USA, 23–24 March (2012)
4. Carroll, J.: Human-Computer Interaction in the New Millennium. ACM Press, Addison-Wesley, New York, Boston (2001)
5. Brod, C.: Technostress: The Human Cost of the Computer Revolution. Addison Wesley, London (1984)

6. Burns, C.M.: Understanding, supporting, and redesigning cognitive work. In: Longo, L., Leva, M.C. (eds.) H-WORKLOAD 2018. CCIS, vol. 1012, pp. 3–12. Springer, Cham (2019). https://doi.org/10.1007/978-3-030-14273-5_1

7. Cain, B.: A review of the mental workload literature. Defence Research and Development Toronto (Canada), Toronto, ON, Canada, pp. 1–35 (2007)

8. Carroll, J.: Completing design in use: closing the appropriation cycle. In: ECIS 2004 Proceedings (2004)

9. Carroll, J., et al.: Identity, power and fragmentation in cyberspace: technology appropriation by young people. In: ACIS 2001 Proceedings (2001)

10. Cascio, W.F.: Psychology of human resource management. In: International Encyclopedia of the Social & Behavioral Sciences, pp. 348–352 Elsevier (2015). https://doi.org/10.1016/B978-0-08-097086-8.73024-8

11. Cascio, W.F., Montealegre, R.: How technology is changing work and organizations. Annu. Rev. Organ. Psychol. Organ. Behav. 3(1), 349–375 (2016). https://doi.org/10.1146/annurev-orgpsych-041015-062352

12. D'Ambra, J., et al.: Application of the task-technology fit model to structure and evaluate the adoption of e-books by academics. J. Am. Soc. Inf. Sci. Technol. 64(1), 48–64 (2013). https://doi.org/10.1002/asi.22757

13. (Mandy) Dang, Y., et al.: Examining the impacts of mental workload and task-technology fit on user acceptance of the social media search system. Inf. Syst. Front. (2018). https://doi.org/10.1007/s10796-018-9879-y

14. Davis, F.D.: Perceived usefulness, perceived ease of use, and user acceptance of information technology. MIS Q. 13(3), 319 (1989). https://doi.org/10.2307/249008

15. Dix, A.: Designing for appropriation. In: Proceedings of the 21st British HCI Group Annual Conference on HCI 2007: HCI…but not as we know it - Volume 2, BCS HCI 2007, University of Lancaster, United Kingdom, 3–7 September 2007 (2007)

16. Dix, A. (ed.): Human-Computer Interaction. Pearson/Prentice-Hall, Harlow, New York (2004)

17. Fischer, T., Riedl, R.: Technostress research: a nurturing ground for measurement pluralism? CAIS 40, 375–401 (2017). https://doi.org/10.17705/1CAIS.04017

18. French Jr., J.R.P., Rodgers, W.L., Cobb, S.: Adjustment as person-environment fit. In: Coelho, G., Hamburg, D., Adams, J. (eds.) Coping and Adaptation, pp. 316–333. Basic Books, New York (1974)

19. Fuglseth, A.M., Sørebø, Ø.: The effects of technostress within the context of employee use of ICT. Comput. Hum. Behav. 40, 161–170 (2014). https://doi.org/10.1016/j.chb.2014.07.040

20. Galy, E.: A multidimensional scale of mental workload evaluation based on Individual – Workload – Activity (IWA) model: validation and relationships with job satisfaction. Tutor. Quant. Methods Psychol. (2019, in press)

21. Galy, E.: Consideration of several mental workload categories: perspectives for elaboration of new ergonomic recommendations concerning shiftwork. Theor. Issues Ergon. Sci. 19(4), 483–497 (2017). https://doi.org/10.1080/1463922X.2017.1381777

22. Galy, E.: Temps de travail pour tant de travail: Vers un modèle intégrant l'individu, la charge mentale et l'activité (ICA). Presses Académiques Franco (2013)

23. Galy, E., et al.: What is the relationship between mental workload factors and cognitive load types? Int. J. Psychophysiol. 83(3), 269–275 (2012). https://doi.org/10.1016/j.ijpsycho.2011.09.023

24. Goodhue, D.L.: Development and measurement validity of a task-technology fit instrument for user evaluations of information system. Decis. Sci. 29(1), 105–138 (1998). https://doi.org/10.1111/j.1540-5915.1998.tb01346.x

25. Goodhue, D.L., Thompson, R.L.: Task-technology fit and individual performance. MIS Q. **19**(2), 213 (1995). https://doi.org/10.2307/249689
26. Hsiao, K.-L.: Compulsive mobile application usage and technostress: the role of personality traits. Online Inf. Rev. **41**(2), 272–295 (2017). https://doi.org/10.1108/OIR-03-2016-0091
27. Hollender, N., et al.: Integrating cognitive load theory and concepts of human–computer interaction. Comput. Hum. Behav. **26**(6), 1278–1288 (2010). https://doi.org/10.1016/j.chb.2010.05.031
28. Isaac, H., et al.: Surcharge informationnelle, urgence et TIC. L'effet temporel des technologies de l'information. Manag. Avenir **13**(3), 149 (2007). https://doi.org/10.3917/mav.013.0149
29. Ivory, M., Hearst, M.: The state of the art in automating usability evaluation of user interfaces. ACM Comput. Surv. **33**(4), 470–516 (2001)
30. Karray, F., et al.: Human-computer interaction: overview on state of the art. Int. J. Smart Sens. Intell. Syst. **1**(1), 137–159 (2008). https://doi.org/10.21307/ijssis-2017-283
31. Kim, J., Gatling, A.: Impact of employees' job, organizational and technology fit on engagement and organizational citizenship behavior. JHTT, JHTT-04-2018-0029 (2019). https://doi.org/10.1108/JHTT-04-2018-0029
32. Kupersmith, J.: Technostress and the reference librarian. Ref. Serv. Rev. **20**(2), 7–50 (1992). https://doi.org/10.1108/eb049150
33. Longo, L.: Experienced mental workload, perception of usability, their interaction and impact on task performance. PLoS ONE **13**(8), e0199661 (2018). https://doi.org/10.1371/journal.pone.0199661
34. Longo, L., et al.: The importance of human mental workload in web design. In: WEBIST 2012: 8th International Conference on Web Information Systems and Technologies, Porto, Portugal, 18–21 April (2012)
35. Mahapatra, M., Pillai, R.: Technostress in organizations. A review of literature. In: 26th European Conference on Information Systems (ECIS2018), Portsmouth, UK (2018)
36. Marchiori, D.M., et al.: Do individual characteristics influence the types of technostress reported by workers? Int. J. Hum.-Comput. Interact. **35**(3), 218–230 (2019). https://doi.org/10.1080/10447318.2018.1449713
37. McCarthy, R., Aronson, J., Mazouz, K.: Measuring the validity of task technology fit for knowledge management systems. In: AMCIS 2001 Proceedings, p. 337 (2001)
38. Mitchell, D.K.: Mental Workload and ARL Workload Modeling Tools (Technical Note ARL-TN-161). U.S. Army Research Laboratory, Aberdeen Proving Ground, MD (2000)
39. Nimrod, G.: Technostress: measuring a new threat to well-being in later life. Aging Mental Health **22**(8), 1086–1093 (2018). https://doi.org/10.1080/13607863.2017.1334037
40. O'Donnell, R., Eggemeier, F.: Workload assessment methodology. In: Boff, K.R., Kaufman, L., Thomas, J.P. (eds.) Handbook of Perception and Human Performance, Cognitive Processes and Performance, vol. 2. Wiley, Hoboken (1986)
41. Orlikowski, W.J.: Using technology and constituting structures; a practice lens for studying technology in organizations. Organ. Sci. **11**(4) (2000). https://doi.org/10.1287/orsc.11.4.404.14600
42. Orlikowski, W.J.: The duality of technology: rethinking the concept of technology in organizations. Organ. Sci. **3**(3), 398–427 (1992). https://doi.org/10.1287/orsc.3.3.398
43. Orru, G., Longo, L.: The evolution of cognitive load theory and the measurement of its intrinsic, extraneous and germane loads: a review. In: Longo, L., Leva, M.C. (eds.) H-WORKLOAD 2018. CCIS, vol. 1012, pp. 23–48. Springer, Cham (2019). https://doi.org/10.1007/978-3-030-14273-5_3

44. Ragu-Nathan, T.S., et al.: The consequences of technostress for end users in organizations: conceptual development and empirical validation. Inf. Syst. Res. **19**(4), 417–433 (2008). https://doi.org/10.1287/isre.1070.0165

45. Riedl, R.: On the biology of technostress: literature review and research agenda. SIGMIS Database **44**(1), 18 (2012). https://doi.org/10.1145/2436239.2436242

46. Rosen, L., Weil, M.: Technostress: Coping with Technology@ Work@ Home@ Play. Wiley, Etobicoke (1997)

47. Salanova, M., et al.: El Tecnoestrés: Concepto, Medida e Intervención Psicosocial [Technostress: Concept, Measurement and Prevention]. Nota Técnica de Prevención. Instituto Nacional de Seguridad e Higiene en el Trabajo, Madrid (2007)

48. Salanova, M., et al.: The dark side of technologies: technostress among users of information and communication technologies. Int. J. Psychol. **48**(3), 422–436 (2013). https://doi.org/10.1080/00207594.2012.680460

49. Sellberg, C., Susi, T.: Technostress in the office: a distributed cognition perspective on human–technology interaction. Cogn. Technol. Work **16**(2), 187–201 (2014). https://doi.org/10.1007/s10111-013-0256-9

50. Sperandio, J.: La charge mentale... au travers de quelques recherches anciennes d'ergonomie. In: 10th Conference on Psychology and Ergonomics EPIQUE Arpege Science Publishing, pp. 57–62 (2019)

51. Sweller, J.: Cognitive load during problem solving: effects on learning. Cogn. Sci. **12**, 257–285 (1988)

52. Sweller, J.: Cognitive load theory, learning difficulty and instructional design. Learn. Instr. **4**, 295–312 (1994)

53. Tarafdar, M., et al.: Impact of technostress on end-user satisfaction and performance. J. Manag. Inf. Sys. **27**(3), 303–334 (2010). https://doi.org/10.2753/MIS0742-1222270311

54. Tarafdar, M., et al.: The impact of technostress on role stress and productivity. J. Manag. Inf. Sys. **24**(1), 301–328 (2007). https://doi.org/10.2753/MIS0742-1222240109

55. Tarafdar, M., Wenninger, H.: Effects of 'fit' on email overload. In: Twenty-Fourth Americas Conference on Information Systems, New Orleans (2018)

56. Tracy, J.P., Albers, M.J.: Measuring cognitive load to test the usability of web sites. In: Annual Conference-Society for Technical Communication, vol. 53, pp. 256–260 (2006)

57. Tomer, G.: Person-technology fit and work outcomes: a study among IT professionals in India. ECIS 2015 Completed Research Papers, Paper 181 (2015)

58. Tomer, G.: Towards developing career technology fit framework and analyzing its influence on work related outcomes among IT professionals (2013)

59. Tu, Q., Wang, K., Shu, Q., Computer stress – related in china. Commun. ACM **48**(4), 77–81 (2008)

60. Vold, M.A.: New technology in the office: attitudes and consequences. Work Stress: Int. J. Work Health Organ. **1**, 143–153 (1987). https://doi.org/10.1080/02678378708258496

61. Wickens, C.D.: Multiple resources and mental workload. Hum. Factors **50**(3), 449–455 (2008). https://doi.org/10.1518/001872008X288394

62. Yan, Z., et al.: Understanding the linkage between technology features and technostress. In: Telemedicine Communication, ECIS 2013 Research in Progress (2013)

63. Yong, J.-H., et al.: A study of person-technology fit in the cloud computing classroom. Int. J. Online Pedagogy Course Des. **7**(3), 1–16 (2017). https://doi.org/10.4018/IJOPCD.2017070101

64. Young, M.S., et al.: State of science: mental workload in ergonomics. Ergonomics **58**(1), 1–17 (2015). https://doi.org/10.1080/00140139.2014.956151

Handling Design Tasks: Effects of Music on Mood and Task Performance

Ying Fang, Ruiqian An, Junxia Wang, and Zhanxun Dong[(✉)]

Shanghai Jiaotong University, Shanghai, China
dongzx@sjtu.edu.cn

Abstract. Listening to music at work seems to be the habit of many designers. As they believe that music can stimulate creativity and help them concentrate. Design tasks mainly fall into two categories: repetitive tasks such as image matting and creative tasks such as brainstorming. In order to study how different types of music actually impact work performance, participants were asked to perform image-matting and brainstorming under three different experimental background conditions (no music, light music, and hard music). Physiological indexes such as heart rate and GSR (Galvanic Skin Response) were recorded to measure the influence of music on mood. Task results were carefully examined to indicate how music impacts task performance. Our analysis reveals that listening to music would lower task performance for repetitive task as well as creative task, and hard music is even worse than light music. However, listening to music doesn't seem to impact anxiety, arousal or mental effort. Plus, arousal improves when competing creative design tasks compared with repetitive design tasks.

Keywords: Design task · Background music · Physiological measure · Work performance

1 Introduction

Listening to music at work is the habit of many designers, as many believe that music could help them concentrate and stimulate creativity. However, very few studies have focused on how music impacts work performance and psychological status of designers.

Current studies have shown that listening to music at work does affect task performance and mood. Music can increase work quality and efficiency for computer information systems developers, and participants reported positive mood as well as better focus on design [1]. Research on computer word processing shows that music can reduce hand motions while typing by 23.2%, and may thus reduce risks of musculoskeletal disorder caused by typing [2]. For driving tasks, music (even negative ones) can bring induced mood which leads to higher energy rate, and this makes dealing with concurrent tasks while driving easier [3]. However, music is rather a double-edged sword that impacts differently depending on the type of task. Researchers pointed out that while music may have facilitating effects on tasks involving great concentration and attention, for tasks such as comprehension tasks music can be

© Springer Nature Switzerland AG 2020
D. Harris and W.-C. Li (Eds.): HCII 2020, LNAI 12186, pp. 31–42, 2020.
https://doi.org/10.1007/978-3-030-49044-7_3

distracting [4]. In a study on reading comprehension of junior high students, performance significantly declined with lyrical music as background [5].

The type of music is a key factor in deciding the influence of music on task completion. Listening to calm music before an examination could reduce physiological measures of anxiety and help achieve better work performance, while listening to harsh music has no such effects and may even increase anxiety [6]. Plus, a study of driving game shows that participants were very negatively affected by high-arousal music selected by experimenter, as they became inaccurate, high-distracted and more anxious, but for low-arousal music the results are more positive [7]. However, light music can also have negative impacts. A study on light music showed that reports of discomfort, earaches and headaches increased and work efficiency for complex mental work significantly decreased with light music as background music [8]. In terms of labor work, a research on 50 machine operators of garment industry showed that relaxation music would hamper work performance and the author suggested that the type of music which doesn't suit work environment may be the reason [9]. It is therefore important to study how different types of music could act on specific tasks to improve concentration, memory and work efficiency, which could help people gain more knowledge and information in a limited period of time [10, 11].

IDEO studio thinks that music can help people transform from the traditional mode of work which is rather rigorous and meticulous to a more relaxed and creative one [12]. Music evokes the more creative perceptual thinking, thus broadening people's horizons and stimulating innovation [13]. Also, music promotes creative processes (e.g. group discussion, brainstorming) by reducing the pressure of breaking silence and creating an environment more suitable for communication, cooperation and creative thinking [14]. However, very few studies have focused on how music impacts work performance and psychological status of designers. In this study, we investigated how different types of music influence completion of repetitive/creative design tasks. Our hypotheses are as follows.

H1: Light music improves work performance of repetitive design tasks while hard music helps with creative design tasks.
H2: Light music can reduce anxiety and arousal while hard music can increase them when people perform repetitive/creative design tasks.
H3: Arousal and anxiety increase, and more mental effort is needed for creative design tasks compared with repetitive design tasks.

2 Method

2.1 Participants

Twenty-two students from the School of Design of Shanghai Jiaotong University participated in the study and 20 copies of valid data (11 female, 9 male) were chosen. The subjects were healthy with no hearing impairment. The age of our participants ranges from 22 to 25, and they have all studied designing for 3 to 5 years.

2.2 Preparations

Repetitive Task and Creative Task. Design tasks could be mainly divided into repetitive tasks and creative tasks. Image matting was chosen as the representation of repetitive tasks, as it is rather monotonous and tedious and a very common task for designers. We prepared three images of single-line closed shape and ensured that they are of the same difficulty. Participants were asked to use pen tool in Adobe Photoshop CC 2018 to complete image matting, and then fill the shape in the same color as that for the original image. As for creative task, brainstorming was selected as it is a fairly common and important task for designers. Specifically, participants were asked to draw a mind map based on a given word, and this is done by writing a second word that relates to the former.

Music Stimuli. Two light music and two hard music were picked by researchers. This was done to ensure that when performing a second task under the same music environment (light music for example), our participant would listen to a different song, which help erase impacts of familiarity.

Light music: With "light music" and "soothing" as keywords, the two music with the highest playback amount.

Hard music: Two hot hard music of the previous year.

Selection criteria: BPM for light music does not exceed 72 times per minute, and BPM for hard music is greater than or equal to 92.

Evaluation Index
Task performance. Repetitive task: Accuracy and efficiency of image matting.

Creative task: Number of words and correlation of words.

Physiological measures. Polygraph (Bio-Trace + Software for Nexus-10) was employed to measure heart rate and GSR. Mean heart rate indicates arousal, while heart-rate variability indicates mental effort. GSR demonstrates changes in the degree of anxiety [15].

Subjective test. Assessment questionnaire.

Tool for Statistical Analysis. IBM SPSS Statistics 21, Python, Bio-Trace + Software for Nexus-10, Natural language analysis tools (Word Relevance).

2.3 Experimental Design

The study employed mixed-subjects design of 2 (repetitive task and creative task) × 3 (no-music, light music, hard music), meaning that each participant has two complete both tasks under three different experimental background conditions. Specific procedure of experiment is as follows:

    Step 1: Listen to instructions carefully.
    Step 2: Practice image-matting to get familiar with the software.
    Step 3: Sit still for 5 min to calm heart rate.
    Step 4: Complete repetitive task in no-music environment.

Step 5: Take a three-minute break.
Step 6: Repeat step 2 under light music environment and hard music environment respectively.
Step 7: Complete creative task, repeat step 2 and step 3.
Step 8: Fill out assessment questionnaire.

3 Results

3.1 Task Performance of Repetitive Task

Accuracy and efficiency are the indicators for performance repetitive task. Accuracy is measured by the number of different pixels between work of our participants and the original image, and the calculation was done by python programming. Efficiency is measured by the time taken to complete the task. Average number of different pixels and time-on-task of all the participants under three music environments are shown in Table 1.

Table 1. Mean number of different pixels and time-on-task under three different experimental background conditions

Music environment	Number of different pixels	Time on image-matting task (s)
No music	82314	148
Light music	94554	119
Hard music	104376	130

We used SPSS to run a one-way ANOVA (Analysis of Variance). We set music type the independent variable, and number of different pixels and time on are the dependent variables. The significance level is 0.05. The results show that background music affects the time required for competing image matting ($p = 0.042 < 0.05$), and the influence on accuracy is very significant ($p = 0.009 < 0.01$) (see Table 2).

Table 2. One-way ANOVA for task performance indicators of repetitive task

		df	F	Sig.
Number of different pixels	Between groups	2	5.093	.009
	Within groups	57		
	Total	59		
Time on image-matting task	Between groups	2	3.365	.042
	Within groups	57		
	Total	59		

Listening to music while performing repetitive tasks would lower the accuracy of performance, and hard music is even worse than light music (see Fig. 1). Plus, hard music would impair would efficiency compared with light music (see Fig. 2). It is noticeable that efficiency seems to be the lowest with no music. The reason may lie in

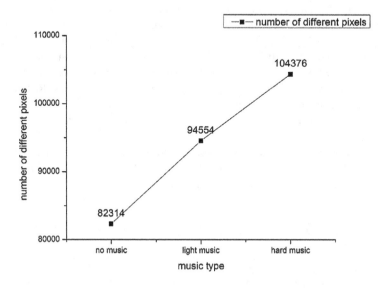

Fig. 1. Number of different pixels under three different experimental background conditions

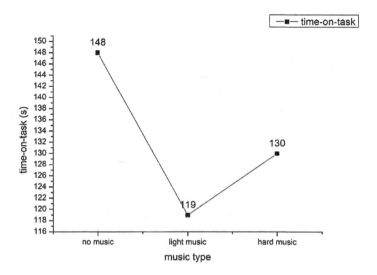

Fig. 2. Time-on-task under three different experimental background conditions

the fact that though having been given time for practice before experiment started, participants weren't familiar enough with the software provided.

3.2 Task Performance of Creative Task

Number of words and correlation of words are the indicators for task performance of creative task. Specifically, correlation of words is measured by Chinese word similarity query. It is based on a base containing 8 million Chinese word vectors. The similarity interval of two words is [0,1], where 0 stands for the lowest similarity and 1 the highest. The Average number of words and correlation of words are shown in Table 3.

One-way ANOVA was run where music type was the independent variable and number of words and correlation of words are the dependent variables. Results show that music environment has influence on correlation of words ($p = 0.043 < 0.05$), however it doesn't seem to have impact on the number of words ($p > 0.05$) (see Table 4).

As is shown in Fig. 3, correlation of words decreases under background music, and hard music is more negative than light music. This shows that music may impair work performance while completing creative tasks.

Table 3. The average number of words and correlation under three different experimental background conditions

	Number of words	Correlation of words
No music	21.85	0.434
Light music	20.85	0.464
Hard music	22.95	0.485

Table 4. One-way ANOVA for task performance indicators of creative task

		df	F	Sig.
Correlation of words	Between groups	2	3.336	.043
	Within groups	57		
	Total	59		
Number of words	Between groups	2	.795	.457
	Within groups	57		
	Total	59		

3.3 Physiological Measures

Mean heart rate, heart-rate variability and GSR are the indicators of physiological status. Mean heart rate indicates arousal, and heart-rate variability shows mental effort. GSR demonstrates changes in the degree of anxiety. Table 5 shows the GSR and mean heart rate statistics under three music environments.

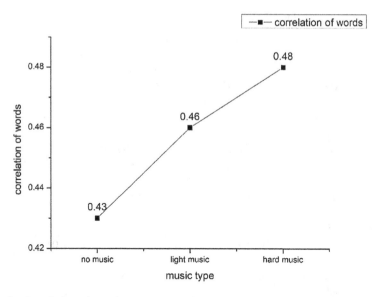

Fig. 3. Correlation of words under three different experimental background conditions

Table 5. GSR and mean heart rate under three different experimental background conditions

Task	Music type	GSR		Mean heart rate (BPM)	
		Mean	Std	Mean	Std
Image matting	No music	4.17	2.80	80.31	10.54
	Soft music	4.39	3.21	80.55	10.30
	Hard music	4.48	3.38	80.35	10.98
Word association	No music	3.48	2.71	87.97	13.51
	Soft music	3.70	2.59	86.49	10.94
	Hard music	4.07	2.72	86.61	11.24

We ran a MANOVA (Multivariate Analysis of Variance) to study how different independent factors interact (see Table 6). Music and task are named the independent variables, while GSR, mean heart rate and heart-rate variability are the dependent variables. Results show that the presence of music has little impact on GSR, mean heart-rate and heart-rate variability ($p > 0.05$). Type of task does act on mean heart rate ($p = 0.02 < 0.05$), however the variable doesn't seem to impact GSR or heart-rate variability. Interaction between music and task exhibits no significance ($p > 0.05$), which means that GSR, mean hear rate and heart-rate variability aren't affected by the interaction between music and task.

Results showed that mean heart rate increases for creative task compared with repetitive task (see Fig. 4). This indicates that the level of arousal increases when performing the probably more challenging creative task. However, the type of task doesn't have much influence on mental effort or anxiety.

Table 6. MANOVA for physiological measures

Source		Type III sum of squares	df	Mean square	F	Sig.
Corrected model	GSR	15.225[a]	5	3.045	.358	.876
	Mean heart rate	1342.513[b]	5	268.503	2.102	.070
	Heart-rate std	36.962[c]	5	7.392	.198	.962
Intercept	GSR	1965.157	1	1965.157	231.263	.000
	Mean heart rate	840960.245	1	840960.245	6585.076	.000
	Heart-rate std	12186.293	1	12186.293	327.217	.000
Music type	GSR	4.024	2	2.012	.237	.790
	Mean heart rate	10.789	2	5.395	.042	.959
	Heart-rate std	.559	2	.280	.008	.993
Task	GSR	10.680	1	10.680	1.257	.265
	Mean heart rate	1314.834	1	1314.834	10.296	.002
	Heart-rate std	29.868	1	29.868	.802	.372
Music type* task	GSR	.522	2	.261	.031	.970
	Mean heart rate	16.890	2	8.445	.066	.936
	Heart-rate std	6.534	2	3.267	.088	.916
Error	GSR	968.715	114	8.497		
	Mean heart rate	14558.596	114	127.707		
	Heart-rate std	4245.621	114	37.242		
Total	GSR	2949.097	120			
	Mean heart rate	856861.355	120			
	Heart-rate std	16468.876	120			
Corrected total	GSR	983.940	119			
	Mean heart rate	15901.110	119			
	Heart-rate std	4282.583	119			

[a]R Squared = .015 (Adjusted R Squared = −.028)
[b]R Squared = .084 (Adjusted R Squared = .044)
[c]R Squared = .009 (Adjusted R Squared = −.035)

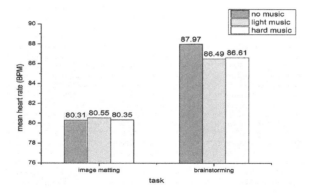

Fig. 4. Mean heart rate of repetitive task and creative task under three different experimental background conditions

3.4 Assessment Questionnaire

Our questionnaire used a 5-point Likert scale. Score for each question ranges from 1 to 5, which corresponds to the five levels of assessment–"very motivating" (1 point), "motivating", "no impact", "disturbing" and "very disturbing" (5 point). After collecting questionnaires from our participants, statistical analysis followed.

Contrary to actual task performance, participants seem to be positive about light music. For both repetitive task and creative task, nearly half think that light music can stimulate task performance and almost no one think it would act negatively. Results showed that participants are more positive about light music when performing repetitive task compared with creative task (see Table 7). When it comes to hard music, participants became more negative about the influence on task completion, especially for repetitive task (see Table 7). It's interesting to note that in terms of the frequency of listening to music while completing design tasks, only 4.6% participants choose perform the design tasks without music, which testifies our observation that music is a common companion for design tasks (Fig. 5).

Table 7. Impact of background music on task completion (1 = "very motivating", 3 = "no impact", 5 = "very disturbing")

	Repetitive task	Creative task
Light music	2.32	2.59
Hard music	3.18	2.86

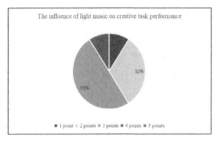
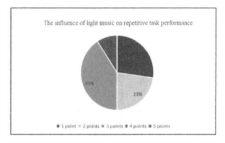

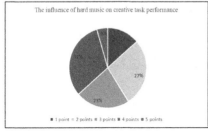
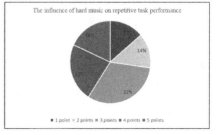

Fig. 5. The influence of two kinds of music on the performance of two kinds of tasks

4   Discussion

We can thus make the following conclusions:

1. Listening to music can lower task performance for both repetitive task and creative task. For repetitive task, listening to music would lower the accuracy of work, and hard music is a worse choice than light music. Plus, work efficiency decreases in the presence of hard music compared with light music. For creative task, listening to music would harm the ability of thinking out of the box, and hard music is still a worse choice than light music.
2. Music doesn't have much impact on anxiety, arousal or mental effort when performing repetitive task and creative task.
3. Arousal increases when performing creative task compared with repetitive task, but there seems little impact on anxiety or mental effort.

Limitations. In this study we chose image-matting and brainstorming as representations for repetitive design task and creative design task. However, it is hard to deny that design tasks are rather complicated and are various in process and perception even within the same type. This poses challenges to the application of our study, namely our result may not be applicable for other repetitive or creative design tasks. Also, sometimes designers need to perform tasks which both involve monotonous work and creative work such as Photoshop and video editing. There are also some other types of tasks such as logical task (e.g. modeling). Plus, participants were design students aged from 20 to 25, who have studied design for 3 to 5 years. Our target participants fall in a quite narrow range, and the impact of music on design tasks might be different for newcomers to design or experienced design professionals. Last but not least, physiological statistics almost demonstrate no significance in our study. The reason may lie in the fact that as an activity where people sit still and remain rather calm, as opposed to activities such as driving or running, the change in those statistics are fairly minor. However, this doesn't mean that music doesn't change psychological status.

Future Work. It is interesting to do research on other repetitive design tasks and creative design tasks, as well as other type of design tasks to figure out whether the impact on music echoes with our result. Also, including participants with different levels of design skills and from different areas of design is an option. It is also advisable to examine at length the impact of music on design tasks using testing methods such as EEG monitoring.

5 Conclusion

In this study we explored the impact of music (no music, light music and hard music) on repetitive design task and creative design task. Contrary to the common knowledge, our findings reveal that listening to music impairs work performance of both repetitive task and creative task. For repetitive task accuracy and efficiency are lowered, and creativity is reduced for creative task. Plus, hard music is an even worse choice than light music. In terms of physiological measures, listening to music while performing tasks doesn't seem to impact anxiety, arousal or mental effort. However, people do have a higher level of arousal when doing the creative task, which is probably more challenging. Assessment questionnaire indicate that participants were generally positive about the impact of light music on task completion, especially for repetitive task. This shows between a gap between how people feel and the actual result.

Further work could lie in exploring the impact of music on different repetitive/creative tasks, as well as other types of design tasks such as logical tasks. Also, including participants with different levels of design skills and from different areas of design is an option. Plus, it is advisable to employ testing methods such as EEG monitoring to dig deep into the physiological impact on music on design tasks.

References

1. Lesiuk, T.: The effect of music listening on work performance. Psychol. Music **33**(2), 173–191 (2005)
2. Sengupta, A.K., Jiang, X.: Effect of background music in a computer word processing task. Occup. Ergon. **12**(4), 165–177 (2015)
3. Van Der Zwaag, M.D., Dijksterhuis, C., De Waard, D., et al.: The influence of music on mood and performance while driving. Ergonomics **55**(1), 12–22 (2012)
4. Dalton, B.H., Behm, D.G.: Effects of noise and music on human and task performance: a systematic review. Occup. Ergon. **7**(3), 143–152 (2007)
5. Anderson, S.A., Fuller, G.B.: Effect of music on reading comprehension of junior high school students. School Psychol. Q. **25**(3), 178 (2010)
6. Lilley, J.L., Oberle, C.D., Thompson Jr., J.G.: Effects of music and grade consequences on test anxiety and performance. Psychomusicol. Music Mind Brain **24**(2), 184 (2014)

7. Cassidy, G., MacDonald, R., et al.: The effects of music choice on task performance: a study of the impact of self-selected and experimenter-selected music on driving game performance and experience. Musicae Scientiae **13**, 357–386 (2009)
8. Li, L., Zhiwei, L., Li, P.: The effect of soft background music on mental performance. Ergonomics (02) (2013)
9. Padmasiri, D.: The effect of music listening on work performance: a case study of Sri Lanka. Int. J. Sci. Technol. Res. **3**(5) (2014)
10. Chen, Y., Wu, Y.: The influence of music type and rhythm on memory. J. Chin. Health Psychol. (02) (2015)
11. Cheng, H.: The influence of background music on the improvement of work attention. Music Time (15) (2015)
12. Shao, L.: Finland: everyone is a designer. Xinmin Wkly. **28**, 82–84 (2012)
13. Li, Y.: IDEO, design changes everything. Design (10), 46–57
14. Li, H.: Analysis on the innovation method and design management of IDEO company. Entrepr. World **04**, 51–53 (2009)
15. Ünal, A.B., de Waard, D., Epstude, K., et al.: Driving with music: effects on arousal and performance. Transp. Res. Part F: Traffic Psychol. Behav. **21**, 52–65 (2013)

Relationship Between Thermal Sensation and Human Cognitive Performance Based on PMV

Jiawei Fu[1], Rui Yan[2], Fulin Wang[2($\boxtimes$)], and Liang Ma[1]

[1] Department of Industrial Engineering, Tsinghua University, Beijing, China
[2] Building Energy Conservation Research Center,
Tsinghua University, Beijing, China
flwang@tsinghua.edu.cn

Abstract. In this paper, the effects on human performance of thermal sensation were investigated. Recruited subjects performed Mackworth clock tests examining vigilance and cognitive performance. The results showed that thermal discomfort caused by elevated air temperature and clothes had a negative effect on performance. A quantitative relationship was established between thermal sensation votes and task performance. The relationship indicates that optimum performance can be achieved slightly below neutral ($PMV_{best} = -0.74$ for group), while thermal discomfort (feeling too warm or too cold) leads to reduced performance. The big individual difference in PMV_{best} implies that occupants have different thermal requirements to keep high performance. Therefore, conventional heating and cooling approaches should work with individual microclimate control approaches to improve individual productivity. The findings can be used for economic calculations pertaining to building design and operation when occupant productivity is considered.

Keywords: Thermal sensation votes · Performance

1 Introduction

Studies [1–3] have identified that human performance bears a close relationship to the indoor environment quality (IEQ). The IEQ covers several factors, including thermal environment, indoor air quality, lighting, and acoustic, etc. Air temperature is one commonly used indicator of thermal environment in IEQ and performance research. Numerous field and laboratory investigations have been conducted to study the relationship between air temperature and human performance. Several studies have proved that air temperature influences performance indirectly through its impact on prevalence of Sick Building Syndrome (SBS) symptoms or satisfaction with air quality [4, 5]. Meanwhile air temperature also directly affects human performance. Berglund et al. [6] and Niemela et al. [7] reported a decrement in performance of call center workers when the temperature was above 25 °C.

Thermal comfort is derived from the actual requirements of people and is influenced by their metabolic heat production, physical activity, clothing, and the four environmental parameters: air temperature, mean radiant temperature, air velocity, and

© Springer Nature Switzerland AG 2020
D. Harris and W.-C. Li (Eds.): HCII 2020, LNAI 12186, pp. 43–51, 2020.
https://doi.org/10.1007/978-3-030-49044-7_4

air humidity. Different combinations of these thermal criteria would result in the same thermal sensation. Wyon et al. [8] showed that subjects performing sedentary work under two conditions and achieving subjectively assessed thermal neutrality by slightly adjusting the air temperature could achieve similar performance results under these two different temperature conditions. Taking these results into account, it seems that using only air temperature to define thermal conditions for optimal performance may not be sufficient when describing the effects of thermal environment on performance. This is in spite the air temperature is the commonly used indicator of thermal environment in IEQ and productivity research. In addition to temperature, it would be necessary to use other parameters describing the thermal environment, such as thermal sensation vote or the predicted mean vote (PMV). Such an approach was used by Loveday et al. [9] who derived a deterministic model relating productivity to three variables: air temperature, relative humidity, and relative air velocity, by Roelofsen [10] who related the loss of performance with PMV and by Jensen et al. [11] who derived the relationship between thermal sensation votes and performance. In addition to these relationships, Kosonen and Tan [12] illustrated how the productivity loss can be minimized through improved thermal comfort design criteria using the PMV index; however, only the effects of feeling too warm on productivity were reported and no relationship between PMV and productivity was created. An attempt to establish a quantitative relationship between thermal environment (thermal sensation votes) and performance was also made in the present study using the data from the experiments examining the effect of thermal discomfort on performance of a wide range of performance tasks.

2   Methods

2.1   Experimental Set-up

The experiment was carried out in a normal office adapted for the experiments. Fourteen volunteers (2 females and 12 males, aged on average 23 ± 2 years) were recruited. Each volunteer was exposed to six different thermal conditions. The intended PMV of six conditions was set between -1 to 1.5. The operative PMV of each condition was calculated according to temperature, dress heat resistance, air speed and air moisture. The order of exposures was balanced. The experiment was carried out from 15, April 2019 to 29, April 2019. Each subject was exposed at the same time of the three successive experimental days (For example, subject 1 at 15:00–17:00 on day 1, day 2 and day 3) to avoid any influence of circadian rhythm on the within-subject difference between conditions.

Each experimental session lasted for 2 h with a 10-min break after approximately 1 h (as shown in Fig. 1). During exposure in the room, subjects performed Mackworth clock test, which has been used in the field of experimental psychology to study the effects of long term vigilance on the detection of signals. The tasks were presented on a PC and were self-paced. The screen has a large red pointer in a large circular background like a clock. The pointer moves in short jumps like the second hand of an analog clock, approximately every second. At infrequent and irregular intervals, the hand makes a double jump, e.g. 12 times every 30 s. The task is to detect when the double jumps occur by pressing a button (as shown in Fig. 2). the reaction time was recorded by the computer clock. Tasks were presented to subjects without feedback, i.e. they did not receive any information on their performance and performed each task until it was completed.

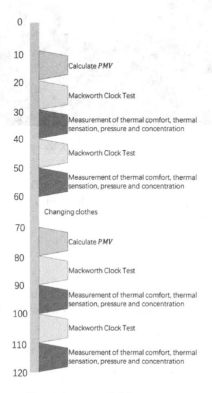

Time during expose (min)

Fig. 1. Experimental procedure.

During experiments, the subjects also evaluated the perceived thermal comfort, thermal sensation, performance, pressure, and concentration several times. Physiological parameters were also measured, including skin temperature and HR, which were recorded using a smart wearable band (cling band) that transmit signals to a PC through Bluetooth connection. The performance index (PI) was calculated as the reciprocal of processing/reaction time. Speed (response time) and accuracy (% correct) were used as measures of performance of tasks completed without feedback. In this case, the performance index (PI) was computed by dividing the mean reaction time by the accuracy of responses.

The SPSS 13.0 (SPSS Inc., Chicago, IL, USA) program was used to conduct the statistical analysis. The measured data were subject to analysis of variance in a repeated measures design; Greenhouse-Geisser statistics were used to adjust the violation of

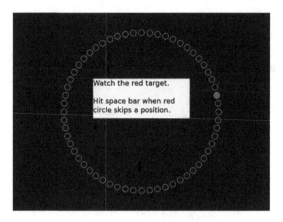

Fig. 2. Mackworth clock test.

sphericity. The significance level was set to be 0.05 ($p < 0.05$). The repeated-measures ANOVA was used as it has been shown to be robust even for data that are not completely normally distributed.

3   Results

The measured physical parameters describing the main parameters of indoor climate in the office under six different exposure conditions as shown in Table 1.

Table 1. *PMV* (mean ± standard deviation) under the six exposure conditions.

Condition (intended PMV)	Operative PMV	Thermal sensation
−1	−0.64 ± 0.14	−0.61 ± 0.47
−0.5	−0.23 ± 0.16	−0.07 ± 0.32
0	0.03 ± 0.23	0.25 ± 0.37
0.5	0.33 ± 0.21	0.25 ± 0.45
1	0.59 ± 0.21	0.32 ± 0.7
1.5	0.92 ± 0.2	0.82 ± 0.52

Figure 3 shows that the self-estimated thermal sensation was significantly different at six thermal conditions ($F(2,76,35.92) = 15.702$, $p < 0.01$). It can be seen that the subjects were thermally neutral (mean thermal sensation votes = −0.07, 0.25, 0.25) at intended PMV = −0.5, 0, 0.5, thermally warm (mean thermal sensation votes = 0.32, 0.82) at intended PMV = 1, 1.5, and thermally cold (mean thermal sensation votes = −0.61) at intended PMV = −1, which indicates that the six thermal conditions were as intended.

The performance of Mackworth Clock tasks decreased significantly when PMV increases (Fig. 4, $F(2.61, 33.91) = 12.18$, $p < 0.01$). The reaction time at the coldest

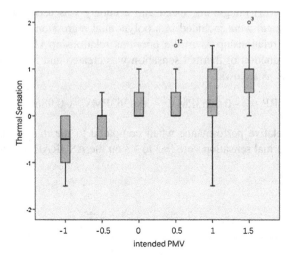

Fig. 3. Subjective thermal sensation under six conditions

conditions (PMV = −1, 0.5)is significantly lower than that at warm conditions (PMV = 1, 1, 5) (post hoc $p < 0.05$).

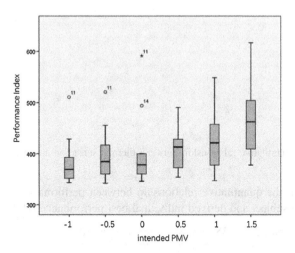

Fig. 4. Performance index (reaction time/accuracy) under six conditions

4   Discussion

4.1   Development of Quantitative Relationship

The performance of Mackworth clock tests was used to derive the quantitative relationship between thermal sensation and performance. As a measure of performance, the performance index (PI) integrating speed and accuracy was used. Standardized data (Relative Performance) were included in a polynomial regression model to reflect the same trend of the relationship as in the previous relationship [11, 13]. The relative performance as a function of thermal sensation was created and it is shown in Fig. 5. This relationship is as follows:

$$RP = -0.0253PMV^2 - 0.0363PMV + 0.9866 \qquad (2)$$

where RP is the relative performance when compared to the maximum performance and PMV is the thermal sensation vote (−3 to +3 on the ASHRAE seven-point thermal sensation scale).

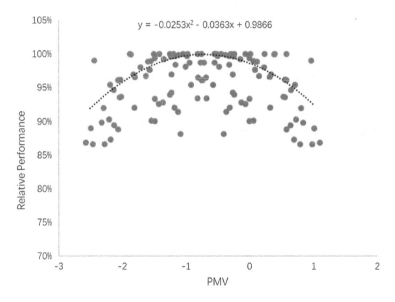

Fig. 5. Quantitative relationship between thermal sensation and performance

We also built the quantitative relationship between performance and thermal discomfort for each subject and derived individual best performance (PMV_{best}) as follow:

No. 1: −0.64; No. 2: −0.35; No. 3: −1.00; No. 4: −0.42;
No. 5: −1.00; No. 6: −1.00; No. 7: −1.70; No. 8: −1.57;
No. 9: −0.55; No. 10: −0.44; No. 11: −0.08; No. 12: −0.38;
No. 13: −0.57; No. 14: −1.00.

It can be seen that the individual difference was very big (Average of all subjects' PMV_{best} is −0.74, and the standard deviation of all subjects' PMV_{best} is 0.48). On average, the optimum performance was at PMV = −0.74 and the further the temperature from PMV = −0.74, the lower the performance. Compared with PMV = 0.74, the performance at PMV = 0 and −1.5 was around 5% and 6% lower, respectively.

4.2 Comparison with Other Models

Measurement of productivity is the essential component of a relationship between IEQ and productivity. Task performance is usually used to reflect productivity and the speed at which the tasks are performed is one of the common measures of performance. Accuracy is another aspect of performance and has been defined as freedom from error in discrete tasks [14]. In IEQ-productivity research, metrics of work speed have more often been used to measure performance than metrics of work accuracy. For many types of task, however there exists a speed-accuracy trade-off [15], meaning that within the capacity of a person he/she can either per- form the task very fast with a large number of errors or very slow with very few errors (the effects on performance mea-sures occur in two opposite directions). Sometimes only one of these two performance indicators is affected by the environmental conditions. In the present work it was decided to integrate speed and accuracy into one measure defining the overall effect of environmental conditions on performance. It should, however, be acknowledged that such an approach may not always be right and further work on this issue is needed [16].

The relationship developed in this study is compared with the relationships developed previously in Fig. 6. In the case of our model, the optimum performance occurs when the thermal sensation vote is −0.74. This is lower than the value predicted from other models showing optimum performance at about −0.5; also some differences can be observed on the cold side of the thermal sensation vote. A potential reason for

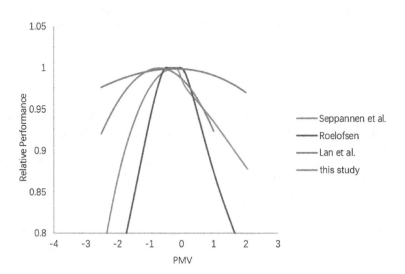

Fig. 6. Comparison of the relationship between thermal sensation and relative performance developed in this work with previously developed relationships by Roelofsen [10], Lan et al. [17], and Seppänen et al. [13].

the observed discrepancy can be the type of data used in the present study and that of Jensen et al. [11]; we used only the performance of Mackworth clock test while in other studies a wide range of tasks including arithmetical calculation, typing, logical reasoning, and tasks examining memory were used. The model of Roelofsen [10] indicates even greater effects of thermal discomfort on productivity compared with the relationships developed in the present work and that of Seppänen et al. [13]. The reason is probably a limited set of data used by Roelofsen to create the relationship.

The individual difference in PMV_{best} was obvious in this study. Similar results were found in previous research [18]. The individual difference in PMV_{best} implies that occupants have different thermal requirements to keep high performance. Therefore, conventional heating and cooling approaches should work with individual microclimate control approaches to improve individual productivity.

5   Conclusions

Experiments with subjects have shown that performance is negatively affected by thermal discomfort caused by elevated air temperature. A quantitative relationship between thermal sensation and performance was established, which can be a useful tool to predict productivity loss due to thermal discomfort in the cost-benefit calculations pertaining to indoor environments. The relationship seems to suggest that the optimum performance is achieved when people feel slightly cool thereby it makes sense to set the PMV limits in workplaces in the range between −1 and 0. The big individual difference in PMV_{best} implies that occupants have different thermal requirements to keep high performance.

Acknowledgements. This work was supported by the National Natural Science Foundation of China [grant number 51578305] and Beijing Municipal Natural Science Foundation [grant number 9172008].

References

1. Roelofsen, P.: The impact of office environments on employee performance: the design of the workplace as a strategy for productivity enhancement. J. Facil. Manag. **1**(3), 247–264 (2002)
2. Woods, J.E.: Cost avoidance and productivity in owning and operating buildings. Occup. Med. **4**(4), 753–770 (1989)
3. Lorsch, H.G., Ossama, A.A.: The impact of the building indoor environment on occupant productivity-part 1: recent studies, measures, and costs. ASHRAE Trans. **100**(2), 741–749 (1994)
4. Seppänen, O.A., Fisk, W.J., Mendell, M.J.: Association of ventilation rates and CO2-concentrations with health and other responses in commercial and institutional buildings. Indoor Air **9**, 252–274 (1999)
5. Fang, L., Wyon, D.P., Clausen, G., Fanger, P.O.: Impact of indoor air temperature and humidity in an office on perceived air quality, SBS symptoms and performance. Indoor Air **14**(Suppl. 7), 74–81 (2004)

6. Berglund, L., Gonzales, R., Gaffe, A.: Predicted human performance decrement from thermal discomfort and ET*. In: Proceedings of the Fifth International Conference on Indoor Air Quality and Climate, Toronto, vol. 1, pp. 215–220 (1990)

7. Niemela, R., Hannula, M., Rautio, S., Reijula, K., Railio, J.: The effect of indoor air temperature on labor on productivity in call centers-a case study. Energy Build. **34**(7), 59–64 (2002)

8. Wyon, D.P., Fanger, P.O., Olesen, B.W., Pedersen, C.J.K.: The mental performance of subjects clothed for comfort at two different air temperatures **18**, 358–374 (1975)

9. Loveday, D., Hanby, V.I., Woodward, W.B.: A software tool for relating the comfort and productivity of occupants in commercial buildings. In: Proceedings of CIBSE National Conference, pp. 69–75 (1995)

10. Roelofsen, P.: The design of the workplace as a strategy for productivity enhancement. In: Clima 2000/Napoli 2001 World Congress, Napoli (2001)

11. Jensen, K.L., Toftum, J., Friis-Hansen, P.: A Bayesian network approach to the evaluation of building design and its consequences for employee performance and operational costs. Build. Environ. **44**, 456–462 (2009)

12. Kosonen, R., Tan, F.: Assessment of productivity loss in air-conditioned buildings using PMV index. Energy Build. **36**, 987–993 (2004)

13. Seppänen, O., Fisk, W., Lei, Q.H.: Room temperature and productivity in office work. In: eScholarship Repository, Lawrence Berkeley National Laboratory, University of California (2006). http://repositories.cdlib.org/lbnl/LBNL-60952

14. Drury, C.G.: Managing the speed-accuracy trade-off. In: Karwowski, W., Marras, W. (eds.) The Occupational Ergonomics Handbook, 1st edn. CRC Press, New York (1999)

15. Wickelgren, W.A.: Speed-accuracy trade-off and information processing dynamics. Acta Psychologica **41**(1), 67–85 (1977)

16. Lan, L.: Mechanism and evaluation of the effects of indoor environmental quality on human productivity. Ph.D. thesis, Shanghai Jiao Tong University (2010)

17. Lan, L., Wargocki, P., Lian, Z.: Quantitative measurement of productivity loss due to thermal discomfort. Energy Build. **43**(5), 1057–1062 (2011)

18. Cui, W., Cao, G., Park, J.H., et al.: Influence of indoor air temperature on human thermal comfort, motivation and performance. Build. Environ. **68**, 114–122 (2013)

The Effect and the Efficiency Balance of Font Size and Font Color Change on the Human Memory in Chinese Vocabulary

Tse-Wei Fu and Jui-Wen Peng[(⊠)]

Department of Industrial Design, National Taipei University of Technology,
Taipei, Taiwan
t106588019@ntut.org.tw, jpeng@mail.ntut.edu.tw

Abstract. The study of reading memory mainly focuses on the language, cognitive and metacognitive skills that people must master, by which people are able to read and study. Little attention has been paid to how words themselves, the perceptual characteristics of words, affect comprehension and memory. In the research, perceptual characteristics were controlled by providing Chinese vocabulary in four different font sizes and two font colors which are experimented in 10 participants. The font sizes were 16pt, 20pt, 28pt, and 40pt; the font colors were red and black, respectively. Experiments were made to compare and measure the two aspects influence on human memory and efficiency balance. The experimental results showed that font size increasing was more effective in increasing memory scores than color-changing. In particular, compared with red words in 16pt, words in black of 40pt significantly increased the memory score. However, there was no significant difference in memory score between 20pt and 28pt black words. According to the experimental results and post-experiment interview, the results were discussed at the end of this paper.

Keywords: Traditional Chinese character · Color psychology · Font size · Memory · Color memory · Reading memory

1 Introduction

This is an era of information explosion, we can easily obtain information through various media by the five senses, among which the visual received the most information. Vision is the easiest way to convey information, which can convey the content of communication to the public directly. But the way information presented is complex and changeable, which makes it not easy to find and identify the information we really want to obtain.

Information transmission is often complex and complicated, our sights then are distracted by various factors, which would affect our reading ability and memory capacity such as the location of information, reading direction, and reading order. In order to highlight the key points in a string of information, the key points are generally bolded, classified, enlarged, or color-coded to guide attention so as to deepen the memory of reading.

© Springer Nature Switzerland AG 2020
D. Harris and W.-C. Li (Eds.): HCII 2020, LNAI 12186, pp. 52–61, 2020.
https://doi.org/10.1007/978-3-030-49044-7_5

Attention enhancement can have a positive effect on memory. Many studies have shown that color can cause people's emotional response and strengthen their attention; therefore, it can be concluded that color can effectively enhance attention and thus improve memory. The larger and bolder the font, the more important the content is, the more attention can be attracted. For example, it can be applied to the marking system to provide user identification function, which can help users get the correct information quickly. If the signs of public places or transportation systems are unclear due to the inconsistent font, it is easy to cause users to identify too slowly and even cause danger [1].

However, there is still a lack of efficiency comparison among these known ways in enhancing memory. Therefore, this study designed a series of experiments to explore and analyze the effects of different sizes in red and black fonts on human memory and find out which has a stronger connection with human memory. Finally, we try to find out the relationship and rules between these two words through experiments.

2 Related Work

2.1 Factors Affecting Memory

Unlike language, the brain does not have a specific limited memory center area for memory. According to the duration of memory, the present research divides it into three stages: 1. Sensory memory 2. Short-term memory 3. Long-term memory. Sensory memory is the memory of a very short time, is the beginning of the memory system, it is most vulnerable to attention and has a short storage time. The memory time is about 0.25 s to 4 s; In short-term memory, the speed of fading to forgetting is quite fast. Most information, if not retrieved repeatedly, is usually retained between 5 s and 20 s, and the longest time is no more than 1 min; Long-term memory is different from the former two. It can be memorized for a long time after the memory is consolidated, which is divided into implicit memory and explicit memory. Implicit memory can be recalled due to some external factors. Explicit memory is conscious memory, and relative implicit memory is specific [2].

Human visual cognition is highly related to many aspects of psychology, such as perception, attention, and memory. Memory is the core of visual cognition. which varies with many factors mutating [3]. In terms of visual communication, there are currently quite a few studies exploring the relationship between memory and the various factors, such as the placement of information and text, segmentation, group classification, word-level proportion and color, and even the reader's emotions and life experience. Among them, color is an important factor that affects vision. Color can be the color of the reading environment, the background color of the information interface, the color annotation of the information picture, or the color of words. These are factors that will affect reading fluency and then affect attention and memory. Taking segmentation as an example, Yingying Wu's research points out that hierarchical segmentation through the structure of context can enable readers to concentrate and increase reading fluency [4].

2.2   Literature Review on the Relationship Between Attention and Memory

Studies have pointed out that attention is the cognitive process of selecting information available in the environment. When we focus on certain pieces of information, our cognitive system has chosen and focused on a certain amount of information for processing. Being aware of certain stimuli increases the likelihood that information will be stored in the brain. In other words, information we pay more attention to is easier to remember than information we don't pay attention to or ignore [3].

There have been many related researches on memory previously. Some scholars have concluded mnemonics that can help people to remember quickly. There are also a lot of relevant teaching books in the market. William James proposed in 1890 that the memory is named short-term memory and long-term memory according to the definition of dualism in the "The Principles of Psychology". At present, sensory memory has been added before short-term memory. The memory time of sensory memory is much shorter than that of short-term memory. Short-term memory is affected by attention, so is sensory memory especially. Attention has also been studied for a long time, with high levels of attention-directed sight having a positive effect on short-term memory and even allowing the brain to store information into long-term memory [8].

The above literature shows that improving attention can effectively improve memory. Therefore, it can be concluded that attention is highly positively correlated with memory capacity in information storage.

2.3   The Effect of Font Size on Reading Memory

At present, some studies have pointed out that in children's reading and learning, the text in the readings takes the font size as a key reminder. The larger font can bring better learning efficiency for children and enable them to remember the key points of the enlarged font more accurately [9].

X Wang found that text on computers and tablets should have a moderate font size. The font sizes of 18pt and 24pt are the most conducive to learning, and meanwhile save learners' time of JOL. At the same time, designers should avoid too small or too large font, such as less than 12pt or more than 36pt, so as not to bring too much cognitive burden to learners and hinder learning [10]. Research by Rhodes and Castel points out the effect of font size of information on JOL (Judgment of Learning). JOL refers to that learners use subsequent memory tests to judge their memory capacity based on the previously read texts. Their study assessed the efficiency of participants' memory of single words presented in large font (48pt) and small font (18pt). The results showed that the information presented with large and small font had no significant effect on the JOL of the overall reading because it impaired the reading fluency. However, the results showed that compared with the small font, the large font can make the participants remember more easily [11].

The above literature can confirm that the font size of information affects the efficiency of memory in visual delivery. Enlargement of key text can effectively enhance the memory, but it does not necessarily allow subjects to fully remember the whole article except that the key points are remembered. In the methods that focus on the character size, there is a range of enlarging and narrowing. It is not necessary to enlarge the character size in a blind way to enhance memory, which may lead to negative effects.

2.4   The Effect of Colored Words on Reading Memory

As evidence of the relationship between attention and memory, color can improve our attention and help us remember certain information. The more attention that is stimulated to focus on information, the more chance it will be transferred to long-term memory storage [6].

The reading of information is affected by various factors, and color annotation is a method to enhance attention and memory. Color marking can make the reading process smoother, and bring a better effect on the reader in the second reading. Moreover, there was a more significant benefit in rereading when highlighting than when reading for the first time, suggesting that highlighting is more effective in memory than other simple cues, such as boldface or typeface serif [20].

Memory is divided into three processes: 1. Sensory memory 2. Short-term memory 3. Long-term memory according to the duration. If any effect of font color on the attention center of the brain is a conservative cause, it has to do with brain area specialized in processing long-term memory [21]. It can be inferred through research literature that colored character is positively related to readers' attention and memory efficiency, which means that compared with colorless character, colored one can effectively enhance attention and memory.

3   Method

The purpose of this experiment was to examine how the perceptual characteristics of the text created by words in red (16pt) and words in black (20pt, 28pt, and 40pt) affected reading memory among participants.

3.1   Design of Vocabulary Samples

Participants were asked to memorize 16 Chinese two-character vocabularies arranged on a computer screen to collect their memory effects for the perceived features of different texts. In the following, we will introduce the rules for word matrix arrangement we designed and word selection for our experiment.

Selections of Vocabulary Samples. In the experiment, since the influence of the participants' familiarity to characters was considered, Chinese vocabulary was selected from Word List with Accumulated Word Frequency in Sinica Corpus [26]. The word frequency of the top 80 vocabularies was filtered out by nouns. The font adopted the figure dragon font of the Etian Chinese system, the reason is that the font is the most common font in the Chinese computer DOS system. The selected regular script is one of the earliest developed and most commonly used Chinese font, which has been formulated and implemented by the Ministry of education.

The experiment took 80 two-character vocabularies listed as an example (see Fig. 1). Under the control environment of fixed light source, line-of-sight distance, and apparent height distance, the fonts in black and red on white background in different font sizes were compared.

序號	詞名	詞性	詞頻
1	我們	名詞	18151
2	他們	名詞	8818
3	問題	名詞	6683
4	台灣	名詞	6414
5	學生	名詞	5523
6	公司	名詞	5418
7	社會	名詞	5282
8	目前	名詞	4867
9	現在	名詞	4236
10	國家	名詞	3550
11	世界	名詞	3518
12	方式	名詞	3362
13	環境	名詞	3276
14	孩子	名詞	3201
15	網路	名詞	3092
16	日本	名詞	3061
17	中心	名詞	3042
18	地方	名詞	2990
19	市場	名詞	2950
20	老師	名詞	2871
21	學校	名詞	2857
22	方面	名詞	2658
23	同時	名詞	2640

Fig. 1. Word list with accumulated word frequency in Sinica Corpus. (The experiment will use the top 80 two-character noun vocabularies from the list.)

The Font Size of Text Samples. In this paper, pt (point) was used as the text font unit in the experiment, and the comparison was made among four types of the font size of 16pt, 20pt, 28pt, and 40pt (see Fig. 2).

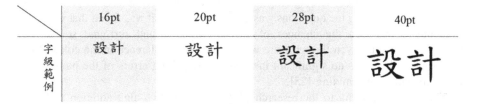

Fig. 2. The font size of vocabulary samples in Chinese.

Arrangement of Vocabulary Samples. In each group, 16 words were arranged in 4 * 4 and displayed in the middle. The traditional reading direction will be caused by the distribution area of the memorized words unbalanced. Therefore, the arrangement of vocabulary samples was designed that each row and each column contained a non-repeating character size (see Fig. 3).

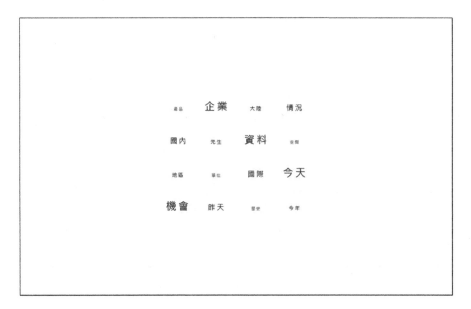

Fig. 3. Arrangement of vocabulary samples.

3.2   Design of Intervals Task

The interval time and behavior of each section in the experiment will also affect the experimental results. For example, the experiment conducted by Sarah R. Allred and

Maria Olkkonen is to study the human memory of the color of objects after the passage of time and to explore whether humans' memory of color is affected by a period of time and life experience. Brightness or chroma in memory may vary because of the passage of time. In order to study the relationship, participants were asked to observe different colors in the same space. The lighting on the object was controlled, then participants' hesitation in answering the questions was observed. Finally, it was found that when the time interval between the changes of different lighting and surround sound was extended to 10 min, even though there was a significant difference in the color of the background, there was no significant increase in the pairing errors of the participants due to memory color mixing [25].

Therefore, according to the research [25], 20 simple two-digit addition and subtraction arithmetic questions were added in the intervals of each section of the experiment in our study (Fig. 4). This interval task was used to enable the participants in each section in the experiment not to be affected by the memory of the previous process so as to guarantee the objectivity of the experimental data, and what representations of the words that remained in the subject's memory after the memory faded was observed.

$$3 + 9 = \qquad 15 + 9 =$$
$$8 + 8 = \qquad 13 + 8 =$$
$$16 - 7 = \qquad 16 - 5 =$$
$$1 + 10 = \qquad 17 + 19 =$$
$$15 + 0 = \qquad 15 + 8 =$$
$$16 - 8 = \qquad 10 + 38 =$$
$$0 + 13 = \qquad 16 + 13 =$$
$$19 - 3 = \qquad 16 - 13 =$$
$$17 + 0 = \qquad 17 + 59 =$$
$$18 - 7 = \qquad 13 + 82 =$$
$$7 + 12 = \qquad 17 + 14 =$$
$$8 + 10 = \qquad 18 + 73 =$$

Fig. 4. Design of intervals task.

3.3   Main Task

Participants. Participants were 10 adults. All participants had rapid naming, reading and verbal abilities in the average range, based on standardized measures.

Materials. In this study, MacBook Pro (15-inch, 2017) was taken as the area sample of the experiment and the corrector (i1 DISPLAY PRO) was used for color correction. The background was white. The color of the text included black and red. The computer was placed in a black portable studio of 60 * 60 * 60(cm) and access to the lab was controlled so that the color, light, and sound of the reading environment would not interfere with reading and memory.

Procedure. The experiment was conducted in 5 rounds. A question contains 16 vocabularies (4 16pt red words, 4 20pt black words, 4 28pt black words, and 4 40pt black words), and the memory time was one minute. At the end of the one-minute memory time, participants would be asked to perform interval tasks, and fill in the two-digit addition and subtraction calculation questions as shown in (Fig. 4). The time of the interval task was open-ended. After the completion of the interval task, the participants were asked to fill in the words remembered in the memory questions mentioned above. The participants were asked to recall as many words as possible. This stage is also not time-limited, if the participants forget how to spell the words they remembered, they can answer and describe in pinyin or by a variety of expressions. After the answer was completed, one round of the test was completed. In the case of memory fatigue, the test participants can request rest in the interval of two rounds. The steps of the 5 rounds are the same, but the questions of the task are different. The experiment is finished when all the questions are completed.

After the experiment is completed, a simple recording interview was conducted for each subject to record the memory pattern and the state of personal recollection during the response to examine whether there was any difference in the responses and feelings of the test subjects, and to examine whether there was any room for improvement of the experiment to provide more precise results to explain our hypothesis.

4 Result

4.1 Experimental Results

The experimental results showed increasing the font size was more effective in increasing the reading memory score than changing the font color, especially compared to vocabulary in red (16pt), when black vocabulary increased to 40pt. However, vocabularies in black, 20pt and 28pt respectively, had no significant difference in memory scores.

After five rounds of experiments, there were 81, 94, 92, and 104 vocabularies are remembered in 16pt, 20pt, 28pt, and 40pt respectively. The final results of the five rounds showed that when the font size was enlarged more than 2 times, it was obviously easier to remember compared with the red ones. One thing that should be noticed was that according to the observations in each round, it was found that the experimental result of the first round alone showed that the number of vocabularies in red (16pt) memorized was greater than the other three larger black ones. In the other four rounds, the number of words remembered in 20pt, 28pt, and 40pt was slightly or significantly more than 16pt words.

4.2   Feedback from Participants

The participants were asked whether they had memory fatigue and whether they needed a rest from the end of each round to the beginning of the next round. Three of the ten participants proposed a pause at the end of the third round, and it was generally considered that no rest was needed during the experiment. However, according to the experimental results, the participants who made the rest request performed significantly better than those who did not. The results showed that memory peaked in the second round, but plummeted to its lowest in the third round. Interviews after the experiment also showed that the participants after the break were less likely to experience memory overlap which most of those who did not request a break began to experience in subsequent experiments. The error rate for answering unselected words also peaked in the third round, and most of the participants themselves did not notice the phenomenon and did not request rest.

Based on the above findings, we believe that in future experiments, we need to extend the interval between each round and add different memory interval tasks. The three participants who proposed to take a break mentioned in the interview that during the process of taking a break, they would perform better in the following memory due to the daily conversations with us had a significant relaxing effect.

Therefore, in future experiments, simple fun games will be added as interval tasks between rounds, in addition to relaxing the participants, but also carrying out memory shuffling to reduce the experimental variables per round. In addition, most participants responded that the vocabularies in each round could be logically connected into sentences for better memory effects. Despite the fact that the result of the experiment showed the different memory effects caused by the change of the font size and the font color; However, in order to more carefully control the experimental variables, the sample presentation of each round in future experiments will be more carefully selected Chinese two-character noun vocabularies that have no relation to each other.

5   Discussion and Conclusion

In this study, a series of Chinese double-character vocabularies with different font sizes and font colors were used to test and compare the methods, to find out which is better for readers to memorize the key points we highlighted in Chinese reading. The results showed that in an ideal environment, the memory effect obtained by enlarging the font size was generally better than coloring the text. Color-coded words could surely achieve a certain effect in attracting attention, but only when the character size was almost the same could they have an obvious effect on the deep storage of information in the brain.

Future experiments would include the following methods, namely, the different typesetting of vocabulary samples, the number of vocabulary samples presented at one time, and the order of vocabulary sample presentation, which can examine the relationship between these ways to improve the effect of vocabulary memory from more aspects.

References

1. The Guide to Public Signage Design. http://lugang.chcg.gov.tw/Codes/Upload/Files/公共標示符碼.pdf. Accessed 15 Jan 2020
2. Tulving, E.: Episodic and semantic memory. Organ. Mem. **1**, 381–403 (1972)
3. Dzulkifli, M.A., Mustafar, M.F.: The influence of colour on memory performance: a review. Malays. J. Med. Sci. MJMS **20**(2), 3 (2013)
4. Wu, Y., Yang, X., Yang, Y.: Importance conveyed in different ways: effects of hierarchy and focus. J. Neurol. **47**, 37–49 (2018)
5. Louis, A., et al.: Discourse indicators for content selection in summarization (2010)
6. Sternberg, R.J.: Cognitive Psychology, 5th edn. Wadsworth Cengage Learning, Belmont (2009)
7. James, W.: The Principles of Psychology. Henry Holt and Company, Inc., New York (1890)
8. Titchener, E.: Lectures on the elementary psychology of feeling and attention (1908)
9. Halamish, V., Nachman, H., Katzir, T.: The effect of font size on children's memory and metamemory. Front. Psychol. **9**, 1577 (2018)
10. Wang, X., Wang, Z., Li, H., Zhou, W.: An eye-movement study on text font size design rules in the digital learning resources (2017)
11. Rhodes, M.G., Castel, A.D.: Memory predictions are influenced by perceptual information: evidence for metacognitive illusions. J. Exp. Psychol. Gen. **137**, 615–625 (2008)
12. Farley, F.H., Grant, A.P.: Arousal and cognition: memory for color versus black and white multimedia presentation. J. Psychol. **94**(1), 147–150 (1976)
13. Birren, F.: Color Psychology and Color Therapy. McGraw-Hill, New York (1950)
14. Greene, T.C., Bell, P.A., Boyer, W.N.: Coloring the environment: hue, arousal, and boredom. Bull. Psychon. Soc. **21**, 253–254 (1983)
15. Morton, J.: Why color matters (2010)
16. Moore, R.S., Stammerjohan, C.A., Coulter, R.A.: Banner advertiser-web site context congruity and color effects on attention and attitudes. J. Advert. **34**(2), 71–84 (2005)
17. White, J.V.: Strathmoor Press, State of Ohio (1997)
18. Ludlow, A.K., Wilkins, A.J.: Case report: color as a therapeutic intervention. J. Autism Dev. Disord. **39**(5), 815–818 (2009)
19. De Giorgio, A.: Enhancing motor learning of young soccer players through preventing an internal focus of attention: the effect of shoes colour. PLoS ONE **13**(8), e0200689 (2018)
20. Yeari, M., Oudega, M., van den Broek, P.: The effect of highlighting on processing and memory of central and peripheral text information: evidence from eye movements. J. Res. Read. **40**(4), 365–383 (2016)
21. Haynes, E.: The effect of text color and text grouping on attention and short-term recall memory (2017)
22. McConnohie, B.V.: A study of the effect of color in memory retention when used in presentation software (1999)
23. Dzulkifli, M.A., Mustafar, M.F.: The influence of colour on memory performance: a review. Malays. J. Med. Sci. MJMS **20**(2), 3 (2012)
24. Wang, H.-W.: The impact of color on short term memory of english letters (2017)
25. Allred, S.R., Olkkonen, M.: The effect of memory and context changes on color matches to real objects. Atten. Percept. Psychophys. **77**(5), 1608–1624 (2015)
26. Word List with Accumulated Word Frequency in Sinica Corpus. http://words.sinica.edu.tw/sou/sou.html. Accessed 17 Oct 2019

A Study on Search Performance and Threshold Range of Icons

Aiguo Lu and Chengqi Xue[⊠]

School of Mechanical Engineering, Southeast University,
Nanjing 211189, China
{luaiguo, ipd_xcq}@seu.edu.cn

Abstract. Icon lightness coding in digital interfaces has a very important impact on interface design. How to reduce visual information confusion and improve cognitive performance through icon feature coding is a subject to be studied in the optimization of interface design. According to previous research, icon composition is subdivided into lines and planes, while background is subdivided into positive background, negative back ground, and no background. The materials used in this study are the positive background of graphical icons, which were represented by color blocks in this study. Employing questionnaire as behavioral experimental tool, the main objective of this paper is to explore threshold range of the graphical icons' Lightness levels in Dark mode and Light mode. In dark mode, the behavioral experiment showed that when the lightness of the color blocks changes within the range of a1(10,-0,-0)–a5(50,-0,-0), the difference between the lightness of the color blocks and the background increases faster, and the difference increases slowly in the range of a6(60,-0,-0)–a10(100,-0,-0). In light mode, the behavioral experiment showed that when the lightness of the color blocks changes within the range of b1(90,-0,-0)–b3(70,-0,-0), the difference between the lightness of the color blocks and the background increases faster, and the difference increases slowly in the range of b4(60,-0,-0)–b10(0,-0,-0). Through this experiment, we can reduce threshold range of icon' background lightness levels of in Dark mode and Light mode. Knowledge gained from the current study could serve as a reference when icons are researched or designed in the future.

Keywords: Computer interface design · Graphical icons · Lightness · Behavioral experiments · Dark mode

1 Introduction

Whether operating graphic users' interfaces for computers or hand-held devices, users need to click visual icons to execute program instructions. (Nasanen and Ojanpaa 2003). It is important and common for users to search for an graphical icon in a complex users interface. The performance of search tasks mainly depend on the difference of the target and the distractors (Duncan and Humphreys 1989; Wolfe 1998). The search time will increases if the difference between the target and the distractors is small (Nasanen et al. 2001). To make it easy for users to find the target icon, designers should ensure that the target is very different from the distractors, coding them with

© Springer Nature Switzerland AG 2020
D. Harris and W.-C. Li (Eds.): HCII 2020, LNAI 12186, pp. 62–68, 2020.
https://doi.org/10.1007/978-3-030-49044-7_6

different distinctive visual characteristics. Among many forms of information encoding, lightness coding in digital interfaces has a very important impact on interface design. How to reduce visual information confusion and improve cognitive performance through icon feature coding is a subject to be studied in the optimization of interface design. The dark mode in 2019 brings challenges to interface design, not only to adapt to the low power consumption mode of smart phones, but also to reduce the visual fatigue caused by light under different lighting environments, so explore the difference in brightness of icon color matching scope is a valuable subject. As early as 1997, Kingdom found that the brightness of the visual space area is not only related to the brightness of the area, but also depends on the luminances of adjacent regions (Kingdom 1997). simultaneous brightness contrast (SBC) is a well-known demonstration for this finding, which usually described as a homogeneous brightness change within an enclosed test patch such that a gray patch on a white background looks darker than an equi luminant gray patch on a black background.

Many domestic and foreign scholars have begun to pay attention to the effects of dark mode and lightness coding on the cognitive performance of digital interfaces. Shieh and Ko (2005) found that icons with a black background and a target color of red had the highest aesthetic preference. After researching the cognitive performance of different colors on a black background, it found that the cognitive performance of yellow is the highest (Zhang et al. 2011). There is still a lot of research on cognitive performance against black background, especially related to study of hue. In addition to the above-mentioned researches on dark mode, research on lightness coding is also common in the field of human-computer interface. WU and other experiments analyzed the cognitive performance of hue, lightness, and saturation in the combination of foreground and background colors. They considered that hue affects cognitive speed and visual preference, but the influence is less than lightness and saturation. Moreover, Ahlstrom and Arend (2005) others use hierarchical lightness coding to improve the design of aviation combat display interfaces.

In this paper, we used background color blocks of graphical icons as experimental materials and made a questionnaire to investigate threshold range of icons' lightness levels by the analysis of questionnaire results. We provided a framework work to improve lightness coding design for icons.

2 Methodology

2.1 Materials

The experimental material are the background color blocks of the graphical icons. The color blocks in behavioral experiment consisted of 20 different background lightness, 10 levels in Dark mode and 10 levels in Light mode. All color blocks used were 32 £ 32 pixels in size and saved in bitmap format with a color depth of 24 bits. As is shown in Fig. 1, In Dark mode, the background lightness of the color blocks varies from 10 to 100 in 10 CIE LAB units with a total of 10 levels. The color of the color blocks is expressed in Lab color mode, which are a1(10,-0,-0), a2(20,-0,-0), a3(30,-0,-0), a4(40,-0,-0), a5(50,-0,-0), a6(60,-0,-0), a7(70,-0,-0), a8(80,-0,-0), a9(90,-0,-0), a10(100,-0,-0), and the background color is expressed as Lab (0, -0, -0). Table 1 shows the rankings of the lightness levels for the Dark mode. As is shown in Fig. 2, In Light mode, the background lightness of the color blocks varies from 0 to 90 in 10 CIE LAB units with a total of 10 levels. The background color of the color blocks is expressed in Lab color mode, which are b1(90,-0,-0), b2(80,-0,-0), b3(70,-0,-0), b4(60,-0,-0), b5(50,-0,-0), b6 (40,-0,-0), b7(30,-0,-0), b8(20,-0,-0), b9(10,-0,-0), b10(0,-0,-0), and the background color is expressed as Lab (100,-0,-0).Table 2 shows the rankings of the lightness levels for the Light mode.

Fig. 1. The lightness of color blocks in Dark mode. The lightness levels of the color blocks from left to right are a1–a10, all color blocks were 32 £ 32 pixels in size and background is 96 £ 96.

2.2   Subjects

Of the 30 students who participated in this experiment, 15 were males and 15 were females. Both participants were university students aged between 22 and 26 years, with normal or corrected vision and without color blindness or color weakness. Before participating in the experiment, all participants were provided informed consent, and their personal details like name, gender, age, grade, major and vision were recorded in a database. During the behavioral experiment, Each student received a questionnaire with 20 questions. There was no time requirement for participants to complete the questionnaire.

2.3   Experimental Equipment and Experimental Procedures

Behavioral experiment were conducted in the form of questionnaires to record behavioral data. First, the study used the questionnaire star and the experimental materials to create a questionnaire, and then send the questionnaire to the mobile phones of the 30 participants. The questionnaire had 20 questions, of which 10 were about dark mode and 10 were about light mode and the pictures that showed in each question were made in Adobe illustrator.

The details of behavioral experimental procedure were as follows. First, the initial brightness value of the matching stimulus was set in Adobe illustrator, 10 levels in Dark mode and 10 levels in Light mode. All stimuli were presented as color blocks. Then these color blocks add a white and black background in Adobe illustrator. After that, a questionnaire was made by questionnaire star (app) and generated a URL link to be send to participants. The completed questionnaire is shown in the Fig. 3.

To begin behavioral experiment, Participants had to open the link of questionnaire, shown on their mobile phones or computers, by pressing URL. Then the explanation given by the researchers about the questions was presented under title. Participants were asked to complete the 20 questions after reading the instructions carefully. When the question was displayed in the center of the screen, the participants need to judge the

Table 1. The lightness levels for the Dark mode

Index	Lightness levels	(L,a,b)	Background (L,a,b)
Lightness	a1	(10,-0,-0)	(0,0,0)
	a2	(20,-0,-0)	
	a3	(30,-0,-0)	
	a4	(40,-0,-0)	
	a5	(50,-0,-0)	
	a6	(60,-0,-0)	
	a7	(70,-0,-0)	
	a8	(80,-0,-0)	
	a9	(90,-0,-0)	
	a10	(100,-0,-0)	

Fig. 2. The lightness of color blocks in Light mode. The lightness levels of the color blocks from left to right are b1–b10. All color blocks used were 32 £ 32 pixels in size and background is 96 £ 96.

Table 2. The lightness levels for the Light mode

Index	Lightness levels	(L,a,b)	Background (L,a,b)
Lightness	b1	(90,-0,-0)	(100,0,0)
	b2	(80,-0,-0)	
	b3	(70,-0,-0)	
	b4	(60,-0,-0)	
	b5	(50,-0,-0)	
	b6	(40,-0,-0)	
	b7	(30,-0,-0)	
	b8	(20,-0,-0)	
	b9	(10,-0,-0)	
	b10	(0,-0,-0)	

difference between the stimulus color blocks and the black or white background. After judging, they had to press the corresponding button. The left-most button indicates that the difference is not obvious and the score record 1. The right-most button indicates that the difference is very obvious and score record 4. The results and the completed questionnaires occurring during this process were recorded by computer. The experiment was conducted in a single session of approximately 5-min duration. Figure 4 depicts that experimental process.

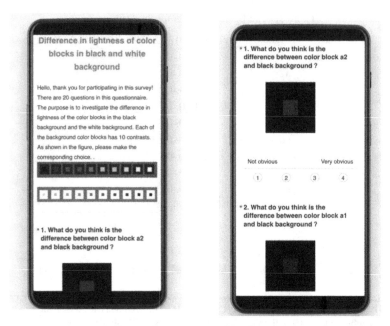

Fig. 3. The questionnaire

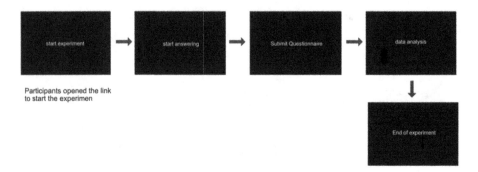

Fig. 4. Experimental flowchart

3   Analysis and Results

3.1   Behavioral Data Analysis

Behavioral data of this study is score of each question. As shown in Fig. 5, A score of 1 indicates that the difference between color blocks and the background is not obvious, a score of 2 indicates that the difference between color blocks and the background is generally obvious, a score of 3 indicates that the difference between color blocks and the background is obvious, and a score of 4 indicates that the difference between color blocks and the background is very obvious.

not obvious   generally obvious   obvious   very obvious

①   ②   ③   ④

Fig. 5. Level 4 scale

Figure 6 illustrates that when the lightness of the color blocks changes within the range of a1(10,-0,-0)–a5(50,-0,-0), the difference between the lightness of the color blocks and the background increases faster, and the difference increases slowly in the range of a6(60,-0,-0)–a10(100,-0,-0) in dark mode. Moreover, in Light mode, the behavioral experiment showed that when the lightness of the color blocks changes within the range of b1(90,-0,-0)–b3(70,-0,-0), the difference between the lightness of the color blocks and the background increases faster, and the difference increases slowly in the range of b4(60,-0,-0)–b10(0,-0,-0).

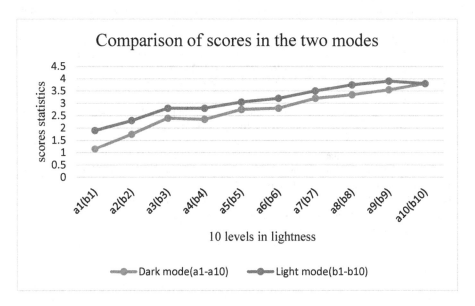

Fig. 6. Comparisons of scores in the two modes

4   Conclusion

There were several findings in the experiment. First, the results showed that when the lightness difference between the color block and the background was the same, the participant thought that the color block was more different from the background in Light mode. Besides, in dark mode, when the lightness of the color blocks changes within the range of a1(10,-0,-0)–a5(50,-0,-0), the difference between the lightness of the color blocks and the background increases faster. That is, the difference in brightness from the background is in the range of 10–50. However, in Light mode, when the lightness of the color blocks changes within the range of b1(90,-0,-0)–b3(70,-0,-0), the difference between the lightness of the color blocks and the background increases faster. That is, the difference in brightness from the background is in the range of 10–30.

Our investigation is the innovation into the statistical study of the recognition and memorability of icons in human-interactive systems. The results would provide valuable guidance to further studies and future design of icons in digital interface.

Acknowledgement. This paper is supported by National Natural Science Foundation of China (No. 71871056).

References

Nasanen, R., Ojanpaa, H.: Effects of image contrast and sharpness on visual search for computer icons. Displays **24**, 137–144 (2003)

Duncan, J., Humphreys, G.W.: Visual search and stimulus similarity. Psychol. Rev. **96**, 433–458 (1989)

Wolfe, J.M.: What can 1 million trials tell us about visual research? Psychol. Sci. **9**, 33–39 (1998)

Nasanen, R., Ojanpaa, H., Kojo, I.: Effect of stimulus contrast on performance and eye movements in visual search. Vition Res. **41**, 1817–1824 (2001)

Kingdom, F.A.A.: Simultaneous contrast: the legacies of Hering and Helmholtz. Perception **26**, 673–677 (1997)

Shieh, K.K., Ko, Y.H.: Effects of display characteristics and individual differences on preferences of VDT icon design. Percept. Mot. Skills **100**(2), 305–318 (2005)

Zhang, J., Gao, Z., Liu, X., et al.: Comparative research on cognitive performance for different color visual targets on black background. J. Shanxi Med. Univ. **42**(7), 542–550 (2011)

Wu, J.H., Yuan, Y.: Improving searching and reading performance: the effect of highlighting and text color coding. Inf. Manag. **40**(7), 617–637 (2003)

Ahlstrom, U., Arend, L.: Color usability on air traffic control displays. In: Proceedings of the Human Factors and Ergonomics Society 49th Annual Meeting, Orlando, FL, United states, pp. 93–97 (2005)

Effects of Digitally Typesetting Mathematics on Working Memory

Francis Quinby[1], Marco Pollanen[1], Michael G. Reynolds[2],
and Wesley S. Burr[1(✉)]

[1] Trent University, Department of Mathematics, Peterborough, ON, Canada
`wesleyburr@trentu.ca`
[2] Trent University, Department of Psychology, Peterborough, ON, Canada

Abstract. Typesetting mathematics proves challenging for many people. In this paper we discuss the results of an experiment comparing working memory interference while transcribing mathematical expressions using both digital applications and handwriting. To measure interference, we used a head-mounted eye-tracker to count the number of times individuals had to refer to a presented expression while transcribing said expression. We also compared the duration of fixations on the mathematical expression being replicated to measure the amount of information encoded into memory during each reference to the stimulus. Results showed a significant difference in the number of transitions necessary to complete the transcription task, such that handwriting the expressions required significantly fewer transitions than digital typesetting. No significant difference was found regarding the average duration of fixation on the stimulus, indicating that participants encoded similar amounts of information during these fixations, regardless of the transcription condition. This supports the hypothesis that digitally typesetting mathematics causes significant working memory interference. These results help to show why typesetting may cause difficulty for individuals, and should be taken into consideration when working to improve the user interfaces of typesetting applications.

Keywords: Eye-tracking · Working memory · Mathematical software · Software user interfaces

1 Introduction

Typesetting mathematics digitally in an efficient and effective manner presents a challenge to many individuals. Though many different programs and applications exist with the capability to effectively transcribe mathematical notation, in most university level mathematics courses handwritten assignment submissions are still the norm [8]. Though a number of studies have shown that university mathematics students feel that the learning and adoption of LaTeX or another digital typesetting tool provides them with benefits in their understanding of the material and ability to communicate content more effectively [2,6,8], these

© Springer Nature Switzerland AG 2020
D. Harris and W.-C. Li (Eds.): HCII 2020, LNAI 12186, pp. 69–80, 2020.
https://doi.org/10.1007/978-3-030-49044-7_7

technologies can present a steep learning curve which many students do not over-come. However, as the world becomes more globally connected, online courses are becoming increasingly prevalent and are important to making education accessible to a greater number of individuals. These types of courses are also becoming mandatory in some places, such as the secondary school system for the province of Ontario, Canada. This makes the effective and efficient digital communication of mathematics more important than ever. Even on university campuses, new teaching methods are being developed which require students to be able to communicate mathematics digitally, such as online office hours, which have been shown to be more beneficial for students than regular office hours [5].

Though there is a clear difference in typesetting mathematics digitally instead of handwriting the material, few studies have directly compared these processes and their different effects on human cognition. Furthermore, a number of digital mathematical typesetting programs and applications employ different models to represent mathematical structure. For example, the fraction, $\frac{1}{2}$, is inserted and represented differently when using the traditional Microsoft Word Equation Editor versus LaTeX code. To examine these differences, we designed a study to compare the effects on working memory and efficiency of typesetting mathematics digitally to those of handwriting similar mathematical material.

The study was also designed to give a baseline comparison of the effects caused by typesetting programs which use different models to represent mathematical structure. Microsoft Equation Editor uses a 1-dimensional model in which mathematical structures inserted from a pallete contain spaces in which to insert the components of the structure, such as the numerator and denominator of an inserted fraction. By comparison, the online application Mathematics Classroom Collaborator, or MC^2 uses a 2-dimensional, free-form model which allows users to insert symbols onto a blank canvas workspace. These symbols can then be dragged and dropped into the appropriate location and resized. We designed this study to give us a comparison of the performance of these two models when transcribing linear expressions, which are handled similarly by both models, to give us a baseline comparison of the interfaces, allowing us to more accurately attribute differences to the underlying models in future studies with expressions containing multi-dimensional mathematical structures.

2   Methods

To provide a baseline comparison of the effects caused by the use of different typesetting models for mathematical structure, Microsoft Equation Editor and an online application, Mathematics Classroom Collaborator (MC^2) were compared to handwriting. Word uses a structure-based input model in which elements in mathematical expressions are input from palettes and represented as nested boxes, e.g., a fraction would be represented as two boxes separated by a fraction bar. By comparison, MC^2 uses a two-dimensional, free-form model which allows users to insert symbols onto a blank canvas workspace with additional moving and resizing.

The experiment was a $2 \times 2 \times 3$ factorial design in which participants ($n = 36$) were instructed to transcribe 12 classes of expressions using one of the following three directed methods: pen and paper, Word, and MC^2. 36 participants were recruited for this experiment from first year service-level mathematics courses at Trent University in Peterborough, Ontario, specifically a statistics course and a calculus course for non-mathematics majors. Results from six participants were excluded due to experimenter error, calibration and equipment failures, and software issues with the server running MC^2.

The presented stimuli varied in length, from two variables ('short condition') to four variables ('long condition'). The presented expressions also varied in complexity: half had variables represented by Roman letters which would be found on a computer keyboard ('simple condition'), and the other half represented by Greek letters which would be found in an interface menu in the two digital models ('complex condition'). The same 36 (12×3) expressions were used in all runs of the experiment, with the order and typesetting application condition which they were to be transcribed with randomly generated for each run. Participants completed three practice trials of the Long-Complex condition prior to the experimental trials, one in each typesetting method condition. Examples of each of these four conditions are shown below in Table 1.

Table 1. Examples of the four general classes of stimuli used in the experiment.

Condition	Example
Short-Simple	$e = -6c$
Short-Complex	$\sigma = 9\gamma$
Long-Simple	$z = -3r - 2e - 2m$
Long-Complex	$\rho = -2\beta + 2\tau + 4\alpha$

Participants wore a ViewPoint EyeTracker®(Biopac Systems, Inc.) head mounted eye-tracker throughout the experiment, allowing assessment to be taken of metrics which should correlate to working memory interference through each experimental trial. Eye-tracking hardware allows for millisecond-level observations of human eye focus, and records two types of events: *fixations* (where the eye fixates for a brief period, $\sim$50 ms, on a location) and *saccades* (where the eye rapidly travels from one location to another) [11]. Extensive cognitive science research has shown that fixation duration and number correlates well with cognitive processing. Thus, for our experiment, we measured the number of transitions of the users' eye fixation location from the presented mathematical expression to the workspace. The efficiency of each model was assessed by measuring the time to complete each experimental trial. Outcome variables of time to completion and number of transitions were standardized to a per variable measurement by dividing by two for the 'short' expressions and four for the 'long' expressions.

3 Results

Prior to analysis, the eye-tracking gaze location video recordings and data had to be interpreted and gaze locations categorized and prepared for quantitative analysis. Each fixation observed was assigned a location, either on the workspace, the stimulus (the mathematical expression to be transcribed) or on an indeterminate location. The start time of each trial was recorded as the exact moment that the presented stimulus appeared on the display monitor and the end time as the start of a three second fixation on a sticker participants were instructed to focus their gaze on when they had completed the trial. Eye-tracking data allowed us to measure eye-gaze transitions from the stimulus to workspace, providing us with a count of the number of times the participant had to refer to the stimulus to replicate it, which in turn provided us with a measure for evaluation of working memory interference: more returns to the stimuli imply more occurrences of "refreshing" working memory with the stimuli to be transcribed. From the eye-tracking data we were also able to obtain a measure of mean duration of fixations on the workspace during experimental trials, which provides us with a measure of the cognitive load of working with the different interfaces, as past research has shown mean fixation duration to increase with increased task load [7]. Finally, efficiency was assessed by measuring the amount of time participants required to complete a trial. All analysis was done in R 3.6.0 [9], with figures created in ggPlot2 [12].

3.1 Working Memory Interference

We found that the digital interface models show a level of interference with users' working memory. The transcription method showed a significant main effect on the number of transitions per variable ($F_{2,28} = 23.47, p < 0.001$). Contrasts showed that digital transcription methods resulted in a significantly higher number of transitions per variable than handwriting ($t_{29} = 5.68, p < 0.001$), with no significant difference found between Word and MC^2 ($t_{29} = 1.62, p = 0.12$). None of the interactions were found to be significant in this analysis, indicating that the effects of transcription method were consistent throughout all other experimental conditions, and a success for our randomization procedure. Summary statistics are shown in Table 2.

Table 2. Number of transitions per variable in presented stimulus, by transcription method. Note the large difference between Pen & Paper and the two digital methods.

Transcription method	Mean	SE
Pen & Paper	1.226	0.053
MC^2	1.588	0.065
Word Equation Editor	1.534	0.075

The main effect of transcription method on the mean dwell time on the stimulus was not significant ($F_{2,28} = 1.137, p = 0.328$), implying that there was no difference in the amount of time participants spent encoding information from the presented stimulus for transfer to their workspace. This leads us to assume that the same amount of information is being encoded into working memory during each fixation on the stimulus during the use of all transcription methods. These results are shown in Table 3.

Table 3. Mean dwell time on stimulus during experimental trials, by transcription method. Note the lack of large difference between the methods.

Transcription method	Mean	SE
Pen & Paper	0.677	0.086
MC^2	0.618	0.093
Word Equation Editor	0.635	0.100

The combination of these two results shows that users, on average, encode the same amount of information into working memory during each fixation on the stimulus when transcribing a mathematical expression, regardless of transcription method. However, due to the fact that more transitions between the stimulus and the workspace are required to complete the transcription task when using the digital interfaces, we can see that there is an element of these interfaces which interferes with what individuals are storing in working memory. This is expected from previous research [1,4].

3.2   Cognitive Load

To examine the question of why this working memory interference may be occurring, we also assessed cognitive load through a comparison of the mean Workspace Fixation Duration (WFD): the average duration of fixations in the location of the workspace. These workspaces were a laptop for the MC^2 and Word Equation Editor trials, and a blank piece of paper for the handwritten trials. Differences were found in this measure, with a significant main effect of transcription method on the mean WFD ($F_{2,28} = 4.57, p < 0.05$) as well as significant main effects of expression length ($F_{1,29} = 35.54, p < 0.001$) and expression complexity ($F_{1,29} = 168.8, p < 0.001$). Contrasts indicate that there was a slight difference in the mean WFD between the digital transcription methods and the handwritten condition ($t_{29} = 2.16, p < 0.05$) but that there was less of a difference between the Equation Editor and the MC^2 conditions ($t_{29} = 2.02, p = 0.052$). The way in which these differences manifest themselves becomes more apparent when we consider the interactions in the next paragraph. Contrasts also indicated that the mean WFD was greater for complex expressions ($M = 1.425, SE = 0.163$) than simple expressions ($M = 1.011, SE = 0.176$) and greater for short expressions ($M = 1.321, SE = 0.172$) than long expressions ($M = 1.115, SE = 0.167$). Descriptive statistics by transcription method are shown in Table 4.

Table 4. Mean Workspace Fixation Duration (WFD) during experimental trials, by transcription method.

Transcription method	Mean	SE
Pen & Paper	1.112	0.171
MC^2	1.307	0.160
Word Equation Editor	1.235	0.177

The most interesting effects in the analysis of WFD may lie in the significant interactions. Interactions were significant between length and transcription method ($F_{2,28} = 7.19, p < 0.01$) as well as between complexity and transcription method ($F_{2,28} = 29.01, p < 0.001$). These two interactions are highlighted in Fig. 1 below.

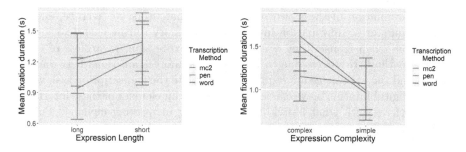

Fig. 1. Interaction plot of length by transcription method (left) and complexity by transcription method (right) on mean Workspace Fixation Duration.

We can see from the interaction plot in Fig. 2 (right) that in the simple trials, the mean WFD was less for the handwritten trials than for the digital trials. As increased fixation duration has been shown to be related to increased cognitive load, this indicates that cognitive load was greater when typing out simple mathematical expressions using the digital transcription applications then when handwriting them. However, this relationship is reversed for the complex trials, in which the mean WFD was higher for the two digital applications than the handwritten condition implying greater cognitive load. While this effect for the complex condition is expected considering the need to navigate the menus to find the Greek variables in Equation Editor and MC^2, it is interesting that it appears cognitive load is less when typing out the simple expressions than when handwriting them. This might be explained by the composition of the simple expressions and the ability for users to find all symbols on the keyboard along with the high level of familiarity most individuals have with typing on a computer.

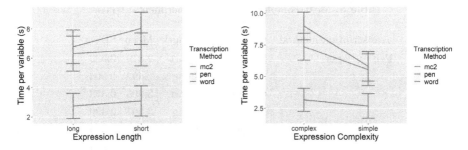

Fig. 2. Interaction plot of length by tool (left) and complexity by tool (right) on time per variable.

The interaction effect between length and transcription method of WFD is also interesting as we can see that for all three methods the mean WFD is lower in long expressions than short expressions. This indicates that there is "start up cost" or an increase in cognitive load at the start of the transcription process for all three methods. Interestingly WFD shows the greatest decrease from short expressions to long expressions for the handwritten condition indicating that this "start up cost" is the greatest when transcribing mathematics with pen and paper. This could be explained by the need to locate an appropriate area on the paper one which to write the expression, but further research is necessary to determine the cause.

3.3 Efficiency

In addition to evaluating the working memory interference which occurs during the transcription of mathematics by the different methods tested, we were also able to evaluate the efficiency of each method. As previously mentioned, efficiency was evaluated as the time taken to complete the transcription process normalized by the number of variables in the transcribed expression, 2 for short expressions and 4 for the long expressions.

Results showed all three main effects, length ($F_{1,29} = 27.33, p < 0.001$), complexity ($F_{1,29} = 156.2, p < 0.001$), and transcription method ($F_{2,28} = 242.1, p < 0.001$) to be significant. Contrasts show mean time for variable was less for long expressions ($M = 5.279, SE = 0.702$) than short expressions ($M = 5.904, SE = 0.667$) as well as less for simple expressions ($M = 4.680, SE = 0.714$) than for complex expressions ($M = 6.502, SE = 0.654$). Furthermore, the two-way interactions of length by transcription method ($F_{2,28} = 11.6, p < 0.001$) and complexity by transcription method ($F_{2,28}, p < 0.001$) were found to be significant as well. Descriptive statistics of completion time for each experimental condition are shown in Table 5 and the significant interactions are shown in Fig. 2, left and right panels.

Table 5. Mean completion time in experimental trials

Transcription method	Complexity	Length	Mean	SE
Pen & Paper	Simple	Short	2.870	0.629
		Long	2.492	0.504
	Complex	Short	3.305	0.573
		Long	3.007	0.499
MC^2	Simple	Short	5.861	0.644
		Long	5.236	0.804
	Complex	Short	7.325	0.679
		Long	7.381	0.581
Word Equation Editor	Simple	Short	6.551	0.691
		Long	5.070	0.682
	Complex	Short	9.511	0.590
		Long	8.487	0.676

3.4   Linear Regression Analysis on Working Memory Interference

While the analysis above shows that there is clearly working memory interference occurring in individuals during the digital transcription of mathematics, it remains unclear exactly how this interference occurs. In this context, it is important to consider how working memory functions during the completion of natural tasks.

Past research has shown that during model replication tasks, individuals do not use working memory to its full capacity and choose to minimize the memory cost. In a classic eye-tracking study in which participants replicated a model of coloured blocks, it was observed that individuals were most likely to follow a

Model → Source to pick up block → Model → Workspace to place block

pattern of fixations where the first fixation on the model would inform the participant of the correct colour of block and the second fixation on the model would inform the participant of where to place the block in the replication [3].

There are many parallels between this task and the transcription task described in this experiment. However, in the handwritten conditions, which represent the most natural method of transcribing mathematics, the mean number of transitions for short expressions was 2.43 and for long expressions, 4.96. Short expressions had 2 variables but 4 to 5 symbols, while long expressions had 4 variables but 8 to 11 symbols. This implies that individuals encode more than one symbol into working memory during each fixation on the stimulus, breaking the expression up into groups of symbols in the process. However, in the digital transcription conditions, on average, more transitions were required to complete the task. Therefore, investigating this effect more closely could provide us with insight into how the memory interference is occurring.

For the following analysis, we examine results from the long, complex experimental condition as there was the greatest variability in the number of symbols in expressions. Although all expressions had four variables present, the total number of symbols ranged from 8 to 11. Therefore, we wanted to assess whether the difference in the number of symbols within this condition had a predictive effect on the total number of transitions required to complete an experimental trial.

The mean number of transitions per variable in the pen and paper condition was closer to 1 ($\bar{x} = 1.226$), than the Equation Editor condition ($\bar{x} = 1.534$) and the MC^2 condition ($\bar{x} = 1.588$) which might provide some insight into how the presented mathematical expressions are encoded into working memory during the transcription process and how these interfaces might interfere with such. We suggest that the expression is naturally broken down into parts based on the variables in the expression, such that the equation $y = -3x - 4z$ would be transcribed in a manner similar to the following, where F indicates a fixation on the stimulus and parts of the equation indicate what is being transcribed in the workspace:

$$F \rightarrow y = \rightarrow F \rightarrow -3x \rightarrow F \rightarrow -4z$$

This expression would then have required 3 transitions to transcribe 8 symbols. If an expression such as $y = 2x - z$, an expression with 6 symbols, were to be transcribed in a similar fashion, such as that shown below,

$$F \rightarrow y = \rightarrow F \rightarrow 2x \rightarrow F \rightarrow -z$$

then this would further support this hypothesis. On the other hand, if it were shown that such an expression required fewer transitions, on average, than this would indicate that encoding into working memory during the transcription process is predicted more by the number of symbols in the expression.

To more closely examine this, we ran an additional post-hoc regression analysis, looking at the relationship between the number of symbols in the presented expressions and the number of transitions from the stimulus for trials within the long, complex condition, assuming a unique intercept for each participant. Data was fit to the model below,

$$y = \alpha + \beta_1 X_p + \beta_2 X_w + \beta_3 X_n + \beta_4 X_p X_n + \beta_5 X_w X_n + \beta_{pt} X_{pt}$$

where y represents outcome variable number of transitions, X_p, X_w represent binary variables for pen and word conditions respectively, X_{pt} represents a binary variable for each individual participant, β_{pt} represents an intercept coefficient for each respective participant, and X_n represents the number of symbols in the mathematical expressions. Fitting this model to the data resulted in the following models for the three tool conditions:

$$
\begin{aligned}
\text{Pen:} \quad y &= 4.00920 - 0.03565 X_n + \beta_{pt} X_{pt} \\
MC^2: y &= 0.27264 + 0.51153 X_n + \beta_{pt} X_{pt} \\
\text{Word:} \quad y &= -0.36935 + 0.56815 X_n + \beta_{pt} X_{pt}
\end{aligned}
$$

This regression model was shown to predict 47% of the variance in the number of transitions ($R^2 = 0.47, F_{34,235} = 8.007, p < 0.001$). The number of symbols was found to be a significant predictor for the MC^2 condition ($\beta = 0.512, p < 0.001$) and the pen condition ($\beta = -0.036, p < 0.05$), though in the pen condition the coefficient was negligible and negative which, though statistically significant, has little practical significance. Interestingly enough, number of symbols was not found to be a significant predictor for the Equation Editor condition ($\beta = 0.568, p = 0.802$), though the coefficient was similar to that of MC^2.

These results show that while writing mathematical expressions in the experiment out by hand, the symbols may be broken up and encoded into working memory in such a way that an increase in the number of symbols does not cause a need for increased fixations on the stimulus, or encoding events. On the other hand, when typesetting the expressions digitally, an increase in the number of symbols within expressions of the same number of variables does cause an increase in encoding events on the stimulus which may indicate that working memory operates in a less efficient manner in terms of what is maintained in memory while navigating the user interfaces.

Unfortunately, the resolution of the scene camera on the head mounted eyetracker used in the study was not fine enough to capture precisely what was transcribed on the paper or on the laptop following each transition from stimulus to workspace. Therefore, though this regression supports our previously stated hypothesis, it is far from conclusive. Further research on this topic should be done to examine exactly what is transcribed in the workspace following each transition to determine individuals break up the expression into parts to transfer from the model expression to the workspace.

4   Conclusion and Future Work

A link between mathematical task performance and working memory has been established in the literature [10], suggesting that using a digital interface to typeset mathematics leaves a user with fewer cognitive resources to devote to the solution of a mathematical problem. This may contribute to the need of most students when asked to typeset a mathematical document to first write out a solution by hand before typing it out using a typesetting program [2].

We can see from the results of this study that digitally typesetting mathematics causes interference with working memory performance. Furthermore, it is apparent that cognitive load is greater when using digital typesetting applications for more complex mathematical expressions which may be the cause for at least part of this interference. It seems that the two typesetting applications tested more closely resemble each other in terms of their cognitive demands than either of them resembles the cognitive load of handwriting. Therefore, if we strive to create a digital typesetting application which is as cognitively demanding or less so than handwriting mathematics there are still improvements which must be made.

The regression analysis performed might give us insight into how mathematical expressions are broken down and stored in working memory. We have shown evidence that when copying polynomial expressions (of degree one) with pen and paper, the expressions seem to be broken up one term at a time. We can also see that, if this is the case, the use of the digital typesetting applications seems to disrupt this strategy of working memory storage. While further research must be done to confirm this effect, this knowledge can prove to be useful in the future user experience design of various mathematical typesetting programs. This finding also directs future research to examine whether this effect is consistent when polynomial terms have degree greater than one.

There are also a number of limitations in this study. While the participants in this study were first-year, non-mathematics majors and can therefore be considered non-experts, we are not able to determine whether these effect are also consistent for expert users of mathematics, expert users of the typesetting applications, or a combination of both. Furthermore, the study only made use of one variation in the mathematical notation in the trial expressions which was the types of variable symbols. Further research should be designed to examine how consistent these effects are across other mathematical notation such as fractions and exponents as these structures are handled differently in various typesetting applications.

We suggest that future design of mathematical typesetting software interfaces should consider working memory interference (which this study has identified) and attempt to optimize the efficiency and usability of the applications in order to counteract the effects. The mathematical sciences have fallen behind other academic disciplines in the development of usable real-time digital tools. Simple tasks, such as transcribing notes and problem solving, are largely still done by pencil-and-paper due to the complexity of mathematical interfaces and the cognitive demands in using them. However, this study shows that with the simplest expressions it appears that the cognitive load can be less when using software rather than pen and paper. This perhaps suggests that there may be room to improve mathematical user interfaces so that they are better for inputting a wider range of expressions. We made a first attempt to quantify the cognitive load improvements required as expressions gain complexity. With the changing face of software interfaces from the keyboard/mouse paradigm to being touch-based, it is perhaps an opportune time to design new interface models that make digital tasks with mathematical expressions as efficient or even more efficient than with traditional pen-and-paper for mathematicians of all experience levels.

References

1. Anthony, L., Yang, J., Koedinger, K.R.: Evaluation of multimodal input for entering mathematical equations on the computer. In: CHI 2005 Extended Abstracts on Human Factors in Computing Systems, pp. 1184–1187. ACM (2005)
2. Bahls, P., Wray, A.: Latexnics: the effect of specialized typesetting software on STEM students' composition processes. Comput. Compos. **37**, 104–116 (2015)
3. Ballard, D.H., Hayhoe, M.M., Pelz, J.B.: Memory representations in natural tasks. J. Cogn. Neurosci. **7**(1), 66–80 (1995)

4. Hayes, J.R., Chenoweth, N.A.: Is working memory involved in the transcribing and editing of texts? Writ. Commun. **23**(2), 135–149 (2006)
5. Hooper, J., Pollanen, M., Teismann, H.: Effective online office hours in the mathematical sciences. MERLOT J. Online Learn. Teach. **2**(3), 187–194 (2006)
6. Loch, B., Lowe, T.W., Mestel, B.D.: Master's students' perceptions of Microsoft Word for mathematical typesetting. Teach. Math. Appl.: Int. J. IMA **34**(2), 91–101 (2015)
7. Marandi, R.Z., Madeleine, P., Omland, Ø., Vuillerme, N., Samani, A.: Reliability of oculometrics during a mentally demanding task in young and old adults. IEEE Access **6**, 17500–17517 (2018)
8. Quinlan, J., Tennenhouse, C.: Perceived utility of typesetting homework in post-calculus mathematics courses. PRIMUS **26**(1), 53–66 (2016)
9. R Core Team: R: A Language and Environment for Statistical Computing. R Foundation for Statistical Computing, Vienna, Austria (2020). https://www.R-project.org
10. Raghubar, K.P., Barnes, M.A., Hecht, S.A.: Working memory and mathematics: a review of developmental, individual difference, and cognitive approaches. Learn. Individ. Differ. **20**(2), 110–122 (2010)
11. Rayner, K.: Eye movements in reading and information processing: 20 years of research. Psychol. Bull. **124**(3), 372 (1998)
12. Wickham, H.: ggplot2: Elegant Graphics for Data Analysis. Springer-Verlag, New York (2016)

Ambiguous Goals During Human-Computer Interaction Induce Higher Mental Workload

Thea Radüntz[1]([✉]), Marion Freyer[1], and Beate Meffert[2]

[1] Federal Institute for Occupational Safety and Health, Unit. 3.4, Mental Health and Cognitive Capacity, Nöldnerstr. 40/42, 10317 Berlin, Germany
raduentz.thea@baua.bund.de

[2] Department of Computer Science, Humboldt-Universität zu Berlin, Rudower Chaussee 25, 12489 Berlin, Germany

Abstract. Digitization of the working environment leads to tasks with high ambiguity, resulting in bad performance. The ambiguity of goals plays an important role in problem solving and could affect mental workload. In our study, we investigated whether goal ambiguity affects mental workload and task performance during planning. We hypothesized that problems with higher ambiguity require higher mental workload and degrade performance.

The Tower of Hanoi (ToH) was employed and 21 participants were instructed to move disks in order to reach a particular goal state. The tower-ending goal state, where all disks had to be arranged on one peg, served as goal state with low ambiguity. The flat-ending goal state, where disks had to be arranged on different pegs, was considered as more ambiguous due to its less apparent goal structure. Mental workload was registered by means of the method of Dual Frequency Head Maps from the electroencephalogram. For evaluating subjects' performance, we registered the error rate and planning time.

Analysis of the electroencephalogram yielded a significantly higher mental workload during the trials with a flat-ending goal state compared to trials with a tower-ending goal. Results of performance data were consistent with findings from the literature, indicating a higher error rate and longer planning time for the flat-ending goal state.

By these, we gained proof that tasks with a higher level of goal ambiguity induce higher mental workload.

Keywords: Mental workload · Problem structure · Goal structure · EEG · Tower of Hanoi · Planning

1 Introduction

Digitization of the working environment leads to tasks with high demands on cognitive capacity and impose high mental workload on employees. The cognitive demands even increase when the tasks include high goal ambiguity that is

© Springer Nature Switzerland AG 2020
D. Harris and W.-C. Li (Eds.): HCII 2020, LNAI 12186, pp. 81–90, 2020.
https://doi.org/10.1007/978-3-030-49044-7_8

particular prominent in the highly digitized working environment. Here, the final goals are often hard to be reached at once and must be split into a series of sub-goals that should be completed in a particular order. Tasks where the correct sequential ordering of the sub-goals is obvious from the final goal are referred to as low ambiguous. However, the complexity of the final goals of present-day's tasks often leads to ambiguous ordering of the sub-goals. This high goal ambiguity may result in a suboptimal sequence of the sub-goals, ineffective and needless steps, or even dead ends, where people have to go back to the initial state and start over again.

In previous studies researchers reported that tasks with high goal ambiguity resulted in suboptimal planning with more steps and bad performance ([11], [2]), more errors [17], and prolonged planning and solution times ([10], [17]). Studies related to the effects of goal ambiguity were conducted with children ([11], [2], [12]) and adults ([10], [17], [1], [9]).

The goal ambiguity could also affect mental workload. Mental workload describes the relation between task demands and personal capacity. Cognitive resources for human information processing are limited and thus, mental workload represents the amount of task demands placed on a person's limited resources [8].

Although research already showed that the degree of goal ambiguity is crucial for planning and problem solving, it remains unclear how the ambiguity influences the mental workload. In our study we aimed to bridge this gap. We investigated whether the goal ambiguity affects mental workload and task performance during planning and problem solving. We hypothesized that problems with higher ambiguity of goal structure require higher mental workload.

2 Methods

2.1 Procedure and Subjects

Our sample consisted of 21 subjects between the ages of 22 and 64 years (2 female, 19 male, mean age 38 ± 11, Table 1). The investigation took place in a non-shielded office. The experiment was fully carried out with each subject in a single day. It consisted of a training phase where the subjects were familiarized with the tasks and the main experiment. The training task was identical with the main task but shorter and easier. It was repeated until the subject reported to be confident in the proceeding. During the training phase, we did not register any data.

Table 1. Sample set.

Age (years)	22–29	30–39	40–49	50–59	60–64	Total
Subjects	6	5	8	1	1	21

During the main experiment, we registered several workload-relevant parameters for consolidating our findings. In particular, we registered the electroencephalogram (EEG), heart rate, errors, and planning time before the first move of the trial. As an error we defined a suboptimal move that led to more steps for reaching the goal state. At the beginning of the task, we conducted a resting measurement as a baseline for the analysis of the heart rate.

The Federal Institute for Occupational Safety and Health (BAuA) in Berlin was in charge of the project. All of the investigations acquired were approved by the local review board of the BAuA and the experiments were conducted in accordance with the Declaration of Helsinki. All procedures were carried out with the adequate understanding and written consent of the subjects.

2.2 Tower of Hanoi

For our study, we employed the Tower of Hanoi (ToH) as a classic task in cognitive science. The ToH was widely used in studies of executive functioning and problem solving, involving various levels of planning to achieve a goal [13].

The computerized version of ToH was realized through the implementation in the E-Prime application suite. Our ToH task consisted of three pegs and three or four discs of graduated size. Subjects were asked to transform a given start state into a goal state (Fig. 1) by taking the upper disk from one peg and placing it on another peg at a time. A larger disc could not be placed on a smaller disc. Subjects were requested to plan their actions before their first movement for reaching the goal with the least number of moves. In case of a non-optimal move, a feedback about the occurrence of an error was shown on the screen and the trial started over again. For avoiding the tendency of a speed and accuracy trade-off, there was no time limit for task solving. In general, subjects needed around 10 min to complete the ToH.

The tower-ending goal state, where all disks were stacked on one peg, provided an unambiguous goal state. Hereby, the disk at the bottom had to be in its goal position first, followed by the second from the bottom, and so on. In contrast, in the flat-ending goal state, where the disks were distributed over the pegs, the prioritization of moves was completely ambiguous. Subjects had to solve six trials with tower-ending and six trials with flat-ending goal state. The task design is presented in Fig. 2.

2.3 EEG, Heart Rate, and Performance Data

The EEG was registered by 25 electrodes placed at positions according to the 10–20 system. It was recorded with reference to Cz and at a sample rate of 500 Hz. For signal registration we used the g.LADYbird/g.Nautilus device by g.tec GmbH and their Matlab interface for the recording. Subsequently, the recorded EEG signal was filtered with a bandpass filter (order 100) between 0.5 and 40 Hz. Independent component analysis (ICA, Infomax algorithm [14]) was used for artifact rejection. For increasing topographical localization a simple Hjorth-style surface Laplacian filter using 8 neighbors [7] was applied. Next, we

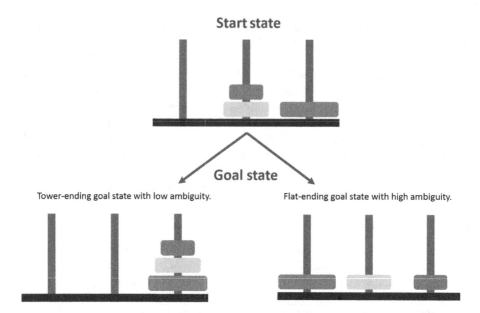

Fig. 1. Tower of Hanoi task. Subjects were required to transform the starting configuration into a goal configuration with low ambiguity (the tower-ending goal state) or
high ambiguity (the flat-ending goal state).

Trial number	1	2	3	4	5	6	7	8	9	10	11	12
Number of discs	3						4					
Number of moves required	5	5	6	6	7	7	7	7	11	11	15	15
Goal state	tower	flat	tower	flat	tower	flat	tower	flat	tower	flat	tower	flat
Goal ambiguity	low	high	low	high	low	high	low	high	low	high	low	high

Fig. 2. Task design for the Tower of Hanoi task.

transformed the artifact-free EEG to average reference and cut it into segments
of 1 s length, overlapping by 0.5 s. We computed the workload relevant frequency
bands (theta: 4–8 Hz, alpha: 8–12 Hz) over the segments by means of Fast Fourier
Transformation (FFT) and generated the Dual Frequency Head Maps (DFHM)
as outlined in the article by Radüntz [15]. For classifying the DFHM of each
subject from the EEG segments as low, moderate, or high workload, we used
the already trained SVM classifiers from the laboratory study [15] and obtained
a value every 0.5 s. Finally, we applied a moving-average time window of 6 s and
adjusted the result in order to gain a DFHM-workload index as percentage value
between 0 (all DFHM classified as low) and 100 (all DFHM classified as high).
For each ToH trial, a mean value of the DFHM-workload index was computed.

We registered the pulse signal by means of a plethysmographic pulse sensor
at the earlobe using g.tec's g.PULSEsensor coupled with g.tec's mobile amplifier

g.Nautilus. The pulse signal was windowed with a Hamming function and filtered with a bandpass filter (order 100) between 0.5 and 3.5 Hz. Peak detection was performed in order to determine the heart rate and the inter-beat intervals. Artifacts were automatically detected by means of statistical analysis, corrected using linear interpolation of the values at neighboring points, and equidistantly resampled with a time resolution of 0.5 s. Heart rate was determined in beats per minute in the time domain. The mean values for each trial were baseline-corrected according to the mean value measured during the resting state.

For performance evaluation, we used the number of errors and the planning time of each trial and person. Biosignal processing and all calculations were done with MATLAB.

2.4 Statistical Analysis

For each dependent variable (i.e., DFHM-workload index, heart rate, number of errors, and planning time), we averaged the corresponding values from the six trials with tower-ending goal state and the six trials with flat-ending goal state, respectively. Based on these values, we calculated the means for the two levels of goal ambiguity (Fig. 2) over the 21 subjects for each dependent variable.

The Shapiro-Wilk test showed a normal distribution for the differences between tower-ending and flat-ending goal state for the DFHM-index and planning time but not for heart rate and errors. Thus, comparisons between tower-ending and flat-ending trials related to the DFHM-workload index and planning time were conducted using paired-sample t-test, while for heart rate and errors we used the Wilcoxon test. Statistical calculations were conducted using SPSS with a significance threshold of 5%.

3 Results

The DFHM-workload index, planning time, and number of errors showed significant differences between the two levels of ambiguity as represented by the tower-ending and flat-ending goal states. Planning time was significantly higher during the flat-ending goal state ($t(20) = -5.544$, $p \leq .001$, $|d| = 1.21$) indicating that the planning demands in tasks with high goal ambiguity ($M = 17.5$ s, $SD = 8.4$ s) were significantly higher than during tasks with low goal ambiguity ($M = 11.4$ s, $SD = 6.0$ s). Accordingly, the number of errors was significantly higher during the flat-ending goal state ($z = -3.63$, $p \leq .001$, $r = 0.56$) indicating that the performance in tasks with high goal ambiguity was significantly worse ($Mdn = 0.67$, range $= [0, 4]$) than in tasks with low goal ambiguity ($Mdn = 0.17$, range $= [0, 1]$).

The DFHM-workload index was significantly higher during the flat-ending goal state ($t(20) = -2.242$, $p = .036$, $|d| = 0.49$) indicating a higher mental workload as assessed by the EEG during trials with high goal ambiguity ($M = 63.13$, $SD = 7.98$) compared to trials with low goal ambiguity ($M = 61.69$, $SD = 7.19$). However, no significant difference could be found for the baseline-corrected heart rate (z = −0.226, p = .821, r = 0.04), although there was the tendency that in flat-ending trials it was higher (Mdn = 2.34, range = [−9.46, 15.87]) than in tower-ending ones (Mdn = 1.66, range = [−12.08, 14.68]). Figures 3 and 4 show the results.

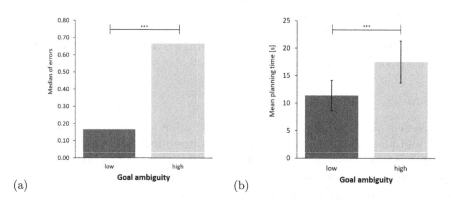

(a)   (b)

Fig. 3. (a) Median of errors and (b) mean value of planning time computed for the ToH-goal states with low and high ambiguity over 21 subjects (∗ ∗ ∗: p < .001; error bars for normal-distributed data indicate 95% confidence interval).

For the sake of completeness, we also looked for possible age differences. Subjects with an age below the median of 40 years were classified as younger (n=11), the remaining as older (n=10). For the normally distributed dependent variables DFHM-workload index and planning time, we calculated two mixed ANOVAs with age as between-subject factor. Results for the DFHM-workload index indicated neither a significant main effect for age (F(1, 19) = 0.347, p = .56, η^2 = .02) nor an interaction effect between age and goal ambiguity (F(1, 19) = 0.545, p = .47, η^2 = .03). Similarly, the planning time did not yield significant differences neither for age (F(1, 19) = 3.415, p = .08, η^2 = .15) nor for the interaction between age and ambiguity (F(1, 19) = 1.193, p = .29, η^2 = .06).

For the non-normally distributed dependent variables heart rate and errors, two Mann-Whitney-U tests were calculated to determine if there were differences between the age groups. As dependent variables for the baseline-corrected heart rate and errors, we employed the gradient between flat-ending and tower-ending goal states, respectively. There was no statistically significant difference in heart rate between younger and older subjects related to goal ambiguity (U = 52.00, Z = −0.211, p = .83, r = −.05). Finally, we were not able to find a statistically significant difference in errors between the age groups (U = 32.00, Z = −1.634, p = .10, r = −.36).

Figure 5 shows the results for the four dependent variables for each age group and goal-ambiguity level.

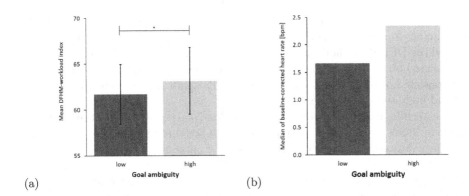

Fig. 4. (a) Mean DFHM-workload index and (b) median of baseline-corrected heart rate computed for the ToH-goal states with low and high ambiguity over 21 subjects (∗: .01 ≤ p < .05; error bars for normal-distributed data indicate 95% confidence interval).

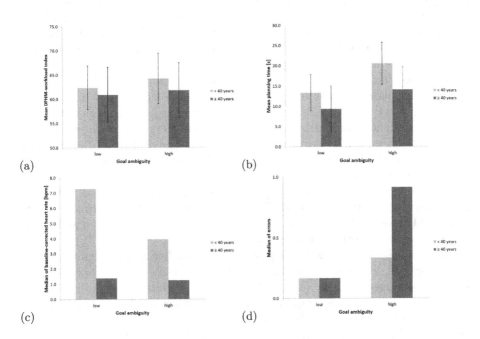

Fig. 5. (a) Mean DFHM-workload index, (b) mean value of planning time, (c) median of baseline-corrected heart rate, and (d) median of errors computed for the ToH-goal states with low and high ambiguity over subjects under (blue) and over (red) 40 years (error bars for normal-distributed data indicate 95% confidence interval). (Color figure online)

4   Discussion and Conclusions

We hypothesized that problems with higher ambiguity of goal states affect task performance and require higher mental workload. For assessing mental workload, we used the DFHM method that was previously developed in a laboratory setting and is based on the EEG. The 21 subjects participating completed the ToH task that consisted of trials with tower-ending and flat-ending goal state representing low and high ambiguity, respectively.

The results indicated that mental workload as registered by the DFHM-workload index from EEG was significantly higher during trials with high goal ambiguity. Moreover, subjects' performance (i.e., planning time and errors) during trials with high goal ambiguity was significantly poorer than during trials with low goal ambiguity. Heart rate did not yield any significant differences. The reason might be the small set size or the lack of time pressure during the task that is particular relevant for heart rate alterations.

In our study, we tried to address a possible interaction between ambiguity and age on workload. We have to admit that our sample set was not very appropriate. From the literature it is well known that working memory and information processing capabilities decrease with age ([3],[5]). Studies indicated significant differences in working-memory capacity and planning abilities for subjects over 60 years ([4], [6], [16]). Thus, a limitation of our study in this context was that our sample set was to young (i.e., median age of 40 years) for revealing age-related trends. Nevertheless, descriptive statistics and inspection of the results showed an increased planning time for younger compared to older subjects resulting in higher mental workload and heart rate but less errors during high-ambiguity trials. This might be an indication that younger subjects were more attached to planning tasks and thus, invested more effort. On the contrary, older subjects seemed less motivated and engaged in high-ambiguity planning tasks as linked to less planning time, decreased mental workload and heart rate, and more errors. Future studies should investigate a possible interaction between goal ambiguity and age with a suitable sample set. Furthermore, time on task and time pressure are notably relevant topics for further research as these could have an additional effect on planning performance and mental workload.

To sum up, we concluded that problems with higher goal ambiguity impose increasingly high demands on cognitive capacity, resulting in not only higher error rates, longer planning time, and suboptimal performance but also in negative consequences of inappropriate workload that could affect human's health and the safety of persons. The issue of goal ambiguity is of particular interest for problem solving in digitized working environments that often comprise suboptimal sequences of sub-goals because of an unclear goal state. Application developers should focus on reducing ambiguity and optimizing the goal structure by predefining and communicating the sequence of the sub-goals to the users through intelligent assistance systems. This way, we await performance at its best whilst simultaneously preserve employee's health.

Acknowledgements. We would like to thank our student assistants Lea Rabe, Emilia Cheladze, and Friederice Schröder for conducting the experiments and the participants for their contribution during the experiments. Furthermore, we want to thank our student assistant Claudia Cao for computational and overall support.

Author contributions. T.R. initiated the project and was responsible for the overall conception of the investigation. T.R. was responsible for the implementation of the tasks and the overall technical support. T.R. was responsible for the signal processing, data analysis, and evaluation. M.F. provided computational support for the data analysis with SPSS and graphic editing. Data interpretation was performed by T.R. The manuscript was written by T.R. Final critical editing was performed by B.M.

References

1. Berg, W.K., Byrd, D.L., McNamara, J.P., Case, K.: Deconstructing the tower: parameters and predictors of problem difficulty on the tower of London task. Brain Cogn. **72**(3), 472–482 (2010)
2. Borys, S.V., Spitz, H.H., Dorans, B.A.: Tower of Hanoi performance of retarded young adults and nonretarded children as a function of solution length and goal state. J. Exp. Child Psychol. **33**(1), 87–110 (1982)
3. Cowan, N., Morey, C.C., Chen, Z., Gilchrist, A.L., Saults, J.S.: Theory and measurement of working memory capacity limits. In: Advances in Research and Theory, pp. 49–104. Elsevier (2008)
4. Dobbs, A.R., Rule, B.G.: Adult age differences in working memory. Psychol. Aging **4**(4), 500–503 (1989)
5. Engle, R.W.: Working memory capacity as executive attention. Cur. Dir. Psychol. Sci. **11**(1), 19–23 (2002). http://www2.psych.ubc.ca/pgraf/Psy583Readings/Engle %202002.pdf. Accessed 02 Apr 2014
6. Gilhooly, K.J., Phillips, L.H., Wynn, V., Logie, R.H., Sala, S.D.: Planning processes and age in the five-disc tower of London task. Thinking Reasoning **5**(4), 339–361 (1999)
7. Hjorth, B.: An on-line transformation of EEG scalp potentials into orthogonal source derivations. Electroencephalogr. Clin. Neurophysiol. **39**(5), 526–530 (1975). http://www.sciencedirect.com/science/article/pii/0013469475900565
8. Kahneman, D.: Attention and Effort. Prentice-Hall, Englewood Cliffs (1973)
9. Kaller, C.P., Rahm, B., Köstering, L., Unterrainer, J.M.: Reviewing the impact of problem structure on planning: a software tool for analyzing tower tasks. Behav. Brain Res. **216**(1), 1–8 (2011)
10. Kaller, C.P., Unterrainer, J.M., Rahm, B., Halsband, U.: The impact of problem structure on planning: insights from the tower of London task. Cogn. Brain Res. **20**(3), 462–472 (2004)
11. Klahr, D., Robinson, M.: Formal assessment of problem-solving and planning processes in preschool children. Cogn. Psychol. **13**(1), 113–148 (1981)
12. Lehto, J.: Are executive function tests dependent on working memory capacity? Q. J. Exp. Psychol. Sect. A **49**(1), 29–50 (1996)
13. Lezak, M.D., Howieson, D.B., Bigler, E.D., Tranel, D.: Neuropsychological Assessment. Oxford University Press, Oxford (2012). https://www.ebook.de/de/product/16114053/muriel_d_lezak_diane_b_howieson_erin_d_bigler_daniel_tranel_neuropsychological_assessment.html

14. Makeig, S., Bell, A.J., Jung, T.P., Sejnowski, T.J.: Independent component analysis of electroencephalographic data. In: Touretzky, D.S., Mozer, M.C., Hasselmo, M.E. (eds.) Advances in Neural Information Processing Systems, vol. 8, pp. 145–151. MIT Press (1996). http://papers.nips.cc/paper/1091-independent-component-analysis-of-electroencephalographic-data.pdf

15. Radüntz, T.: Dual frequency head maps: A new method for indexing mental workload continuously during execution of cognitive tasks. Front. Physiol. 8, 1019 (2017). https://www.frontiersin.org/article/10.3389/fphys.2017.01019

16. Rönnlund, M., Lövdén, M., Nilsson, L.G.: Adult age differences in tower of Hanoi performance: Influence from demographic and cognitive variables. Aging Neuropsychol. Cogn. 8(4), 269–283 (2001)

17. Ward, G., Allport, A.: Planning and problem solving using the five disc tower of London task. Q. J. Exp. Psychol. Sect. A 50(1), 49–78 (1997)

Research on Interface Complexity and Operator Fatigue in Visual Search Task

Keran Wang$^{(\boxtimes)}$ and Wenjun Hou

School of Digital Media and Design Arts, Beijing University of Posts and Telecommunications, Beijing 100876, China
evakrl206@163.com

Abstract. During the visual search task performed by the operator, if the magnitude of the information flow does not match the design of the boundary, it will directly affect the information decoding process of the operator's brain. The resulting visual load will easily cause the operator to fatigue, Which seriously affects the performance of visual search tasks. Most of the literature focuses on the relationship between the physiological data of the eyes and operator fatigue, and the research on the relationship between the source of fatigue and the design of interface elements is insufficient. This research focuses on the eye movement data during the visual search task, and studies the relationship between the objective complexity of the human-machine interface and operator fatigue, and provides design suggestions and empirical evidence for interface designers when designing interfaces. The study found that in visual search tasks, the number of interface elements exceeding 20 and the interface element size being too small can significantly induce operator fatigue.

Keywords: Visual search task · Fatigue · Interface complexity · Eye tracking data

1 Introduction

With the rapid development of information technology and control technology, computer digital interface [1], due to its multiple sources of information, large amounts of information content, and intricate information structure relationships, has become an important carrier and medium for human-computer information interaction.

In airborne complex human-computer system, if the magnitude of information flow does not match the design of boundary when the operator performs visual search tasks, it will directly affect the decoding process of brain information, and the visual load generated thereby will easily make the operator fatigued.

In a complex human-machine system, if the operator does not perform a task, if the magnitude of the information flow does not match the design of the boundary, it will directly affect the information decoding process of the operator's brain. The resulting visual load will easily make the operator Generate fatigue, which severely affects the performance of visual search tasks [2].

D. Harris and W.-C. Li (Eds.): HCII 2020, LNAI 12186, pp. 91–99, 2020.
https://doi.org/10.1007/978-3-030-49044-7_9

Most of the literature focuses on the relationship between the physiological data of the eyes and operator fatigue, and the research on the relationship between the source of fatigue and the design of interface elements is insufficient.

This study selects the most typical feature search experiments in visual search [3], through questionnaire research and eye tracking experiments:

1. Questionnaire study: Explore the relationship between the complexity of different design interface elements (the density of the interface, the number of interface elements, the size, the number of colors, etc.) and the subjective perceived fatigue of the user, and the visual complexity for subsequent eye tracking research The degree independent variable provides experimental materials and the results are compared and verified.
2. Eye tracking research: explore the impact of different visual complexity interfaces on user behavior and cognitive load, and explore the inherent relationship of different indicators by combining related theories.

This research focuses on the eye movement data during the visual search task, and studies the relationship between the complexity of the airborne complex human-machine system interface and the operator's fatigue, and provides design suggestions and empirical evidence for interface designers when designing the interface.

2 Related Studies and Remaining Issues

2.1 Visual Fatigue

Visual fatigue is a phenomenon of reduced visual performance caused by tension or discomfort in visual work [4]. There are many factors that affect visual fatigue, including objective factors such as the operating environment, tasks, and individual conditions of the operator, as well as the operator's Subjective psychological factors [5].

Na [6] and others, from the Department of Industrial Engineering of Tsinghua University investigate the effect of display movement speed on visual fatigue in dynamic visual search tasks, they found visual fatigue increased as search time increased within 40 min search time.

Zagermann [7], from the University of Konstanz in Germany, et al. presents a refined model, embedding the characteristics of HCI into the relation of eye tracking data and cognitive load, they found fixations average gaze time, saccade speed, saccade width, pupil diameter and other eye movement indicators can indicate fatigue. Krzyszt of Krejtz [8], University of Clemson, proposed to use pupil diameter oscillation frequency to evaluate and explain cognitive load.

Abdulin et al. from the university of Texas, used the fixation and scanning eye movement behavior data to quantitatively evaluate the user's visual fatigue. During the task execution, the short-term fixation-related qualitative score was more sensitive to the assessment of visual fatigue than the indicators based on the scanning behavior [9].

The number of fixations indicates the number of times a user views an area of interest (AOI). Rudmann et al. found that the direction of gaze represents an interface element related to current cognitive activity. It can be interpreted as repeated interest in a certain area.

2.2 Interface Visual Complexity

Visual complexity can be defined as "diversity in stimulation patterns." [10] Geissler et al. believe that stimulus complexity should include the number of elements, the degree of dissimilarity between elements, and the degree of unity between elements [11].

Deng and Poole divide visual complexity into two dimensions: visual diversity (referring to the variety of design element types), visual richness (including the number of design elements, the content of text, the number of graphics, and the links and layout of pages Wait) [12]. And Lee et al. [13] found that the number and type of elements on the car screen have a significant impact on perceived visual complexity.

Tuch et al. [14] found that the increase of visual complexity will lead to the increase of task completion time and the decrease of recognition rate in behavior, and lead to the negative evaluation of emotional arousal and emotional valence in emotion.

These studies provide important ideas and methods for the following experimental design, and also provide theoretical support for the interpretation of experimental results

3   Methodology

3.1   Experimental Materials

At present, most researches usually use subjective evaluation as the index to measure the interface complexity, that is, the user perception complexity replaces the objective visual complexity, but the subjective evaluation has unstable factors and needs a lot of time to collect. Based on previous studies on interface complexity, this study selected three effective indicators that objectively affect interface complexity: the number of interface elements, the size of interface elements, and the number of color types of interface elements.

As control variables, the selection on geometric rules more round as interface core elements, with the number of interface elements (12, 24, 36), the size of the interface elements (radius $r = 0.5$ cm, 1 cm, and 1.5 cm), the types of interface elements color number (3, 6, 9) as a variable, in the same area in the plane (15 cm × 18 cm) using python - the random module randomly generated 100 different pictures as the experimental material, complexity and random automatically label on each interface elements. Take Fig. 1 and Fig. 2 and Fig. 3 for example.

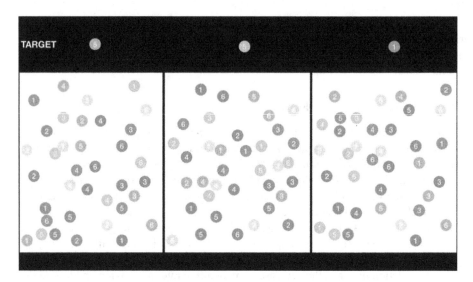

Fig. 1. The number of interface elements is 36, the size of the interface elements (radius $r = 1.5$ cm), the types of interface elements color number is six. The target of visual search task is above the picture.

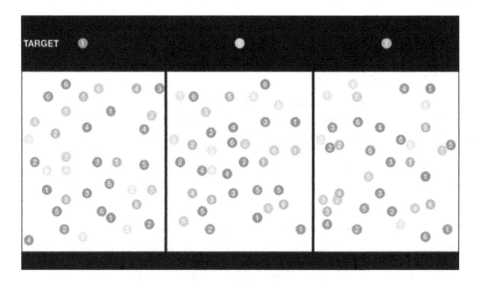

Fig. 2. The number of interface elements is 36, the size of the interface elements (radius $r = 1$ cm), the types of interface elements color number is six. The target of visual search task is above the picture.

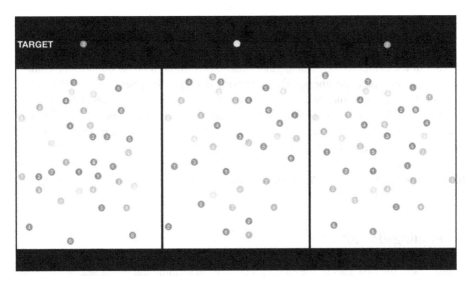

Fig. 3. The number of interface elements is 36, the size of the interface elements (radius $r = 0.5$ cm), the types of interface elements color number is six. The target of visual search task is above the picture.

3.2   Subjects

12 (male: 6, female: 6) graduate students from Beijing University of Posts and Tele-communications, with an average age of 23.1 years (SD = 0.6), all of whom had no impaired eyesight, good mental and physical health, and no experience of taking special drugs (antibiotics, narcotic drugs, etc.), smoking, drinking and other vices.

3.3   Experimental Tasks

In this study, based on the human brain processing information module and the user interaction behavior in the real situation, the subjects were asked to perform visual search tasks within 2500 ms as best as they could, and to find objects with the same color and number in the given page, and to press the space as soon as possible to respond when they found the objects. This task simulates the information search behavior that occurs on the interface when a real user is interested in certain keywords.

3.4   Experimental Process

First of all, the subjects signed the information book of eye movement tracking experiment and understood the general contents of the experiment. Then they sat down before the test machine smoothly and adjusted the seat back and height to a comfortable position. Before the experiment, the subjects were required to calibrate the position of their eyeballs using a 9-point correction procedure to ensure the accuracy of the experiment.

The subjects first had to go through two exercises to get familiar with the experimental process of the task, and then entered the formal experimental procedure. The process of each experimental task is described as follows: firstly, the fixation point appears in the center of the screen, guides the user's initial fixation point to the center, and then enters the experiment. When experimental stimuli appear on the screen, the user needs to quickly find the asterisk on the page and press the spacebar to continue the next search task. After performing the visual search task as best as they could within 2 min 30 s, the subjects were required to fill in a 7-point likert scale to assess their subjective fatigue level.

Meanwhile, Tobii Pro Glasses 2 eye tracker was used to extract eye movement data of these users, including pupil diameter, scanning speed, average eye hop frequency, fixation hotspot and trajectory.

4   Findings/Results

In this study, the first 15 s of the entire process of performing the visual search task was defined as state 1, and the last 15 s was defined as state 2. Combining the current status of fatigue-related eye movement indicators and the eye movement behavior recorded by the Tobii Glasses 2 eye tracker, This article extracts left and right pupil diameter, saccades, saccade speed and other eye movement index data related to fatigue.

Figure 4 is the average value of the experimenters of the above indicators at Fig. 1.

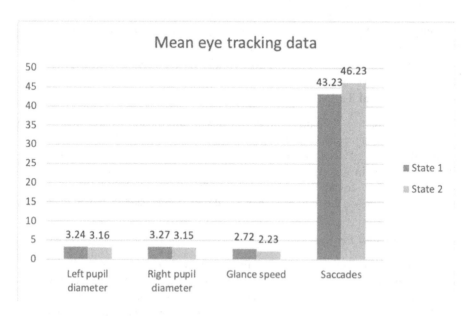

Fig. 4. The average value of the experimenters of the above indicators at Fig. 1.

Table 2 is the homogeneity test result of the above-mentioned eye movement index of the experimenter in state 1 (initial normal state) and state 2 (fatigue state), where df is the degree of freedom. According to the homogeneous analysis of the variance of the fatigue-related eye movement indexes selected initially, the significance of all eye movement indexes is greater than 0.05, indicating that there is no significant difference in the variance of the above eye movement indexes at the level of $\alpha = 0.05$ (Table 1).

Table 1. Homogeneous analysis of variance.

Heading level	Levene statistics	df1	df2	Saliency
Left pupil diameter	0.260	1	10	0.621
Right pupil diameter	2.500	1	10	0.145
Glance speed	0.030	1	10	0.866
Saccades	0.074	1	10	0.792

According to the ANOVA single-factor analysis of variance of eye movement indicators in Table 1, the significant probability values of left and right pupil diameters and saccade speeds are less than 0.05, indicating that the working state has significant effects on pupil diameter, average gaze time, and saccade speed. The F value is the ratio of the mean square error between the groups to the mean square error within the group, and the P value is the significance probability. In particular, the results of Table 2 indicate that the pupil diameter of the right eye is more affected by the working condition.

Table 2. ANOVA analysis results.

Eye track index	F	P
Left pupil diameter	0.260	0.001
Right pupil diameter	2.500	0.003
Glance speed	0.030	0.012
Saccades	0.074	0.021

1. When the number of interface elements is 12 and 24, the user's left and right pupil diameters and saccades change slightly. The average value of the difference between the left pupil diameter and the right pupil is 0.083 mm under normal and fatigue conditions. However, when the number of interface elements is compared with 24 and 36, the average saccade speed of the user is significantly slowed down, the average value of the difference of the left pupil diameter is 0.167 mm, and the right pupil is 0.201 mm. In order to investigate this reason, I interviewed the subjects, and at the same time, I measured the visual acuity of the experimenters and found that all the subjects had better vision than the left eye. From this reasoning: In the visual search task, due to the difference in the vision of the left and right eyes,

the subject was more dependent on the right eye, so the right eye was more affected by fatigue.

2. The user's fatigue experience decreases significantly with the increase in the size of the interface elements, the user's average gaze time decreases, the saccade speed increases, and the pupil diameter decreases to a minimum.

3. The user's fatigue experience increases significantly with the increase in the color of the interface elements. The average gaze time of the user gradually increases, the saccade speed slightly decreases, and the pupil diameter remains basically unchanged.

5   Conclusions

Visual search task is a process of recognition and matching, which only needs a little calculation. The experiments in this paper show that the number, size and color of interface elements are significantly related to the source of operator fatigue of visual search task. The diameter of left and right pupil, average fixation time and scanning speed are significantly related to the fatigue of user's subjective feelings. In visual search tasks, the number of interface elements exceeding 20 and the interface element size being too small can significantly induce operator fatigue. We should fully consider the complexity of interface when designing human-computer interface.

References

1. Chenqi, X.: Significant achievements of human-computer interface research in the digital age ——a review of human factor reliability in digital nuclear power plants. J. Saf. Environ. **19** (05), 1860 (2019). (in Chinese)

2. Makri, D., Farmaki, C., Sakkalis, V.: Visual fatigue effects on steady state visual evoked potential-based brain computer interfaces. In: 2015 7th International IEEE/EMBS Conference on Neural Engineering (NER), pp. 70–73. IEEE, Montpellier (2015)

3. Yantis, S., Jonides, J.: Abrupt visual onsets and selective attention voluntary versus automatic allocation. J. Exp. Psychol. Hum. Percept. Perform. **16**(1), 121–134 (1990)

4. Lambooij, M., Fortuin, M., Heynderickx, I., et al.: Visual discomfort and visual fatigue of stereoscopic displays: a review. J. Imaging Sci. Technol. **53**(3), 030201 (2009)

5. Ostrovsky, A., Ribak, J., Pereg, A., et al.: Effects of job-related stress and burnout on asthenopia among high-tech workers. Ergonomics **55**(8), 854–862 (2012)

6. Na, L., Yunhong, Z., Ruifeng, Y., Gaoyan, Z.: Effect of display movement speed on visual fatigue in dynamic visual search tasks. Ind. Eng. **20**(2), 51–55 (2017). (in Chinese)

7. Zagermann, J., Pfeil, U., Reiterer, H.: Studying eye movements as a basis for measuring cognitive load. In: Extended Abstracts of the 2018 CHI Conference on Human Factors in Computing Systems (CHI EA 2018). Association for Computing Machinery, New York, NY, USA, pp. 1–6 (2018)

8. Krzysztof, K., Katarzyna, W., Izabela, K., et al.: Dynamics of emotional facial expression recognition in individuals with social anxiety. In: Proceedings of the 2018 ACM Symposium on Eye Tracking Research & Applications, pp. 1–9. ACM, Warsaw (2018)

9. Evgeniy, A., Oleg, K.: User eye fatigue detection via eye movement behavior. In: Proceedings of the 33rd Annual ACM Conference Extended Abstracts on Human Factors in Computing Systems, pp. 1265–1270. Chi, EA (2015)
10. Berlyned, E.: Conflict, Arousal, and Curiosity. McGraw-Hill Book Company, New York (1960)
11. GeisslerG, L., Zinkhan, G.M., Watson, R.T.: The influence of home page complexity On consumer attention, attitudes, and purchase intent. J. Advert. **35**(2), 69–80 (2006)
12. Deng, L., Poole, M.S.: Affect in web interfaces: a study of the impacts of web page visual complexity and Order. MIS Q. **34**(4), 711–730 (2010)
13. Lin, Y.C., Yeh, C.H., Wei, C.C.: How will the use of graphics affect visual aesthetics a user-centered approach for web page design. Int. J. Hum. Comput. Stud. **71**(3), 217–227 (2013)
14. Tuch, A.N., Bargas-Avila, J.A., Opwis, K., et al.: Visual complexity of websites: Effects On users' experience, physiology, performance, and memory. Int. J. Hum Comput Stud. **67**(9), 703–715 (2009)

Assessment of Mental Workload Using Physiological Measures with Random Forests in Maritime Teamwork

Yu Zhang[1], Yijing Zhang[2(✉)], Xue Cui[1], Zhizhong Li[1],
and Yuan Liu[3]

[1] Department of Industrial Engineering, Tsinghua University, Beijing 100084,
People's Republic of China
yu-zhang18@mails.tsinghua.edu.cn, cuix41@163.com,
zzli@tsinghua.edu.cn
[2] Department of Industrial Engineering, Beijing University of Civil Engineering
and Architecture, Beijing 100044, People's Republic of China
zhangyijing@tsinghua.edu.cn
[3] China Institute of Marine Technology and Economy,
Beijing 100081, People's Republic of China
liuyuan_cimtec@126.com

Abstract. Assessment of mental workload plays an important role in adaptive systems to perform dynamic task allocations for teamwork onboard. In our study, workload assessment models were established based on EEG, Eye movement, ECG, and performance data, respectively. The data were collected from team subjects operating maritime target identification and coping device allocation tasks collaboratively in a computer simulation program. Physiological measures were collected from wearable sensors, and the team workload was self-assessed using the Team Workload Questionnaire (TWLQ). Mental workload models were trained by the random forests algorithm to predict team workload with self-reported TWLQ measure as reference and physiological measures and objective performance measures as inputs. The low levels of MAPE (Mean Absolute Percent Error) suggested that these measures can be used to provide accurate assessment of operator mental workload in the tested type of maritime teamwork. This study demonstrates the possibility to assess operator status according to physiological measures, which could be employed in adaptive systems.

Keywords: Mental workload · Physiological measures · Random forest · Maritime tasks · Teamwork

1 Introduction

Modern work in safety-critical systems is often characterized by teamworking with a high level of informatization and simultaneous multitasking. During the work process, an operator needs to maintain high concentration of attention for a long time, which may cause high level of mental workload. Prolonged high mental workload can

© Springer Nature Switzerland AG 2020
D. Harris and W.-C. Li (Eds.): HCII 2020, LNAI 12186, pp. 100–110, 2020.
https://doi.org/10.1007/978-3-030-49044-7_10

degenerate the operator's status, which will further lead to degraded performance and errors, even serious accidents sometimes. According to the report released by the American Bureau of Shipping about maritime accidents, approximately 80–85% of accidents involved human errors and about 50% were initially caused by human errors (Baker and Seah 2004). Human error has been identified as a major contributor to maritime accidents just like in some other industries. To this end, assessing mental workload of operators accurately is crucial to the implementation of an adaptive system to support maritime teamwork onboard.

Workload has been defined as a set of task demands, as effort, and as activity or accomplishment (Hartman and McKenzie 1979). Similarly, mental workload could be defined as the amount of mental effort that an individual uses to perform tasks in essence (Gao et al. 2013). Unfortunately, up to now, mental workload is still an ill-defined construct (Rizzo et al. 2016). There is still not a generally accepted definition for the 50-year old construct. Mental workload is induced not only by cognitive demands of tasks but also by other factors, such as stress, fatigue and the level of motivation (Sheridan and Stassen 1979). There have been hundreds of studies on the measurement of mental workload in the past half-century. Many different techniques have been applied in mental workload measurement, including physiological measurements, dual-task (or secondary task) methods, primary task measurements, attention allocation and subjective measurements (Sheridan and Stassen 1979). Continuous and objective measurements instead of overall subjective measurements are desired in adaptive systems (Rusnock and Borghetti 2018), which puts forward a higher requirement to measurement techniques and data collection. Fortunately, neurological and physiological measurements have great potential for assessing workload in complex tasks (Parasuraman and Wilson 2008).

Until now, the state-of-the-art computational models for mental workload are mainly theory-driven instead of data-driven (Moustafa et al. 2017). However, inductive research methodologies can also be applied to create mental workload models from data and produce alternative inferences just like in other scientific fields. Regarding computational models, a model could be trained directly with measurable variables by finding patterns and relationships between these variables and corresponding performance measures (Dearing et al. 2019). At present, one of the most popular research fields devoted to the development of inductive models is machine learning, which is a sub-field of Artificial Intelligence (AI). Because of the multifaceted characteristics of mental workload, the ambiguity and uncertainty associated with many non-linear variables shaped this construct. Moreover, the difficulties associated with the aggregation of these variables and the development of computational models made the need to use machine learning to arise.

With the acceleration and spread of machine learning in the past two decades (Bishop 2006), researchers initiated to investigate mental workload using inductive data-driven research methodologies (Borghetti et al. 2017; Dearing et al. 2019; Lee and Tan 2006; McKendrick et al. 2019; Moustafa et al. 2017; Rusnock et al. 2015; Zhang et al. 2004). Moustafa et al. (2017) presented a study investigating the capability of a selection of supervised machine learning classification techniques to produce data-driven computational models of mental workload, which tended to outperform the NASA Task Load Index and the Workload Profile in the prediction of human performance (concurrent validity). Besides, McKendrick et al. (2019) investigated two machining learning algorithms for classifying mental workload based on fNIRS signals, and the Rasch model labeling paired with a random forest classifier led to the best model fits and showed an evidence of both cross-person and cross-task transfer. Instead of considering workload categorically, finer-grain automation decisions are possible by representing workload numerically. Besides using machine learning algorithms for classification, some studies have been conducted by regression analysis. Smith et al. (2015) examined the efficacy of using two regression-tree alternatives (random forests and pruned regression trees) to decrease workload estimation cross-application error based on a remotely piloted aircraft simulation. Borghetti et al. (2017) used Random forests algorithms on electroencephalogram (EEG) input to infer operator workload based upon IMPRINT (Improved Performance Research Integration Tool) workload model estimates.

The vast majority of previous studies for mental workload based on machine learning have measured single one physiological variable, such as EEG (Borghetti et al. 2017; Heger et al. 2010). Meanwhile, the mental workload assessment or prediction was performed for individuals instead of teams. In this paper, we reported our initial effort to induce data-driven workload models for maritime teamwork with different physiological variables as inputs. We chose the Random forests algorithm based on past modeling success (Borghetti et al. 2017; Moustafa et al. 2017; Smith et al. 2015).

2 Methodology

2.1 Dataset

In this study, we utilized an existing dataset consisting of multiple physiological measures collected as part of a previous study conducted by the authors' lab. In total 108 male university students were recruited as the participants, who were divided into 54 two-operator teams randomly in the experiment. The two-operator teams were asked to complete maritime target identification and coping device allocation tasks collaboratively in a computer simulation program. All the participants were undergraduates or graduates in STEM programs with normal or correct-to-normal vision.

While the participants performing tasks in teams, physiological measures (see Table 1) were recorded, including eight channels of brain electrical activities (by NeuroFlight Cognitive Assessment Training Analysis System at a rate of 512 Hz), eye movement (by SMI IviewX RED Eye Tracker), and heart measures (by Bioharness Physiology Monitoring System at a rate of 250 Hz). Furthermore, the objective performance data in the experiment were also recorded automatically.

Each team should complete all the required operations within a time limit, otherwise, the target disappeared and the platform would record it as an uncompleted target. Each team performed 24 trials for each of the six scenario complexity levels at a certain level of time pressure and filled out the Team Workload Questionnaire (TWLQ) (Sellers et al. 2014) after each treatment to measure their subjective team workload. The subjective scores served as the outputs to train our machine learning model.

Table 1. Summary of dataset variables.

Physiological measures and objective performance	
EEG	**Eye Movement**
Frequencies:	Blink Rate
Delta(1–3 Hz)	Averaged Blink Duration
Theta (4–7 Hz)	Fixation Rate
Low Alpha (8–9 Hz)	Averaged Fixation Duration
High Alpha (10–12 Hz)	Saccade Rate
Low Beta (13–17 Hz)	Average Saccade Duration
High Beta (18–30 Hz)	
Low Gamma (31–40 Hz)	
Mid Gamma (41–50 Hz)	
ECG	**Objective Performance**
Heart Rate	Response Latent Time
Heart Rate Variability	Judgment Time
	Completion Time
	Miss Rate
	Completion Rate
	Operation Accuracy
Subjective workload measures	
TWLQ	
Overall Score, Task Workload, Team Workload, Task-Team Balancing	

2.2   Workload Model Training

In this study, we pursued to predict operator mental workload from physiological data using a machine learning method. The Random Forests method was selected as the machine learning algorithm due to its resistance to overfitting, its simplicity, and its ability to model nonlinear data (Breiman 2001). It is a commonly used and effective algorithm in the field of machine learning and data analysis. Random forests are integrated decision trees, which can reduce the variance of the overall model. Furthermore, by constructing a multitude of decision trees at the training stage and outputting the mean prediction of the individual trees for regression, random forests can correct decision trees' issue of overfitting to their training data set (Breiman 2001). Compared with one single decision tree, random forests usually have better generalization performance. Besides, random forests are not sensitive to outliers in the data set and do not require excessive parameter tuning.

Taking into account the characteristics above, the supervised random forest algorithm was selected to train the workload model. We optimized only two tuning parameters: the number of features randomly sampled as candidates at each internal split node and the number of trees in the forests (Liaw and Wiener 2001). The number of features to set the size of the random subset of features was considered when splitting a node. The lower the number, the more variance is reduced whereas bias is also increased. The number of trees in the forest controls how many decision trees the forest contains. Added trees generally mean better, but would increase computation time and lead to performance plateaus as the number increases. Regression Model can be trained and tested using the randomForest() functions (Liaw and Wiener 2001) that are available in the randomForest R package. In our study, Random forests algorithm was employed on different physiological measures as inputs to infer subjective operator workload measured by TWLQ. All the variables in the physiological data set were averaged across trials of every scenario complexity level for each team before used to train the model. The MAPE (Mean Absolute Percent Error) of the regression model was selected to test the utility of the model (Rayer 2007; Swanson et al. 2011).

3   Results and Discussion

3.1   EEG-Based Assessment Model

After removing the missing values from the EEG data set, the data set was divided into a training set and a test set according to a ratio of 7 to 3, resulting in 158 training data records and 70 test data records. The first step of modeling was to find the optimal parameter mtry, that is, the optimal number of variables for the binary tree in every specified node. By calculating the increase of the random forest error after each variable was removed, the value corresponding to the smallest error is found. The second step was to find the optimal parameter ntree, which is number of decision trees, as shown in Fig. 1. When ntree was 200, the error in the model tended to be stable.

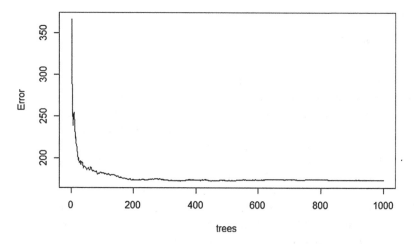

Fig. 1. Error in model and number of trees in the forest

The important() function can induce the importance of the variables filtered by the random forest model, where %IncMSE indicates the Out-of-bag error when this variable is excluded (The Bootstrap method cannot draw all the samples, but only about 2/3 of the samples can be drawn each time. The remaining 1/3 is used as an out-of-bag observation to test the model). IncNodePurity represents the total amount of node impurity reduction caused by this variable, which is shown in Fig. 2.

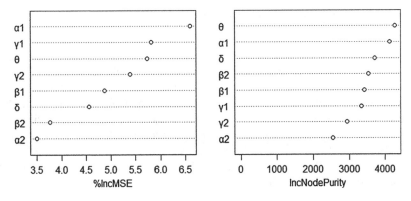

Fig. 2. The Out-of-bag error (%IncMSE) and the total amount of node impurity reduction (IncNodePurity)

By observing Fig. 2, the three most important variables can be found to be Alpha1 relative power, Gamma1 relative power, and Theta relative power. The random forest generated 200 trees, and the three variables were randomly selected at each split.

The third step of modeling is to set the model parameters and build a random forest model. According to the calculation results above, the random forest generated 200 trees, and 3 variables were selected at each split. To test the utility of the model, we input the test set into the model, and the output is shown in Fig. 3. We called the regr. eval() function in the DMwR package and got that MAPE of the EEG-based assessment model was 19.34%.

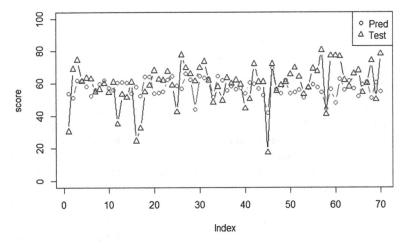

Fig. 3. Comparison of test and predicted values from the EEG-based assessment model

3.2   Eye Movement-Based Assessment Model

After removing the missing values from the eye movement data set, the data set was divided into a training set and a test set according to a ratio of 7:3, resulting in 65 training data records and 25 test data records. Firstly, we found the optimal parameter mtry was 5. The second step was to find the optimal ntree. When ntree was 400, the error in the model tended to be stable. By calling the important () function, the three most important variables we obtained were total blink time, average pupil size of X, and the number of saccades. The random forest generated 400 trees, and 5 variables were randomly selected at each split. The test set was input to the model, and the output was shown in Fig. 4. By calculation, we found that the MAPE of the eye movement-based assessment model was 16.03%.

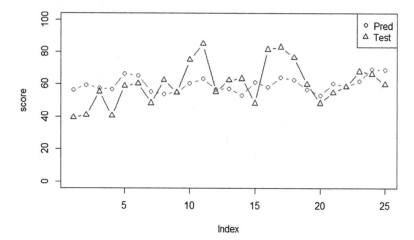

Fig. 4. Comparison of test and predicted values from Eye movement-based assessment model

3.3   ECG-Based Assessment Model

Similar to the operation above, the data set was divided into a training set and a test set according to a ratio of 7 to 3 after removing the missing values from the collected ECG data set. We obtained 116 training data records and 52 test data records. The first step was to find the optimal parameter mtry. By calculating the increase of the random forest error after each variable was removed, the value with the smallest corresponding error was found, and mtry was 3. The second step was to find the optimal parameter ntree. When ntree was 400, the error in the model tended to be stable. The three most important variables obtained were HF, LF, and TP by calling the important () function. The random forest generated 400 trees, and 3 variables were randomly selected at each split. The test set was input to the model, and the output was shown in Fig. 5. The calculation result of MAPE was 15.87% for the ECG-based assessment model.

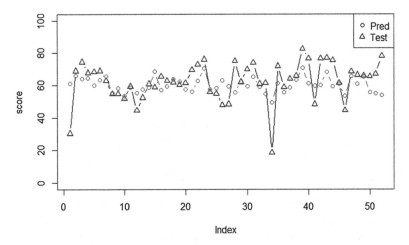

Fig. 5. Comparison of test and predicted values from the ECG-based assessment model

3.4 Performance-Based Assessment Model

After removing missing values from the performance data set, the data set was divided into a training set and a test set according to a ratio of 7 to 3. We obtained 214 training data records and 91 test data records. The first step was to find the optimal parameter mtry. By calculating the increase of the random forest error after each variable was removed, the value corresponding to the smallest error was found, and the mtry was 8 by calculation. The error in the model tends to be stable when ntree was 300. By calling the important () function, we found the three most important variables were the total completion rate, device allocation time of operator 2, and identity judgment time of operator 1. The random forest generated 300 trees, and 8 variables were randomly

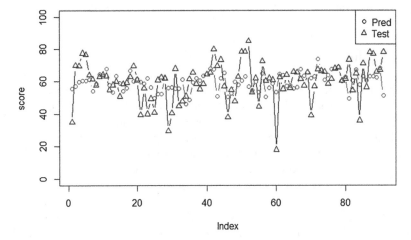

Fig. 6. Comparison of test and predicted values from the performance-based assessment model

selected at each split. The test set was input to the model, and the output was shown in Fig. 6. The calculation results of MAPE was 17.09% for the performance-based assessment model.

4 Discussion

Mental workload assessment models using different physiological measures were established, including assessment models based on EEG, eye movement, ECG and objective performance. From the criterion of MAPE, the error of the ECG-based assessment model was the minimum, followed by the eye movement-based assessment model, and then the performance-based assessment model. The EEG-based assessment model owned the maximum error. Observing the number of variables as the inputs of each model, we found that the performance-based assessment model had the most number of best variables (mtry) for the binary tree in every specified node, the eye movement-based assessment model used the least number of variables, accounting for only about 1/7. For the eye movement-based assessment model, the sample size provided by the original data set was 324, but there were many missing values and outliers. Totally 90 valid samples remained after data cleaning, and the training set and test set were divided according to a 7 to 3 ratio. Thus only 65 samples were left for the training of the model, so the MAPE of the model was relatively large. The number of decision trees in the random forests constructed by the four models was not much different. A detailed comparison of four models was shown in Table 2.

Table 2. Detailed comparison table of four models.

Assessment model	Training set	Test set	ntree	mtry	MAPE (%)
EEG	158	70	200	3	19.34
Eye movement	65	25	400	5	16.03
ECG	116	52	400	3	15.87
Performance	214	91	300	8	17.09

5 Conclusion

Focusing on the effective assessment of mental workload, many researchers have carried out various work, including the classification of workload levels, determination of workload values, monitoring of continuous workload during work, and also evaluation of the workload level from designated tasks. This study was conducted based on the data under different time pressure and scenario complexity levels. We attempted to construct an effective workload assessment model through the random forests algorithm. Three types of physiological data were measured concurrently including EEG, ECG, and eye movements. Assessment models were compared from MAPE (Mean Absolute Percent Error), optimal number of variables (mtry), and number of decision trees (ntree). We found that the MAPE of the ECG-based model was the minimal, the performance-based assessment model employed the most number of best variables (mtry) for the binary tree in every specified node, while the eye movement-based assessment model used the least number of variables, accounting for only about 1/7. This study demonstrated the possibility to assess the mental workload of operator onboard according to different physiological measures in maritime teamwork.

Acknowledgment. This study was supported by Project No. JCKY2016206A001 and No. 6141B03020604.

References

Baker, C.C., Seah, A.K.: Maritime accidents and human performance: the statistical trail. In: MarTech Conference, Singapore (2004)

Bishop, C.M.: Pattern Recognition and Machine Learning. Springer, New York (2006)

Borghetti, B.J., Giametta, J.J., Rusnock, C.F.: Assessing continuous operator workload with a hybrid scaffolded neuroergonomic modeling approach. Hum. Factors **59**(1), 134–146 (2017)

Breiman, L.: Random forests. Mach. Learn. **45**(1), 5–32 (2001). https://doi.org/10.1023/A:1010933404324

Dearing, D., Novstrup, A., Goan, T.: Assessing workload in human-machine teams from psychophysiological data with sparse ground truth. In: Longo, L., Leva, M. (eds.) H-WORKLOAD 2018. CCIS, vol. 1012, pp. 13–22. Springer, Cham (2019). https://doi.org/10.1007/978-3-030-14273-5_2

Gao, Q., Wang, Y., Song, F., Li, Z., Dong, X.: Mental workload measurement for emergency operating procedures in digital nuclear power plants. Ergonomics **56**(7), 1070–1085 (2013)

Hartman, B., Mckenzie, R.E. and Advisory Group for Aerospace Research Development Neuilly-Sur-Seine: Survey of Methods to Assess Workload (1979)

Heger, D., Putze, F., Schultz, T.: Online workload recognition from EEG data during cognitive tests and human-machine interaction. In: Dillmann, R., Beyerer, J., Hanebeck, U.D., Schultz, T. (eds.) KI 2010. LNCS (LNAI), vol. 6359, pp. 410–417. Springer, Heidelberg (2010). https://doi.org/10.1007/978-3-642-16111-7_47

Lee, J., Tan, D.: Using a low-cost electroencephalograph for task classification in HCI research. In: Proceedings of the 19th Annual ACM Symposium on User Interface Software and Technology, pp. 81–90 (2006)

Liaw, A., Wiener, M.: Classification and regression by RandomForest. Forest **23** (2001)

Mckendrick, R., Feest, B., Harwood, A., Falcone, B.: Theories and methods for labeling cognitive workload: classification and transfer learning. Front. Hum. Neurosci. **13**, 295 (2019)

Moustafa, K., Luz, S., Longo, L.: Assessment of mental workload: a comparison of machine learning methods and subjective assessment techniques. In: Longo, L., Leva, M.C. (eds.) H-WORKLOAD 2017. CCIS, vol. 726, pp. 30–50. Springer, Cham (2017). https://doi.org/10.1007/978-3-319-61061-0_3

Parasuraman, R., Wilson, G.: Putting the brain to work: neuroergonomics past, present, and future. Hum. Factors: J. Hum. Factors Ergon. Soc. **50**(3), 468–474 (2008)

Rayer, S.: Population forecast accuracy: does the choice of summary measure of error matter? Popul. Res. Policy Rev. **26**(2), 163 (2007). https://doi.org/10.1007/s11113-007-9030-0

Rizzo, L., Dondio, P., Delany, S.J., Longo, L.: Modeling mental workload via rule-based expert system: a comparison with NASA-TLX and workload profile. In: Iliadis, L., Maglogiannis, I. (eds.) AIAI 2016. IAICT, vol. 475, pp. 215–229. Springer, Cham (2016). https://doi.org/10.1007/978-3-319-44944-9_19

Rusnock, C., Borghetti, B., McQuaid, I.: Objective-analytical measures of workload – the third pillar of workload triangulation? In: Schmorrow, D., Fidopiastis, C. (eds.) AC 2015. LNCS (LNAI), vol. 9183, pp. 124–135. Springer, Cham (2015). https://doi.org/10.1007/978-3-319-20816-9_13

Rusnock, C.F., Borghetti, B.J.: Workload profiles: a continuous measure of mental workload. Int. J. Ind. Ergon. **63**, 49–64 (2018)

Sellers, J., Helton, W., Näswall, K., Funke, G., Knott, B.: Development of the team workload questionnaire (TWLQ). Proc. Hum. Factors Ergon. Soc. Ann. Meet. **58**(1), 989–993 (2014)

Sheridan, T.B., Stassen, H.G.: Definitions, models and measures of human workload. In: Moray, N. (ed.) Mental Workload: Its Theory and Measurement, pp. 219–233. Plenum Press, New York (1979)

Smith, A.M., Borghetti, B.J., Rusnock, C.F.: Improving model cross-applicability for operator workload estimation. Proc. Hum. Factors Ergon. Soc. Ann. Meet. **59**(1), 681–685 (2015)

Swanson, D.A., Tayman, J., Bryan, T.M.: MAPE-R: a rescaled measure of accuracy for cross-sectional subnational population forecasts. J. Popul. Res. **28**(2), 225–243 (2011). https://doi.org/10.1007/s12546-011-9054-5

Zhang, Y., Owechko, Y., Zhang, J.: Driver cognitive workload estimation: a data-driven perspective. In: Proceedings of the 7th International IEEE Conference on Intelligent Transportation Systems (IEEE Cat. No.04TH8749), pp. 642–647 (2004)

The Effect of Time Pressure and Task Difficulty on Human Search

Qianxiang Zhou[1,2], Chao Yin[1,2], and Zhongqi Liu[1,2(✉)]

[1] Key Laboratory for Biomechanics and Mechanobiology of the Ministry of Education, School of Biological Science and Medical Engineering, Beihang University, Beijing 100191, China
lzq505@163.com

[2] Beijing Advanced Innovation Centre for Biomedical Engineering, Beihang University, Beijing 102402, China

Abstract. In order to study the effect of time pressure and task difficulty on human search performance, an experimental study was conducted. Thirteen people participated in the experiment. The time pressure is changed by changing the presentation time of the target information on the display. The task difficulty is changed by changing the number of the target information displayed on the computer display. Subjects' reaction time (RT), accuracy rate and eye movement data were recorded during the experiment. The results show that the subjects' reaction time decrease and accuracy rate increase. The subjects' reaction time increased and accuracy rate decreased when the task difficulty increase. Eye movement data also changed with the change of time pressure and task difficulty. When time pressure increased, the pupil diameter increased, number of fixation points decreased, fixation frequency increased and average fixation time decreased. Pupil diameter increased when task difficulty increase. So the conclusion could be made that eye movement behavior and search performance change with the change of task difficulty and time pressure. During Human-machine Interface Design, the task difficulty and time pressure need to be scientifically designed to optimize the performance of the operator.

Keywords: Time pressure · Task difficulty · Visual search · Performance · Eye movement

1 Introduction

People obtain information from the Human-machine Interface by various senses during human machine interaction. Among these senses, the vision is the most important. According to statistics, more than 80% of the information is obtained visually [1]. Visual search is the most common way to obtain information visually. Visual search performance is affected by multiple factors, such as the lighting of the environment, the color coding of the search target, the contrast between the search target and the background, the degree of concentration of people, time pressure and the difficulty of the search task. Among them, time pressure and task difficulty are the two most common factors affecting search performance, the rational design of two factors is crucial to the efficiency of Human-machine interaction. Time pressure refers to the

© Springer Nature Switzerland AG 2020
D. Harris and W.-C. Li (Eds.): HCII 2020, LNAI 12186, pp. 111–124, 2020.
https://doi.org/10.1007/978-3-030-49044-7_11

pressure caused by time constraints and the resource demand to complete the task with limited time [2]. In Human-machine Interface, many tasks have time requirements to be completed in limited time. Under different time pressure, operators may change their target search strategy or process. Wang Yuan et al. studied the effects of time pressure on search behavior in sequential decision, the results showed that time pressure has a effect on search depth. They found that excessive search will be made under low time pressure and insufficient search under high time pressure. Time pressure also significantly affects decision-makers' search strategy preferences, that the search is based option under low time pressure, while it is based attribute under high time pressure. They found that the time pressure will significantly improve search speed [3]. Task difficulty is a complex factor that is related to the density, size and contrast of the target.

Much research has been done on time pressure and task difficulty for target search performance. For example, Walrath et al. studied the effects of time pressure, target density, target color and other factors on visual search, and found that time pressure has a significant effect on search time. High time pressure took an inhibition effect to other factors on search time. While time pressure had no significant effect on accuracy rate [4]. Tsimhoni et al. studied the driving behavior of drivers, and found that time pressure can affect the gaze behavior of drivers. The results showed that the total time of gaze is unchanged, the number of gazes increase and the time of each gaze become shorter, so improving the efficiency of visual search [5]. Guo Xiaoyu studied effects of time pressure on search performance during dynamic visual search. High time pressure significantly improved search performance at low moving speed, lower time pressure also improved search performance, but it was not significant. Higher time pressure and lower time pressure had no significant effect on performance at high moving speed. However, search performance decreased appreciably under high time pressure. The experiment found that low time pressure has no significant effect on search performance, while high time pressure will appreciably reduce search performance [6]. Wang Zhuang studied effects of task complexity and time pressure on the operational performance of computerized procedures. The results showed that under different time pressure, task complexity had a significant effect on performance, and its relation model presented different rules that can be used to predict performance [7]. Fan Xiaoli et al. studied the effects of time pressure and information complexity on human visual search performance. The results showed that both factors have a significant effect on the performance of target search, and both factors affect the target search accuracy rate. Only both factors matched best can get higher reaction accuracy rate [8]. Yang Lindong et al. studied the effects of target appearance probability, target existing or not, and time pressure on visual search performance. The results showed that time pressure and low target appearance probability significantly reduce search time, fixation time and number of fixations [9]. Yang Lindong et al. also studied the effects of personality and task difficulty on search performance, and found that the search time of low task difficulty is significantly lower than that of high task difficulty, and the search accuracy rate of low task difficulty is also higher than that of high task difficulty. So, search performance was better for low task difficulty [10]. Chan et al. studied the effects of different task difficulty on visual search performance. They also found that the search time of low task difficulty is shorter than high task difficulty [11].

Although there are many studies on the effect of time pressure and task difficulty on search performance, there still lack of further research on visual search behavior. The cognitive mechanism of the effect of time pressure and task difficulty on visual search behavior can be further studied if eye tracking technology was used to study the person's target search behavior.

2   Method

2.1   Participants

Thirteen university students participated in the experiment. Their ages were between 21 and 27, the average age was 23 years. Subjects' uncorrected visual acuity or corrected visual acuity was all over 1.0, and they were right-handed.

2.2   Experimental Equipment

The equipment used in the experiment was a RED500 desktop eye tracker from German's SMI Corporation. The composition of the eye tracker mainly included a 15-inch laptop, a 22-inch display, an eye tracking module fixed below the display and some data line, etc. The eye tracker's sampling frequency was 120 Hz, resolution was 0.03° and accuracy was 0.4°.

2.3   Pre-experimental

Experimental Tasks. The target search was made in the pre-experiment, and the subjects' eye movement data were not measured in the experiment. The purpose of the pre-experiment was to screen out the tasks that the accuracy rate is over 85% and these screened tasks will be performed with the measurement of their eye movement data.

The display content of the target search task was realized by programming in Visual C++ 6.0. The task difficulty was changed by increasing the number of information displayed on the monitor. Time pressure was changed by changing the length of time that information is displayed on the monitor.

In Fig. 1, 2, 3, 4, 5 and 6, the graphic symbol were displayed in 1 row and 1 column (1R1C), 2 rows and 2 columns (2R2C), 3 rows and 3 columns (3R3C), 4 rows and 4 columns (4R4C), 5 rows and 5 columns (5R5C), 6 rows and 6 columns (6R6C) in the center area of the display. Only one of them was a target, the others were interference targets. The shape of the target is fixed (Fig. 2, the oblique line in the circle points to the upper tight 45°). The display time of information on the monitor was 1 s, 2 s, 3 s, 4 s, 5 s and 6 s, which mean different time pressure. The longer the display time, the lower time pressure, conversely, the higher time pressure. The number of display information corresponding to each display time was shown in Table 1. Only one target was displayed in 1 row and 1 column that the task is the least difficult. The difficulty of target search task increased When the displayed information increase. It was the most difficult task with 6 rows and 6 columns display symbol. In the experimental task, the display information was randomly changed according to the display time. The target symbol might be appeared or not appeared on display. The subjects should press the left or right mouse When they are sure of the target appearing. If there was no a target, they should press right mouse. The subjects didn't not press mouse if they don't sure the targets appearing and the display information would automatically switch to the next screen of display content when the display time is over. The program automatically recorded the results of each operation to a text file that include the reaction time, the right press and wrong press. The number of experiments corresponding to the amount of display information for each display time was 30.

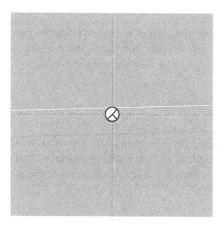

Fig. 1. Display information of 1R1C

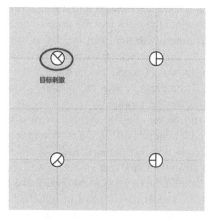

Fig. 2. Display information of 2R2C

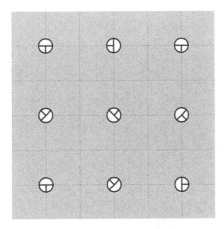

Fig. 3. Display information of 3R3C

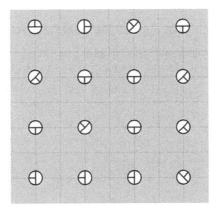

Fig. 4. Display information of 4R4C

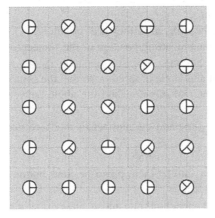

Fig. 5. Display information of 5R5C

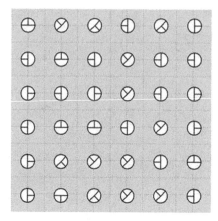

Fig. 6. Display information 6R6C

Table 1. List of display time and number of display symbol

The number of display	Display time					
	1 s	2 s	3 s	4 s	5 s	6 s
1R1C	√	√	√	√	√	√
2R2C		√	√	√	√	√
3R3C		√	√	√	√	√
4R4C			√	√	√	√
5R5C					√	√
6R6C						√

Experimental Procedure. The experiment was conducted as follow procedures:

(1) The experimental task was introduced to the subjects so that they understand the experimental content.
(2) Subjects practiced the tasks until they are proficient.
(3) Adjusted the position and sitting position of the subject to keep the subject's eyes from monitor about 55 cm.
(4) Adjusted the height and angle of monitor so that the center of the subject's line of sight fall on the center of monitor, and the angle between the line of sight and monitor keep 90°.
(5) Inputted the display time and the number of interface items for each experiment on the display.
(6) Read the experimental instructions to the subjects, asked them to focus on the stimulus display area, and let them press the Start button to start the experiment.

Experimental Result. The reaction time and accuracy rate of subjects' perform with 6 kinds of time pressure and 6 kinds of task difficulty were listed in Table 2.

Table 2. Results of reaction time and accuracy rate under 6 kinds of time pressure and 6 kinds of task difficulty

Display time	The number of display	Reaction time (s)	Accuracy rate
1 s	1R1C	0.653	96.67%
2 s	1R1C	0.825	90.00%
	2R2C	1.347	96.67%
	3R3C	1.200	40.00%
3 s	1R1C	0.982	90.00%
	2R2C	1.686	83.33%
	3R3C	1.822	83.33%
	4R4C	2.097	66.67%
4 s	1R1C	0.958	93.33%
	2R2C	1.814	96.67%
	3R3C	2.337	73.33%
	4R4C	2.423	46.67%
5 s	1R1C	1.050	96.67%
	2R2C	1.759	93.33%
	3R3C	2.290	80.00%
	4R4C	3.327	66.67%
	5R5C	3.324	53.33%
6 s	1R1C	1.483	96.67%
	2R2C	2.001	93.33%
	3R3C	2.968	93.33%
	4R4C	3.714	80.00%
	5R5C	3.596	66.67%
	6R6C	3.390	53.33%

According to the data in Table 2, the curves could be fit out of Fig. 7 and Fig. 8. Figure 8 was subject's target search time data under different task difficulty and time pressure. It reflected the overall change of the subject's target search time with the change of task difficulty and time pressure. As shown in Fig. 8, under 6 kinds of time pressure, the overall trend of the subject's target search time was great accordant, it increased with the increasing of the task difficulty. With the same task difficulty, the overall trend of the subject's target search time with the change of time pressure was also great accordant, it decreased with the increasing of time pressure. In other words, under the same time pressure, with the increasing of task difficult, the target search speed of the subject decreased. Under the same task difficulty, with the increasing of time pressure, target search time decreased. Figure 9 was subject's accuracy rate data with different task difficulty and time pressure. It reflected the change of the subject's accuracy rate of the target search with the change of task difficulty and time pressure. As shown in Fig. 9, under 6 kinds of time pressure, the overall trend of the subject's target search accuracy rate was great accordant, it decreased with the increasing of task difficult. The result of one-way ANOVA showed that time pressure has a significant effect on reaction time and accuracy ($P<0.05$). Under the same task difficulty, the overall trend of the subject's target search accuracy rate with the change of time pressure was great accordant, it decreases with the increasing of time pressure.

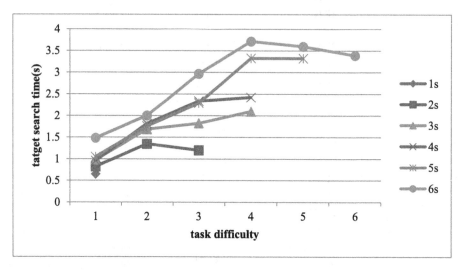

Fig. 7. The change of target search time under 6 kinds task difficulty and 6 kinds time pressure

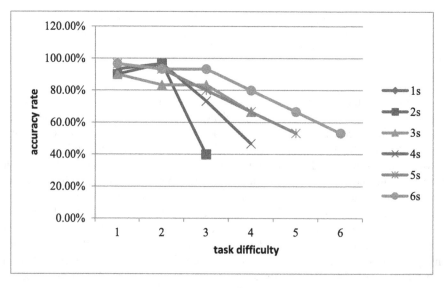

Fig. 8. The change of target search accuracy rate under 6 kinds task difficulty and 6 kinds time pressure

The results of one-way ANOVA showed that task difficulty has a significant effect on both reaction time and accuracy rate ($P < 0.05$).

2.4   Eye Movement Experiment

Experimental Tasks. Based on the data of the pre-experiment, time pressure remained unchanged, and the difficulty of the display task whose removal accuracy is less than 85% is the task of this experiment (Table 3).

Table 3. List of target display time and number of displays

The number of displays		1R1C	2R2C	3R3C	4R4C
Display time	1 s	✓			
	2 s	✓	✓		
	3 s	✓	✓		
	4 s	✓	✓		
	5 s	✓	✓	✓	
	6 s	✓	✓	✓	✓

The experimental task was edited and implemented in Experiment Center 3.6. Stimulus materials were static pictures. In the way of computer screenshot, the screenshot of the stimulus picture in the Visual C++ program in the pre-experiment was saved as picture and inserted into the Experiment Center. There were thirty stimulus pictures for each task difficulty. There was a buffer picture unrelated to the experimental content following each stimulus picture. If the subjects were sure in advance that there is or no target in the picture, press the spacebar to switch to the buffered picture. Then they reported in voice whether the targets appear or didn't appear. If the subjects were not sure whether there is or not, do not pressed the spacebar, the picture automatically switched to the buffer picture when picture reached display time. The buffer picture had not the display time limit; the subjects pressed the spacebar to switch the next stimulus picture. After completing 30 stimulus pictures, the subjects needed to rest for a while to relieve the tension and fatigue of their eyes. Subjects' eye movement data was recorded during the experiment.

Experimental Procedure. The experiment was conducted as follow procedures:

(1) Introduced the experiment process to the subjects. After the subjects clearly understood the experiment process, they started to train.
(2) Adjusted the position and sitting position of the subject to keep the subject's eyes from monitor about 55 cm.
(3) Adjusted the height and angle of monitor so that the center of the subject's line of sight fall on the center of monitor, and the angle between the line of sight and monitor keep 90°.

(4) made the calibration of eye movement measurement.
(5) Subjects started the target search task. After completing 30 stimulus pictures, the subjects needed to rest for a while to relieve the tension and fatigue of their eyes.

3   Results and Discussion

The 4 indexes of average pupil diameter, the number of fixation points, fixation frequency, and average fixation time were often used to analyze people's reading and searching behaviors on the interface, and the results of the 4 indexes were as follows.

3.1   The Effect of Time Pressure on Eye Movement

In the experiment, 6 kinds of time pressure were designed. The average pupil diameter under the 6 kinds of time pressure was shown in Table 4. It could be seen that pupil diameter increase with the increasing of time pressure. The results of one-way ANOVA showed that the effect of time pressure on pupil diameter have a significant effect ($P < 0.05$).

Most studies considered that the size of the pupil diameter reflect the degree of effort of the subjects. In this experiment, due to the increasing of time pressure, the subjects spent more attention resources, and needed to work harder to find the target, which lead to pupil dilation.

Table 4. Average pupil diameter under different time pressure

Time	Average pupil diameter (mm)
1 s	5.16 ± 0.76
2 s	5.08 ± 0.79
3 s	5.01 ± 0.70
4 s	4.98 ± 0.77
5 s	4.86 ± 0.81
6 s	4.94 ± 0.79

The number of fixation points with 6 kinds time pressure were shown in Table 5. It could be seen that with the change of time pressure, the number of fixation points didn't show a monotonic increase or decrease. However, the overall trend was that when the time pressure is high, the number of fixation points was less, and when the time pressure is low, the number of fixation points was more. It might be that the subjects are relatively nervous under the time pressure. So, searching for the target harder could find the target faster, which lead to the number of fixation points decrease. The results of one-way ANOVA test showed that the effect of time pressure on the number of fixation points reached a significant effect ($P < 0.05$).

Table 5. The average number of fixation points under different time pressure

Time	The average number of fixation points (number)
1 s	1.59 ± 0.4
2 s	1.45 ± 0.4
3 s	1.82 ± 0.5
4 s	1.53 ± 0.4
5 s	1.59 ± 0.5
6 s	1.57 ± 0.5

The fixation frequency with 6 kinds time pressure was shown in Table 6. It could be seen that with the increasing of time pressure, the fixation frequency increased. It might be that under the high time pressure, subjects accelerated the frequency of eye attention switching in order to find the target faster in a relatively short time, which lead to the increasing of the fixation frequency. The results of the one-way ANOVA test showed that the effect of time pressure on the fixation frequency reached a significant effect ($P < 0.05$).

Table 6. Average fixation frequency under different time pressure

Time	Average fixation frequency (number/s)
1 s	3.04 ± 1.2
2 s	3.03 ± 1.2
3 s	3.20 ± 1.0
4 s	2.97 ± 0.9
5 s	2.72 ± 0.7
6 s	2.74 ± 0.6

The average fixation time with 6 kinds time pressure was shown in Table 7. It could be seen that except the distortion at 2 s and 6 s, the overall trend was that the average fixation time decrease with the increasing of time pressure. It might be that under the time pressure, the subjects spent more mental resources in order to find the target faster. So accelerated the pace of information identification and extraction, which lead to the decreasing of the average fixation time. The results of the one-way ANOVA test showed that the effect of time pressure on the average fixation time reached a significant effect ($P < 0.05$).

Table 7. Average fixation time under different pressures

Time	Average fixation time (mm)
1 s	280.8 ± 149
2 s	297.8 ± 153
3 s	262.4 ± 140
4 s	330.8 ± 146
5 s	359.0 ± 177
6 s	280.8 ± 126

3.2   The Effect of Task Difficulty on Eye Movement

Results from current researchers suggested that pupil diameter is a sensitive index of task difficulty, which generally increase with the increasing of task difficulty. The average pupil diameter data for the 4 kinds of task difficulty was obtained in this experiment (Table 8). It could be seen that with the increasing of task difficulty, the pupil diameter increased. It might be when task difficulty increase, the subjects spent more resources, including more attention resource, more visual resource and more cognitive resource to complete the task harder, which lead to the increasing of the pupil diameter. It could also be seen that when task difficulty increase, the subjects need to process more information, so pupil enlargement will help the subjects to process more information. The results of the one-way ANOVA test showed that the effect of task difficulty on pupil diameter reached a significant effect ($P < 0.05$).

Table 8. Average pupil diameter with different task difficulty

Task difficulty	Average pupil diameter (mm)
1R1C	4.67 ± 0.78
2R2C	4.87 ± 0.86
3R3C	5.01 ± 0.82
4R4C	5.12 ± 0.84

The result of the number of fixation points with 4 kinds task difficulty was shown in Table 9. It could be seen that with the increasing of task difficulty, the number of fixation points increased significantly. It might be that due to the increase of task difficulty, the number of targets to be searched increased, so more targets needed to be identified, which lead to the number of fixation points increase. The results of the one-way ANOVA test showed that the effect of task difficulty on the number of fixation points reached a significant effect ($P < 0.05$).

Table 9. Average the number of fixation points with different task difficulty

Task difficulty	The average number of fixation points (number)
1R1C	1.57 ± 0.02
2R2C	4.32 ± 0.23
3R3C	7.06 ± 0.61
4R4C	13.02 ± 0.78

The results of fixation frequency with 4 kinds task difficulty was shown in Table 10. It could be seen that with the increasing of task difficulty, fixation frequency increased significantly. The explanation for this was the same as that for the number of fixation points. The results of the one-way ANOVA test showed that the effect of task difficulty on fixation frequency reached a significant effect ($P < 0.05$).

Table 10. Average fixation frequency under different task difficulty

Task difficulty	Average fixation frequency (number/s)
1R1C	2.87 ± 0.7
2R2C	3.80 ± 1.1
3R3C	3.83 ± 1.1
4R4C	4.00 ± 0.9

The results of the average fixation time with 4 kinds task difficulty was shown in Table 11. The average fixation time was generally used to study the readability of the interface. The interface with good readability had less average fixation time, and vice versa. The task stimulus in this experiment was not a readability study, so data results had no correlation with the changes of task difficulty.

Table 11. Average fixation time with different task difficulty

Task difficulty	Average fixation time (mm)
1R1C	319.5 ± 126
2R2C	160.7 ± 46
3R3C	166.3 ± 38
4R4C	163.4 ± 24

4   Conclusion

In this paper, an experimental study on the effects of time pressure and task difficulty on target search performance and eye movement behavior was conducted, so the conclusions can be made as follows:

(1) Time pressure and task difficulty have significant effects on reaction time and accuracy rate. When time pressure increase, there is a tradeoff effect between reaction time and error rate, that is, reaction time becomes faster, error rate increases to a certain extent. When task difficulty increase, the reaction time and error rate will become worse, the reaction time will increase and the accuracy rate will decrease.

(2) The change of time pressure can be well reflected by the pupil diameter, the number of fixation points, fixation frequency and average fixation time. When time pressure increase, pupil diameter increases, the number of fixation points decreases, fixation frequency increases and average fixation time decreases. The change of pupil diameter can also well reflect the change of task difficulty. With task difficulty increase, pupil diameter increases.

References

1. Liu, W., Yuan, X.G., Yang, C.X., et al.: Comparison study between biological vision and computer vision. Space Med. Med. Eng. **14**(4), 303–307 (2001)
2. Chan, A.H.S., Tang, N.Y.W., Donez, O.L., Benson, L.: Decisions under time pressure: how time constraint affects risky decision making. Organiz. Behav. Hum. Decis. Proces. **71**(2), 121–140 (1997)
3. Wang, Y., Qin, J.C., Bian, W.J.: An experimental study based on the moderating effect of expectation between sequential search behavior and time pressure. Manag. Rev. **30**(8), 182–193 (2018)
4. Wal, L.C., Backs, R.W.: Time stress interacts with coding, density, and search type in visual display search. In: Proceedings of the Human Factors and Ergonomics Society Annual Meeting, pp. 1496–1500. SAGE Publications, Thousand Oaks (1989)
5. Tsimhoni, O., Green, P.: Time-sharing of a visual in-vehicle task while driving: the effects of four key constructs. In: Driving Assessment 2003: The Second International driving Symposium on Human Factors in Driver Assessment, Training and Vehicle Design, pp. 113–118. University of Iowa, Iowa (2003)
6. Guo, X.J.: The effect of speed, time stress and number of targets upon dynamic visual search. Master's thesis of Tsinghua University (2011)
7. Wang, Z.: Effects of task complexity and time pressure on operation performance of computerized procedures. Master's thesis of Tsinghai University (2012)
8. Fan, X.L., Zhou, Q.X., Liu, Z.Q.: Principle of plane display interface design based on visual search. J. Beijing Univ. Aeronaut. Astronaut. **41**(2), 216–221 (2015)
9. Yang, L.D., Song, F.X., Zhang, Y.H., Yu, R.F.: Effects of target prevalence, target presence and time pressure on visual search performance. Ind. Eng. J. **19**(6), 83–89 (2016)
10. Yang, L.D., Yu, R.F.: A study of personality and effect of task difficulty on visual search performance. Ind. Eng. J. **19**(1), 91–96 (2016)
11. Chan, A.H.S., Tang, N.Y.W.: Visual lobe shape and search performance for targets of different difficulty. Ergonomics **50**(2), 289–318 (2007)

Human Physiology, Human Energy and Cognition

Human Energy in Organizations: Theoretical Foundations and IT-Based Assessment

Michael Fellmann[1]([✉]) [iD], Fabienne Lambusch[1] [iD],
and Oliver Weigelt[2] [iD]

[1] Business Information Systems, University of Rostock, Rostock, Germany
{michael.fellmann, fabienne.lambusch}@uni-rostock.de
[2] Organizational and Personnel Psychology, University of Rostock,
Rostock, Germany
oliver.weigelt@uni-rostock.de

Abstract. Today's working world can be characterized by an increase in flexibility, complexity and speed. For employees and organizations as a whole, it is challenging to keep pace to market requirements while at the same time retaining health, productivity and well-being of their employees. In this paper, we focus on human energy as one particularly relevant aspect of employee well-being. The experience of human energy has been described in terms of vitality, vigor, or enthusiasm. Tracking human energy in employees continuously over time does not only provide insights into the contingencies of energetic well-being at the individual level. It is also likely very relevant to promote and sustain well-being at the team level, department level, or the organizational level. However, up to now, it is largely unexplored how IT-supported measurements of human energy can contribute to the assessment of organizational energy. Hence we analyze the prospects of applying such measurements to infer levels and dynamics of human energy over time at the organizational level. We draw on existing approaches of organizational energy from the literature as a frame for analysis and outline how IT-supported continuous tracking might assist human resource management and leadership to promote and sustain the well-being of organizations.

Keywords: Human energy · Organizational energy · Self-assessment · IT-supported measurement · Energy-aware information systems

1 Motivation and Introduction

The digital transformation has changed and continues to change the world of work tremendously. On the bright side, telecommuting [1] and other types of flexible work arrangements [2] provide employees with additional autonomy with regard to how, when, and where they do their jobs. On the dark side, employees face new challenges upon managing the boundaries between work and other life domains [3]. Furthermore, intensification of work has contributed to increase pressure on employees for decades [4] and continues to do so. Hence, promoting and sustaining (mental) health and well-being of employees has become a major issue for organizations [5] – besides

© Springer Nature Switzerland AG 2020
D. Harris and W.-C. Li (Eds.): HCII 2020, LNAI 12186, pp. 127–140, 2020.
https://doi.org/10.1007/978-3-030-49044-7_12

performance per se. Recent research in organizational behavior has emphasized the role of job design [6], leadership [7, 8] and interventions [9] for promoting employee health and well-being. In this paper, we explore ways to leverage technology and in particular continuous tracking of individual experiences over time as a tool to promote employee well-being. Our emphasis will be on applications in organizations that reach beyond the individual level and may be relevant for well-being at the macro-level, that is teams, departments, or whole organizations. As we will outline below, well-being measured at the individual level can be aggregated to higher levels of the organization, such as project teams or departments [10]. Studying well-being within a multilevel framework provides the opportunity to examine differences between departments and dispersion or consensus within departments as focal outcomes. Keeping track of some individual level well-being indicator over time even provides the opportunity to model phenomena of team development (e.g., team members becoming more similar over time) [11]. While studying annual survey responses through a multilevel lens in nothing new, research that considers changes over time across levels is currently in nascent state. In this paper, we focus on one specific aspect of employee well-being, namely *human energy*, because energy is a common theme in numerous research streams of psychology, organizational behavior, and management [12, 13]. Furthermore, energy plays a prominent role in common sense and lay theories about (employee) well-being. The remainder of our contribution is structured as follows. In Sect. 2, we elaborate on the background regarding human energy, both in regard to the individual and organizational contexts. In the next Sects. 3 and 4, we show how human energy can be mapped and analyzed on an organizational level and provide use cases for human energy-aware business process management. After this, we conclude the paper and discuss future research directions in Sect. 5.

2   Background

2.1   Human Energy and Its Assessment

Research on occupational stress analyzes a broad range of aspects of well-being that can be described by states such as *fatigue* [14], *thriving* [15], *work-related rumination* [16, 17], *engagement* and *burnout* [18]. Studying these phenomena is important to understand issues related to health and quality of life [5]. One recurrent theme in the literature on employee well-being and self-regulation is individual resource status [19]. Conceptualizing employee well-being in terms of individual resource status is conducive to studying phenomena like strain and recovery as processes unfolding over time [20], i.e. tracking increases or decreases in available resources [21]. In the literature, two resources have attracted considerable attention: *time* and *energy*. Recently, QUINN and colleagues have proposed to integrate different conceptualizations and streams of research on energetic well-being under the umbrella of *human energy* [22]. In their review QUINN and colleagues distinguish between two aspects: *Physical energy* and *energetic activation*, whereby they define the former broadly as the capacity to do work. In this paper, we will focus on the *energetic activation* aspect of human energy, that is, the subjective experience of energy as reflected in high levels of vitality, vigor, enthusiasm, and zest. Accordingly, momentary states of vigor, subjective vitality, exhaustion, or fatigue reflect aspects of energetic activation [13, 23, 24].

Valid measures integrated in everyday information systems could act as facilitators to keep track of the own resource status and desirable behaviors. There is a broad range of verbal scales available to capture individual energetic activation (e.g. [14, 25, 26]). Given that most of the established verbal scales have been developed mainly for application in cross-sectional research, these scales have to be adapted considerably for application in experience sampling research [27]. For instance, typical survey items like "I look forward to each new day" from Ryan and Frederick's scale of vitality neither make sense nor work, when surveying participants multiple times a day – as is typical in experience sampling research. At the same time, measurement instruments in experience sampling research need to be brief to minimize participant burden [28]. Recently, a single-item pictorial scale of energetic activation has been introduced to facilitate keeping track of individual resource status over time [29, 30]. It refers to the common metaphor of a battery in order to capture momentary resource status ranging from an almost empty to a fully charged battery (cf. Fig. 1).

Fig. 1. Battery pictorial scale

This measure can be used for assessments of the momentary state of a person in regard to energy. The scale was introduced presenting a study with empirical results from a diary study across twelve days among 57 employees. The pictorial scale was highly correlated with momentary ratings of subjective vitality and fatigue at the intraindividual and the interindividual level (i.e., has convergent validity). The scale is easy to understand and use. Especially if fluently integrated into existing information systems of organizations, assessments via the brief scale imply a low effort. Drawing on these results, continuous assessment of human energy becomes now possible as a practice for (organizational) self-reflection.

2.2   Organizational Energy: Concepts and History

While human energy is important for the individual, it is also important for organizations such as a company, institution, association or administrations. Since organizations typically involve multiple persons, the construct of human energy hence has to be viewed on a more coarse-grained level leading to a construct of *organizational energy* [31]. In this regard, SCHIUMA (2007) [32] states that "the energy of employees is recognized as an important factor in their performance and in maximizing their overall contribution to the organization. Organizational energy is dynamic in nature; it is more than just the sum of the energy of its employees. It also includes the interaction and dynamics of teams and the organization as a whole" [32]. In more detail, SMIT (2017) [33] summarizes a definition from the Office of Government Commerce (OGC) and states that "organisational energy is the extent to which an organisation has mobilised the full available effort of its people in pursuit of its goals. It is the collective effort of the people and the effective management of that effort within the organisation that are the deciding factor on whether or not the strategic objectives are achieved. Organisational energy refers to a collection of critical organisational elements such as leadership, direction, management, motivation,well-being, effective communication, teamwork, skills and experience, all of which must come together in order to release the organisation's collective energy in the right direction" [33].

The construct of organizational energy is not new. According to SRIRUTTAN [34], it has been formed in the 1980s where it was inspired from the studies of DRUCKER (1959, 1999) in regard to management and leadership. Early theories on organizational energy were formed by SMITH and TOSEY in 1999 [35], according to SRIRUTTAN [34]. These theories provide a holistic framework for dynamic, learning organizations in which organizations are also characterized as an energy system. Later on in the early 2000s, the importance of organizational energy was acknowledged in the highly cited article "Unleashing organizational energy" from BRUCH and GHOSHAL [36]. The authors state that "without a high level of energy, a company cannot achieve radical productivity improvements, cannot grow fast and cannot create major innovations". In the following twenty years, researchers explored (amongst others) energy aspects in regard to the workplace [37, 38], to practices performed on the workplace e.g. in the context of knowledge work [39], the impact of organizational energy on performance [34], its role in change management [40] and relation to leadership styles [41].

2.3   On the Relation Between Human Energy and Organizational Energy

In order to explore the relation between human energy and organizational energy, scales of organizational energy have to be analyzed in regard to the items of these scales that refer to human energy. To do so, a recent analysis of organizational energy concepts by BAKER (2019) [42] is useful. In this analysis, special attention is given to ATWATER and CARMELI's (2009) eight-item feelings of energy scale [42] as well as to COLE et al.'s (2012) 14-item productive energy scale to measure energy in work environments [43]. Both scales strongly relate to human energy. For example, in ATWATER and CARMELI's scale [42] the items such as "I feel energetic and active at work" or "I have high energy to complete my work" directly relate to the level of human energy. Also in COLE's et al. scale [43] some items in the affective dimension (e.g. "People in my work group feel energetic in their job") relate to human energy.

However, although concepts of organizational energy relate to human energy, they are not the same. As it was derived by an in-depth studying of the previously cited works, traditional concepts of organizational energy are more encompassing and additionally comprise e.g. interpersonal relations. Thus, organizational energy may be harder to assess and to measure in comparison to our single-item scale of human energy. Due to this, it is questionable whether concepts of organizational energy are applicable for continuous measurement and feedback that would be highly sought for sustainable management of organizations. For example, "squeezing out" the maximum of performance from employees may increase productivity for a short time, but it could ruin the enterprise in the long run due to high rates of burn-out and fluctuation. In this regard, periodical assessments of organizational energy conducted from time to time may be too slow to detect sudden changes in the energy e.g. due to new management practices or restructuring of the organization. Consequently, assessing the level of human energy on a more regular (if not daily) basis could be a valuable information on its own being highly useful for managers. This is in line with BAKER (2019) [42] who states that "these individual-level measures can be aggregated to the group level as the sum, mean, or variability of individual members' energy". In this way, aggregated human energy measured on a frequent basis could complement traditional concepts of organizational energy that can be assessed from time to time using energy assessments [32] or other methods proposed in the literature on organizational energy [44].

In the next two Sects. 3 and 4, we focus on the measurement and application of aggregated human energy inside organizations. A simple form of human energy application is to map the current energy level. This could support reflective processes as well as fostering energy-related improvements (Sect. 3). Furthermore, human energy could play a role in the design, configuration, execution and analysis of the business process that underlie the organizations' day to day operations (Sect. 4).

3 Organizational Human Energy Mapping

Mapping organizational energy first requires an aggregation of individual measurements of human energy as implied by BAKER (2019) [12]. The author states that individual levels of energy could be "aggregated to the group level as the sum, mean, or variability of individual members' energy" [12]. It has to be added that this aggregation can be performed at various level of details (e.g. group level, department level, organization-wide). Following the aggregation step, the next step is mapping the current human energy level in the organization since it might be relevant in order to reflect on the current state and to find ways of sustainably improving the energy level. In the simplest form, an aggregated level of team energy could be mapped along the typical phases of a project showing the average level of energy of all involved team members (cf. Fig. 2). This will give insights into the current state of the project and the increase and decrease of energy. The latter might depend on the project or on factors outside the project which in either case have to be analyzed in case of lacking energy.

In order to capture human energy, self-assessment instruments could be used that could be triggered if employees work for a part of their time (or full-time) on a project. Self-assessment using our single-item pictorial scale for low-effort assessment can be complemented by automated IT-based data analysis tools that try to infer the level of energy. This may be accomplished e.g. based on the duration of concentrated uninterrupted work on single content objects such as documents or code.

A slightly more advanced application of human energy mapping at the organizational level would be to map energy according to its intensity and significance for the pursuit of organizational goals. This would allow managers to detect changes which might be useful for reflection of their activities. In this direction, an established model for classifying different types of energy is the matrix of BRUCH and VOGEL [46]. The basic idea is to classify organizational energy according to its intensity (high or low) and quality (negative or positive). The matrix has four quadrants, being corrosive

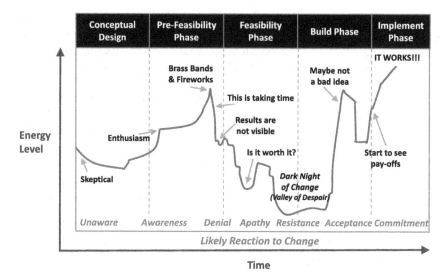

Fig. 2. Mapping of energy along typical phases of projects (source: [45], p. 154)

energy, productive energy, resigned inertia and comfortable energy (cf. Fig. 3). Intuitively, high levels of energy can facilitate the pursuit of enterprise' goals (productive energy) or they can be directed against the enterprise (corrosive energy). If there is only a small level of energy, this could hint to structural problems of the work or management issues. It could also hint to a comfortable situation where no challenges or threats are experienced. The ultimate goal of human energy management at the organizational level might be to increase productive energy and to decrease corrosive energy. For an increase of productive energy, BRUCH and VOGEL [46] suggest two strategies which they roughly describe as "slaying the dragon" or "winning the princess". Whereas the former implies a challenging situation (e.g. being threatened by a

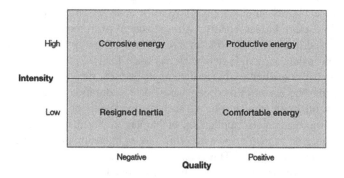

Fig. 3. BRUCH and VOGEL's types of organizational energy [46]

competitor), the latter could translate to the creation of completely new and futuristic products that contribute to the long-term success of the organization.

In order to semi-automatically capture the quality of energy devoted to work, IT-supported tools might assist the user. This could possibly be achieved with a combination of automated stress and sentiment detection. In regard to stress detection, state-of-the-art approaches for stress detection using unobtrusive sensors that can be used in office environments achieve roughly 80–90% of accuracy in detecting stress. Some of these approaches require a smartphone (e.g. [47], the model is able to predict self-reports with an accuracy of 72%), whereas others complement smartphone data with activity trackers to detect stress (e.g. [48] with accuracy of 82% on test data) or use wrist devices (e.g. [49] with accuracy on 55 days of real-life data for a 2-class problem of 92%). Furthermore, detecting work stress in offices by combining data of different sensors as described by KOLDIJK et al. (2018) [50] is quite promising and achieved 90% accuracy for classifying neutral and stressful working conditions on a test data set. Another measurement method is sensing stress of computer users by increased typing pressure on the keyboard. This feature can be leveraged since it was found out that during the stressful conditions, the majority of computer users show significantly increased typing pressure (>79%) and more contact with the surface of the mouse (75%) [51]. Also, chairs and floor equipped with sensors for pressure contribute to stress measurement [52]. Regarding sentiment detection and mood tracking, a recent overview on mood detection using various information input methods is given from LIETZ et al. (2019) [53]. In addition, another possibility is to leverage the output that results from information input processes. Most notably, text-based information that a user creates can be analyzed e.g. using sentiment mining, which is widespread particularly to analyze the sentiment expressed in customer reviews [54]. While a combination of highly stressful working days in combination with negative sentiment could signify high levels of corrosive energy, experiencing stress with positive sentiments could signal high levels of productive energy (e.g. managing to submit a paper on time).

Whereas the former suggestions of human energy application were focused on mapping the level of energy and its quality, it might also be valuable to map energy in relation to department-spanning collaborations. This could be used e.g. to spot energy

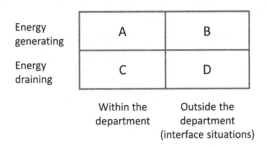

$$A + B + C + D \ = \ 100 \text{ units of energy}$$

Fig. 4. DE's box for characterizing mental energy (source: [55])

losses due to inefficient work organization. In this regard, a matrix suggested by NITISH R. DE [55] may provide a basic frame (cf. Fig. 4).

This matrix divides the mental energy used by managers into four quadrants according to changes (Energy generating or Energy draining) and relation to the organizational structure (Within the department or Outside the department). An advanced feature of matrix of DE in comparison to the matrix introduced before is that it is meant both to classify human energy spending as well as to quantify it. Regarding the latter, a restriction is given that the sum of all entries in the quadrants must be 100 points.

In order to semi-automatically populate such a matrix, various IT-based tools could be leveraged. Quite analogously to the matrix presented before, the quality of energy in terms of an increase or decrease could be assisted by approaches for automated stress detection and sentiment recognition. Moreover, observing the stakeholders involved in activities could be leveraged to discriminate between activities within the department or interface situations. For example, if an employee sends an e-mail to a colleague working in a different department or to an external customer, this could be classified as an interface situation. Regarding the idea of quantifying the amount of energy that is spent in each of the quadrants, the duration of activity execution could be leveraged. For example, in the field of knowledge work state-of-the-art tools such as RESCUETIME allow for a user-controlled, unobtrusive time tracking. Activity duration in conjunction with information about the quality of energy that is devoted to these activities can be used to quantify the units of energy for each quadrant. With this, managers or team leaders would have a valuable instrument to reflect on the major sources and drains of information inside their own department and department-spanning and can act accordingly.

4   Human Energy-Aware Business Process Management

Business process management is concerned with the design, enactment, execution and analysis of business processes. As of now, current approaches for the design and execution of processes are agnostic to the current level of human energy. In the following, we will discuss preliminary use cases how human energy could be integrated into the design and execution of business processes (see Table 1) both regarding the design (build time) and execution (runtime) of processes.

In Use Case 1, the focus is on measurement during execution, which is used to obtain information about the energy state of the employees. In Use Case 2, preventive measures are also included. In Use Case 3, the amount of energy required should already be considered at the level of process design through annotations. In Use Case 4, the focus is on obtaining (ex post) best practices with regard reduce energy waste. Finally, Use Case 5 focuses on the design (ex-ante) of low-energy consumption processes through appropriate process models. Since human energy has not been

Table 1. Use cases, motivation and research question for human energy-aware BPM

Use Case	Motivation	Possible research question(s) (RQ)
Runtime (process execution)		
Human energy measurement	Human energy assessment is important for the correct execution of processes and to prevent wellbeing/health issues	RQ1. How can the energy demand be reliably detected and quantified? RQ2. Which control and correction mechanisms in case of low energy are desirable from the user's point of view?
Human energy management	Managing human energy is important e.g. to prevent burnout-effects	RQ3. How can relevant individual behavioural clues e.g. to save human energy be generated? RQ4. How can managers promote the use of self-management approaches?
Buildtime (Process Design)		
Annotation of process models with human energy-related data	There are often documented models in organisations, but little is known about the profile of processes regarding human energy demands	RQ5. How can energy measurement be used to annotate existing models? RQ6. How can an annotation for new models be performed ex ante? RQ7. Which analyses or optimizations are possible based on the annotations?
Process patterns for human energy preservation	Experienced employees have (implicitly) developed good working methods and sequences	RQ8. How can human energy measurement be used to identify energy demand-reducing patterns of experienced employees?
Human energy-saving process models	Reduce energy consumption at the source	RQ9. How can human energy-related data be used for process model design? RQ10. Which mechanisms for human energy preservation can be integrated into processes?

considered so far in business process management, we give a short motivation as well as further research questions related to the use cases.

In addition to the research questions given in the Table 1, there is also the question of which approvals and consents the analysis and measurement procedures in companies require and how the works council can be effectively involved. In principle, control over the use of data should be given to the individual/employee. Ethical and legal aspects should also be taken into account.

5 Final Reflection and Outlook

In the current working world, there is an increase in flexibility, complexity, and high time pressure that together lead to work intensification and blurring boundaries between life domains. Thus managing human energy is highly important for long-term productivity, health and well-being, both on the level of the individual and the organization. Towards this goal, we first reviewed different notions of human energy at the individual level including the latest development of a pictorial scale for low-effort human energy measurement that can be used for continuous management of this important resource. After this, we reviewed concepts of organizational energy and elaborated on the relation between human energy and organizational energy.

Regarding the relation between human energy and organizational energy, it was found out that concepts vary greatly, since organizational energy is not seen by most of the previous works as a mere aggregate of team member energy. Hence, the continuous measurement and aggregation of human energy in organizations during daily operations could complement the more established notions of organizational energy that comprise additional aspects such as "team spirit".

Independent of this distinction, established classifications of human energy could be leveraged for visualizing the current energetic state of the organization. Using a brief measure such as the battery scale in conjunction with classifications of the quality of energy e.g. as being *productive* or *corrosive*, one can support individuals, teams or departments inside organizations to recognize which behaviors and strategies are favorable or dysfunctional in the long term when dealing with the requirements. On this basis, organizations can develop in such a way that they operate a more sustainable human resource management geared to unfold, promote and develop human potential in the long term – instead of pushing their employees to the limit. Future research could clarify the role of human energy mappings on the organizational level, e.g. in regard to identify the need for interventions from the leadership and management perspective. In this regard, also more network-oriented visualizations of energy flows inside the organization could provide an alternative for visualization as suggested by BAKER [42]. Further, it could be explored how questions such as "Which managers manage their department sustainably?" could be answered, based on mapping approaches of human energy at organizational level. Moreover, the existence of positive feedback loops should be investigated, i.e. if visualizations of high levels of productive energy can help to foster proactive work behavior and prevent the perception of work overload and feelings of overstrain.

In regard to business process management approaches, advanced process management and workflow systems could record the respective energy state of the employees either via self-assessment using the low-effort pictorial scale or via unobtrusive ways e.g. by analyzing typing behavior or keyboard pressure. Future research could focus on the question how work units could be assigned in such a way that a global energy maximum of the workforce is reached. In the best case, the consequence should be that the system learns "personal favorite tasks" of the employees, i.e. tasks that one employee likes to do, but that all others do not particularly like to do. This will probably presuppose that there are human energy-relevant attributes of work to which employees respond differently. As a simple example, in customer service this could be regional proximity or personal acquaintance with the customer, which can make the work easier and more pleasant.

All in all, it should be in the interest of companies to take preventive measures to reduce a waste of human energy (especially corrosive energy) in order to reduce stress and burnout and loss of overall effectiveness. Future research regarding human energy application should focus on the one hand on the acceptance of different instruments and methods for human energy measurement by employees. To this end, empirical analyses are to be carried out in order to obtain scientifically validated statements on how large the scope for human energy-aware approaches in concrete work scenarios really is. On the other hand, research should also focus on analyzing data in such a way that it is helpful to inform managers or leaders for better coping with work practices in the sense of "professional human energy-management".

Despite of the manifold opportunities of human energy management on the organizational level, ethical questions as well as data privacy and protection have to be explored in more detail in the future. Even if human energy-related data is anonymized, there is the danger of identification of individuals based on the actual value. Hence it has to be explored how classical approaches for ensuring privacy in survey techniques could be adapted for the measurement of human energy at various levels of detail on the individual, team, department or organizational level. Nevertheless, we hope to stimulate further research on human energy in organizations since the fact that mental illnesses and burnout are on the rise and call for action in regard to develop approaches that help to maintain long-term productivity, health and wellbeing.

References

1. Allen, T.D., Golden, T.D., Shockley, K.M.: How effective is telecommuting? Assessing the status of our scientific findings. Psychol. Sci. Public Interest 16, 40–68 (2015)
2. Shore, L.M., Coyle-Shapiro, J.A.-M., Tetrick, L.E.: The Employee–Organization Relationship: Applications for the 21st Century. Routledge/Taylor & Francis Group, New York (2012)
3. Barber, L.K., Jenkins, J.S.: Creating technological boundaries to protect bedtime: examining work–home boundary management, psychological detachment and sleep. Stress Health: J. Int. Soc. Invest. Stress 30, 259–264 (2014)
4. Green, F., McIntosh, S.: The intensification of work in Europe. Labour Econ. 8, 291–308 (2001)
5. Reilly, N.P., Sirgy, M.J., Gorman, C.A. (eds.): Work and Quality of Life: Ethical Practices in Organizations. Springer, Dordrecht (2012). https://doi.org/10.1007/978-94-007-4059-4
6. Grant, A.M., Parker, S.K.: Redesigning work design theories: the rise of relational and proactive perspectives. Acad. Manag. Ann. 3, 317–375 (2009)
7. Inceoglu, I., Thomas, G., Chu, C., Plans, D., Gerbasi, A.: Leadership behavior and employee well-being: an integrated review and a future research agenda. Leadersh. Q. 29, 179–202 (2018)

8. Skakon, J., Nielsen, K., Borg, V., Guzman, J.: Are leaders' well-being, behaviours and style associated with the affective well-being of their employees? A systematic review of three decades of research. Work Stress **24**, 107–139 (2010)

9. Tetrick, L.E., Winslow, C.J.: Workplace stress management interventions and health promotion. Ann. Rev. Organ. Psychol. Organ. Behav. **2**, 583–603 (2015)

10. Kozlowski, S.W.J., Klein, K.J.: A multilevel approach to theory and research in organizations: Contextual, temporal, and emergent processes. In: Klein, K.J., Kozlowski, S.W.J. (eds.) Multilevel Theory, Research, and Methods in Organizations: Foundations, Extensions, and New Directions, pp. 3–90. Jossey-Bass, San Francisco (2000)

11. Lang, J., Bliese, P.D., de Voogt, A.: Modeling consensus emergence in groups using longitudinal multilevel methods. Pers. Psychol. **71**, 255–281 (2018)

12. Baker, W.E.: Emotional energy, relational energy, and organizational energy: toward a multilevel model. Ann. Rev. Organ. Psychol. Organ. Behav. **6**, 373–395 (2019)

13. Quinn, R.W., Spreitzer, G.M., Lam, C.F.: Building a sustainable model of human energy in organizations: exploring the critical role of resources. Acad. Manag. Ann. **6**, 337–396 (2012)

14. Frone, M.R., Tidwell, M.-C.O.: The meaning and measurement of work fatigue: development and evaluation of the Three-Dimensional Work Fatigue Inventory (3D-WFI). J. Occup. Health Psychol. **20**, 273–288 (2015)

15. Kleine, A.-K., Rudolph, C.W., Zacher, H.: Thriving at work: a meta-analysis. J. Organ. Behav. **40**, 973–999 (2019)

16. Weigelt, O., Gierer, P., Syrek, C.J.: My mind is working overtime—towards an integrative perspective of psychological detachment, work-related rumination, and work reflection. Int. J. Environ. Res. Public Health **16**, 2987 (2019)

17. Wendsche, J., Lohmann-Haislah, A.: A meta-analysis on antecedents and outcomes of detachment from work. Front. Psychol. **7**, 2072 (2017)

18. Crawford, E.R., LePine, J.A., Rich, B.L.: Linking job demands and resources to employee engagement and burnout: a theoretical extension and meta-analytic test. J. Appl. Psychol. **95**, 834–848 (2010)

19. Ragsdale, J.M., Beehr, T.A.: A rigorous test of a model of employees' resource recovery mechanisms during a weekend. J. Organ. Behav. **37**, 911–932 (2016)

20. Zijlstra, F.R.H., Cropley, M., Rydstedt, L.W.: From recovery to regulation: an attempt to reconceptualize 'recovery from work'. Stress Health: J. Int. Soc. Invest. Stress **30**, 244–252 (2014)

21. Sonnentag, S., Venz, L., Casper, A.: Advances in recovery research: what have we learned? What should be done next? J. Occup. Health Psychol. **22**, 365–380 (2017)

22. Crain, T.L., Brossoit, R.M., Fisher, G.G.: Work, Nonwork, and Sleep (WNS): a review and conceptual framework. J. Bus. Psychol. **33**, 675–697 (2018)

23. Fritz, C., Lam, C.F., Spreitzer, G.M.: It's the little things that matter: an examination of knowledge workers' energy management. AMP **25**, 28–39 (2011)

24. Zacher, H., Brailsford, H.A., Parker, S.L.: Micro-breaks matter: a diary study on the effects of energy management strategies on occupational well-being. J. Vocat. Behav. **85**, 287–297 (2014)

25. Ryan, R.M., Frederick, C.: On energy, personality, and health: subjective vitality as a dynamic reflection of well-being. J. Pers. **65**, 529–565 (1997)

26. Shirom, A.: Feeling vigorous at work? The construct of vigor and the study of positive affect in organizations. In: Perrewe, L.P., Ganster, C.D. (eds.) Emotional and Physiological Processes and Positive Intervention Strategies, vol. 3, pp. 135–164. Emerald Group Publishing Limited, Bingley (2003)

27. Beal, D.J.: ESM 2.0: state of the art and future potential of experience sampling methods in organizational research. Ann. Rev. Organ. Psychol. Organ. Behav. **2**, 383–407 (2015)

28. Gabriel, A.S., et al.: Experience sampling methods: a discussion of critical trends and considerations for scholarly advancement. Organ. Res. Methods **22**, 969–1006 (2019)

29. Weigelt, O., Wyss, C., Siestrup, k., Fellmann, M., Lambusch, F.: Ein Bild sagt mehr als tausend Worte - Entwicklung und Überprüfung einer Piktogramm-Skala zu Human Energy [A picture is worth a thousand words - development and validation of a single-item pictorial scale of human energy]. In: Neue Formen der Arbeit in der digitalisierten Welt: Veränderungskompetenz stärken. Braunschweig (2019). https://doi.org/10.13140/RG.2.2.33862.22082

30. Lambusch, F., Weigelt, O., Fellmann, M., Siestrup, k.: Application of a pictorial scale of human energy in ecological momentary assessment research. In: Harris, D., Li, W.-C. (eds.) HCII 2020, LNAI 12186, pp. 171–189, 2020. Springer, Heidelberg (2020)

31. Hannah, S.T., Avolio, B.J., Cavarretta, F.L., Hennelly, M.J.: Conceptualizing organizational energy. In: Organization Science Winter Conference XVI (2010)

32. Schiuma, G., Mason, S., Kennerley, M.: Assessing energy within organisations. Measur. Bus. Excell. **11**, 69–78 (2007)

33. Smit, M.J.: Development of a project portfolio management model for executing organisational strategies: a normative case study. PM World J. **6** (2017)

34. Siruttan, B.: Organisational energy and performance: relevance and implications among knowledge workers (2012)

35. Smith, P.A.C., Tosey, P.: Assessing the learning organization: part 1-theoretical foundations. Learn. Organ. **6**, 70–75 (1999)

36. Bruch, H., Ghoshal, S.: Unleashing organizational energy. MIT Sloan Manag. Rev. **45**, 45–52 (2003)

37. Barker, B.: Manufacturing best practice and human intellectual energy. Integr. Manuf. Syst. **12**, 7–14 (2001)

38. Dutton, J.E.: Energize Your Workplace: How to Create and Sustain High-Quality Connections at Work. Wiley, Hoboken (2003)

39. Fritz, C., Lam, C.F., Spreitzer, G.M.: It's the little things that matter: an examination of knowledge workers' energy management. Acad. Manag. Perspect. **25**, 28–39 (2011)

40. Land, M., Hex, N., Bartlett, C.: Building and aligning energy for change. A review of published and grey literature, initial concept testing and development (2013)

41. Abualhamael, Z.W.H.: The power of productive organisational energy in relation to leadership style and job satisfaction: the context of Saudi Arabian universities (2017)

42. Atwater, L., Carmeli, A.: Leader–member exchange, feelings of energy, and involvement in creative work. Leaders. Q. **20**, 264–275 (2009)

43. Cole, M.S., Bruch, H., Vogel, B.: Energy at work: a measurement validation and linkage to unit effectiveness. J. Organiz. Behav. **33**, 445–467 (2012)

44. Schippers, M.C., Hogenes, R.: Energy management of people in organizations: a review and research agenda. J. Bus. Psychol. **26**, 193–203 (2011)

45. Matthias, T.M.: A conceptual model of information system implementation within organisations (2009)

46. Bruch, H., Vogel, B.: Strategies for creating and sustaining organizational energy. Employ. Relat. Today **38**, 51–61 (2011)

47. Hovsepian, K., al'Absi, M., Ertin, E., Kamarck, T., Nakajima, M., Kumar, S.: cStress: towards a gold standard for continuous stress assessment in the mobile environment. In: UbiComp 2015, pp. 493–504 (2015)

48. Lawanont, W., Mongkolnam, P., Nukoolkit, C., Inoue, M.: Daily stress recognition system using activity tracker and smartphone based on physical activity and heart rate data. In: Czarnowski, I., Howlett, R.J., Jain, L.C., Vlacic, L. (eds.) KES-IDT 2018 2018. SIST, vol. 97, pp. 11–21. Springer, Cham (2019). https://doi.org/10.1007/978-3-319-92028-3_2

49. Gjoreski, M., Gjoreski, H., Luštrek, M., Gams, M.: Continuous stress detection using a wrist device. In: Lukowicz, P. (ed.) Proceedings of the 2016 ACM International Joint Conference on Pervasive and Ubiquitous Computing Adjunct, pp. 1185–1193. ACM, New York (2016)
50. Koldijk, S., Neerincx, M.A., Kraaij, W.: Detecting work stress in offices by combining unobtrusive sensors. IEEE Trans. Affect. Comput. **9**, 227–239 (2018)
51. Hernandez, J., Paredes, P., Roseway, A., Czerwinski, M.: Under pressure. In: Jones, M. (ed.) Proceedings of the SIGCHI Conference on Human Factors in Computing Systems, pp. 51–60. ACM (2014)
52. Nakashima, Y., Kim, J., Flutura, S., Seiderer, A., André, E.: Stress recognition in daily work. In: Serino, S., Matic, A., Giakoumis, D., Lopez, G., Cipresso, P. (eds.) MindCare 2015. CCIS, vol. 604, pp. 23–33. Springer, Cham (2016). https://doi.org/10.1007/978-3-319-32270-4_3
53. Lietz, R., Harraghy, M., Calderon, D., Brady, J., Becker, E., Makedon, F.: Survey of mood detection through various input modes. In: Makedon, F. (ed.) PETRA 2019. The 12th ACM International Conference on Pervasive Technologies Related to Assistive Environments, Conference Proceedings, 05–07 June 2019, Rhodes, Greece, pp. 28–31. ACM, New York (2019)
54. Salehan, M., Kim, D.J.: Predicting the performance of online consumer reviews: a sentiment mining approach to big data analytics. Decis. Support Syst. **81**, 30–40 (2016)
55. De, N.R.: Organisational energy: a diagnostic exercise. Economic and Political Weekly, M95-M102 (1981)

Computer-Based Neuropsychological Theory of Mind Assessment: A Validation Study

Gilberto Galindo-Aldana[1]([⊠]), Alberto L. Morán[2],
Cynthia Torres-González[1], Lesdly Cabero[3], and Victoria Meza-Kubo[2]

[1] Laboratory of Neuroscience and Cognition, Mental Health, Profession
and Society Research Group, Autonomous University of Baja California,
Mexicali, Mexico
gilberto.galindo.aldana@uabc.edu.mx
[2] Faculty of Sciences, Research Group of Techlogies for Intelligent
Environments Research Group, Autonomous University of Baja California,
Ensenada, Mexico
[3] CETYS Universidad, Mexicali, Mexico

Abstract. Theory of mind is defined as the ability to attribute mental states to oneself and others, this psychological capacity is referred like one of cornerstones of efficient social interaction. The goal of the present study was to create an adapted computer-based version of the Reading the Mind in the Eyes Test (RMET), and to measure its validity in a sample of children Mexico. Method. Instruments: *Computer-Based RMET preparation.* The test consisted on a 28 trial of visual stimuli based on the Baron-Cohen, et al. version [1], containing eyes section from the face of male and female. The application was developed in Java, following a modular, iterative and incremental software development process based on requirements gathering, analysis, design, construction and testing. *Participants.* A total of 54 participants mean age 10.78 years old (SD = .74), voluntarily accepted to respond the test. The results were divided into three sections: i) comparative findings between paper and computer-based versions for total scores, showed no significant differences; ii) KR20 showed strong internal consistency inter-item correlations; iii) general scores also showed a significant correlation between the versions. Discussion. Other studies in Latin-America, had shown internal validity of the paper-based RMET. However, this is the first to demonstrate the internal validity of a computer-based variant of this test for the adult version. Future statistical analysis should include mean difference analysis for data dispersion, and normative data for the target children population for clinical purposes validity.

Keywords: Assessment · Clinical-Neuropsychology · Executive-Function

1 Introduction

1.1 Theory of Mind

Theory of mind (ToM) is a term proposed by Premack and Woodruff in 1978 [1] to refer to a psychological function that allows making inferences about the other's

© Springer Nature Switzerland AG 2020
D. Harris and W.-C. Li (Eds.): HCII 2020, LNAI 12186, pp. 141–149, 2020.
https://doi.org/10.1007/978-3-030-49044-7_13

oneself mental states such as thoughts, feelings, etc. Further, this ability is considered an executive function as it is involved in the comprehension, prediction and explanation of the conduct of others, their intentions, beliefs and their knowledge about a situation [2–5]. The importance of this mental function is that it allows establishing functional social interactions since we can evaluate and adjust our behavior to specific social conditions [6]. A better development of theory of mind makes it more likely to have successful social interactions, since it is possible to understand not only the intentions of the others, but the way in which the other people represents the world and it can be different from the own way [2, 7].

Assessing cognitive processes is a remarkable need in clinical neuropsychology, but it represents a constant improvement and precision of instruments development. According to Björngrim [8], during 80's and 90's decades, the analysis carried out around the subject favored to the traditional test that were made on paper; afterwards, at the beginning of the year 2000, it is possible to observe a significant reduction in the differences between the traditional methods versus the digital ones. The instruments that require a computer are now more frequently used in clinical practice; therefore, the research field has been paying more attention to it. In general terms, the American Academy of Clinical Neuropsychology and the National Academy of Neuropsychology emphasize and point out the benefits of computerized tests: they allow to evaluate in a quicker way, a greater number of patients; measure time reaction more precisely; notably reduces the costs and time in evaluations; exports data automatically; and they are capable of integrate and automate interpretative algorithms, as well as the rules which determine the disability or changes in a trustworthy, statistic manner [9]. In the clinical work, computerized tests have been so useful to detect and diagnose a wide variety of conditions or neuropsychologist alterations that have been exhaustive studied in different populations. Evaluating cognitive process through digital mechanism allows getting more information, including sensitive and multidimensional information. In the case of adults with Alzheimer's disease or other neurodegenerative disorders, besides reaching the status information about cognitive deterioration and functional changes, it is possible to detect subtle changes and patterns of movement. On the other hand, in patients with deficit in working memory, and visual integration alterations, commonly observed in people with attention-deficit disorder, autism disorder, and Down's syndrome, the use of digital tools makes possible to stimulate and measure cognitive and motor skills [10].

In clinical neuropsychology, it can be found that ToM can be assessed since early stages of life. From the classical Piagetian perspective, children develop ToM from its fundamentals of egocentric preoperational perspective that progressively moves towards a world's decentered and independent understanding [11]. Recently, neurocognitivist perspective, suggest that the social cognition (that includes ToM) develops as a brain network that becomes increasingly fined tuned to relevant stimuli and events in activity-dependent manner, including abilities like the perception and processing of faces, joint attention, and empathy [12, 13].

2 Method

Participants. 54 participants were included in the sample; two were excluded due to incomplete data on paper-based assessment. The final two related-sample groups consisted of 52 children 10–12 years old, 10.78 mean age 0.74 SD, 57% were male, while 43% were female. The protocol was successfully approved by the Ethics Committee, indicating no need for informed consent, the application took place in elementary school's facilities, and parents and school's authorities gave proper authorization for participation.

Instruments and Procedures. The paper-based assessment consisted of the 28 item version for children [6]. Each item was presented in a single paper sheet, displaying the visual black and white colored stimuli; all stimuli presented the eyes section of a human face, 58% male, 42% female. At each corner of the paper sheet four options were displayed in times new roman 14 pt. text defining one correct response for the displayed visual stimuli, showing the closest description to the eyes intentions representation, and three incorrect options. Neither time limit nor feedback was given for each stimuli or test in general, once the participant made a selection the next paper-sheet stimuli was administered.

The computer-based RMET test consisted on a 28 trial of visual stimuli based on the Baron–Cohen, et al. original version [6], containing eyes section from the face (Fig. 1) of male 58% and female 42% displayed in black and white over a gray background screen. Four options were presented simultaneously to the eyes picture in text boxes under it. The application was developed in Java, following a modular, iterative and incremental software development process based on requirements gathering, analysis, design, construction and testing. All logic and interaction flow were directly programmed in the application modules. The modules included are: *Patient registration*, where the patient captures his/her full name and age, and based on these a profile is created. *Instructions*, as its name implies, gives the patient information about what the test consists of and how they should respond. *Test questions*, present each of the questions conforming the test, each of which include the image to be identified and the options that the user may choose to respond; the order of the questions and the answers is random in each instance of the test. *Transition between questions*, patients have 10 s to answer each question, in case it is not answered at that time, the question is omitted and the next question is presented. At the end of each question, the patient is presented with a distracting stimulus (black cross on a white background) for a short time, at the end of which the next test question is presented. *Save results*, the results of each test (i.e. question number, answer and score) are stored temporarily and at the end of the test they are exported to a file that is identified with a name based on the patient's name, date and time of test application.

Fig. 1. Sample stimuli from the computer-based version of the theory of mind test.

Each child was assessed individually in a quiet room, 50% of the participants received the paper-based application first, followed by the computer-based assessment, while the other 50% of the participants received the inverse sequence of administration of the instruments.

Data Analysis. Analyses were performed in three moments. First we employed related-samples T-test to compare the total score which compiles a sum of correct responses from the 28 items of each application format. Secondly, percentages of selection for each one of the four options for each stimulus were calculated and mean percentages were compared. Finally a Pearson's coefficient was obtained between the percentages of selections and KR20 inter-item coefficient for correct and error responses on the computer-based RMET data.

3   Results

For each application format we first calculated the sum of correct responses from each item. Afterwards related-samples two-tailed T-test was calculated, which showed no difference between the correct responses means ($t = -.036$, DF = .51, $p = .971$).The linear regression showed non-proportional bias coefficient ($B = -.135$, $p = .243$) among the samples. Results from errors and correct responses for each item were used to estimate inter-item internal consistency reliability test with Kuder-Richardson 20 (KR20) test for dichotomic variables from the computer-based version. A strong internal consistency coefficient was found in this analysis ($KR20 = .805$). Differences between selections for each expression are listed in Table 1. Selections in the target column indicate the number of correct responses made by each participant and are represented in percentage in the table. However, proportion of errors when making any of the other selections was also analyzed to compare if significant differences were presented.

Table 1. Percentage per selection made for each stimulus. P = paper-based application, C = computer-based application

Item	Target	%P	%C	Foil 1	%P	%C	Foil 2	%P	%C	Foil 3	%P	%C
1	Amable (kind)	17	40	Cara de odio (hate)	44	30	Enojado (cross)	23	15	Sorprendido (surprised)	16	15
2	Triste (sad)	56	58	Enojado (cross)	4	19	Sorprendido (surprised)	4	11	Pocoamable (unkind)	36	12
3	Simpático (friendly)	31	42	Preocupado (worried)	38	29	Triste (sad)	21	21	Sorprendido (surprised)	10	8
4	Disgustado (upset)	75	68	Tranquilo (relaxed)	10	10	Sorprendido (surprised)	4	11	Entusiasmado (excited)	11	11
5	Persuasivo (making somebody do something)	37	48	Arrepentido (feeling sorry)	25	19	Tranquilo (relaxed)	25	23	Bromista (joking)	13	10
6	Preocupado (worried)	61	58	Aburrido (bored)	27	25	Cara de odio (hate)	4	9	Cruel (unkind)	8	10
7	Interesado (interested)	65	71	Aburrido (bored)	17	10	Arrepentido (feeling sorry)	13	11	Bromista (joking)	5	8
8	Recordandoalgo (remembering)	82	75	Enojado (angry)	8	8	Simpático (friendly)	3	9	Contento (happy)	8	8
9	Pensativo (thinking)	66	69	Molesto (annoyed)	10	12	Cara de odio (hate)	9	4	Sorprendido (surprised)	15	15
10	Incrédulo (not believing)	50	43	Amable (kind)	16	17	Tímido (shy)	19	29	Triste (sad)	15	11
11	Cara de querer algo (hoping)	70	63	Cara de asco (disgusted)	13	10	Dominante (bossy)	10	25	Enojado (angry)	7	2
12	Serio (serious)	69	64	Confundido (confused)	19	15	Triste (sad)	4	2	Bromista (joking)	8	19
13	Pensativo (thinking about something)	67	64	Disgustado (upset)	11	19	Entusiasmado (excited)	11	10	Contento (happy)	11	7
14	Pensativo (thinking about something)	54	69	Contento (happy)	14	14	Amable (kind)	17	8	Entusiasmado (excited)	15	9
15	Incrédulo (not believing)	21	31	Tranquilo (relaxed)	42	42	Ganas de jugar (wanting to play)	14	10	Simpático (friendly)	23	17
16	Decidido (made up her mind)	68	72	Aburrido (bored)	11	13	Sorprendido (surprised)	11	13	Bromista (joking)	10	2
17	Un poco preocupado (a bit worried)	77	67	Pocoamable (unkind)	10	12	Enojado (angry)	11	8	Simpático (friendly)	2	13
18	Pensandoenalgo triste (thinking about something sad)	60	73	Enojado (angry)	6	6	Dominante (bossy)	23	11	Simpático (friendly)	11	10
19	Interesado (interested)	56	56	Soñador (daydreaming)	27	21	Enojado (angry)	7	15	Triste (sad)	10	8
20	No satisfecho (not pleased)	79	73	Amable (kind)	0	6	Sorprendido (surprise)	10	8	Entusiasmado (excited)	11	13
21	Interesado (interested)	59	63	Tranquilo (relaxed)	24	25	Contento (happy)	11	6	Bromista (joking)	6	6
22	Pensativo (thinking about something	56	64	Sorprendido (surprised)	11	15	Amable (kind)	25	11	Divertido (playful)	8	10
23	Seguro (sure about something)	75	63	Sorprendido (surprised)	11	15	Contento (happy)	6	8	Bromista (joking)	8	14
24	Serio (serious)	51	64	Avergonzado (ashamed)	25	11	Confundido (confused)	16	19	Sorprendido (surprised)	8	6
25	Preocupado (worried)	51	48	Tímido (shy)	25	23	Culpable (guilty)	14	23	Soñador (daydreaming)	10	6
26	Nervioso (nervous)	56	46	Arrepentido (sorry)	29	19	Bromista (joking)	4	16	Tranquilo (relaxed)	11	19

(continued)

Table 1. (*continued*)

Item	Target	%P	%C	Foil 1	%P	%C	Foil 2	%P	%C	Foil 3	%P	%C
27	Incrédulo (not believing)	44	42	Avergonzado (ashamed)	21	14	Satisfecho (pleased)	16	27	Entusiasmado (excited)	19	17
28	Contento (happy)	53	46	Aburrido (bored)	27	27	Cara de asco (disgust)	10	12	Cara de odio (hate)	10	15
	Mean	57.3	58.6		18.7	17.3		12.3	13.4		11.6	10.7
		P_r = **.835**, p < **.001**			P_r = **.829**, p < **.001**			P_r = **.437**, p **.020**			P_r = .315, p = .103		

P_r = Pearson's correlation coefficient between percentage of responses in paper and computer-based applications for each selection, p = sig. p value for the correlations, significant correlations are marked in bold

Table note: selection tagnames correspond to original spanish and english test nomenclatures.

4   Discussion

The objective of this study was to present the first validation results of a computer-based neuropsychological instrument for assessing ToM, towards its comparison with the original paper-based version in a sample of children. The use of computer-based versions of clinical assessment instruments offers several advantages, although they require to be evaluated to determine their equivalence with traditional (paper and pencil) versions [14], they result attractive because computer-based versions have features that are not possible in traditional versions of the instruments [15].

Evaluating the comparability of paper-based and computer-based test is crucial before introducing computer-aided assessment [16]. In the scientific literature, there are many comparability studies that have examined the impact of transferring a test from paper to computer among which are testsfor assessing cognitive processes such as language [17], memory [18, 19] and working memory [20]. However, this is the first work which applies and validates a social cognition assessment task for Mexican children in a computer-based format.

Several studies have described that major advances in ToM understanding occur between 3 and 6 years old, and they are linked to improvements in both, cognitive and social functioning in neurotypical population. Nevertheless, ToM is a developmental phenomenon with important advances over early to middle childhood [5, 21]. In atypical development, ToM abilities fail to fully develop, such as neurodevelopmental disorders (like autistic spectrum disorders) or in acquired brain damage [6]. For this reason it is important to have validated assessment instruments that can be used both in typical and atypical development populations.

Our study demonstrates no significant differences between the paper version and the digital one, as the comparative analysis shows, suggesting that the test properties are very similar. However, it has been reported that the use of digital tools for the assessment could have some advantages over the traditional ones, such as: to capture and engage the person's interest, to assess a greater number of individuals quickly, reduced cost because the administration and the scoring are carried out automatically, automated data exporting for research purposes, likewise it is more suitable for repeated evaluations as it allows for the randomization of stimuli producing alternative forms of the test, increased accessibility to patients in areas in which neuropsychological services are scarcer and more accurate measurement of time indicators such as the reaction time [22–24].

In clinical populations, such as people diagnosed with autism spectrum disorders (ASD), researchers have suggested that technological tools may be particularly promising as clinical mechanisms for some children with this disorder given that many of them exhibit greater strengths in understanding the "physical" world compared with the social one, and they respond well to technological feedback and show interest in technology [25]. Gillespie-Lynch, Kapp, Shane-Simpson, Shane &Hutman [26] carried out a study whose aim was to evaluate whether computer-mediated communication or use of the internet to interact with others, benefits to ASD participants. The sample was conformed by 291 participants with ASD diagnosis and 311 without it. The perception of communicative benefits of internet use was explored through a digital survey. Results showed that the ASD participants reported significantly less face-to-face interactions compared with the control group. The perceived benefits of computer mediated communication that showed greater difference between ASD and non-ASD groups, were: express true self, practice interaction, time to think, choose who talk to, find people like you and written down. This findings are particularly important in the case of our instrument because children with autism spectrum disorders present as one of their main characteristically symptoms, alterations in the development of social cognition and therefore in the theory of mind.

The computerized tests have been used in other fields of child neuropsychology, such as the motor disabilities populations [27], patients with sequelae of sport's concussion [15] or from treatments for serious diseases such as cancer. Heitzer, Ashford, Harel and collaborators [28] reported the results of their study in 73 children diagnosed with medulloblastoma and treated with radiotherapy, who completed a neuropsychological assessment using the Cogstate test before and three months after of the radiotherapy. At the baseline, standard neuropsychological measures of similar cognitive domains were administered. Pearson´s correlation coefficient showed moderate correlations between Cogstate measures (as reaction time) and well-validated neuropsychological measures. Further, analysis revealed that the performance at the baseline was within age expectations but following radiation therapy, there was a decline in performance, suggesting that the test is sensitive to acute cognitive effects of medulloblastoma treated with radiation therapy.

Future validation results of this neuropsychological assessment instrument proposal, should include exploratory and confirmatory factor analysis, and mean difference comparison based on confidence intervals ranks, finally normative data with regression methods including a comprehensive sample controlling variables, such as family, socio-cultural, and clinical features.

References

1. Premack, D., Woodruff, G.: Does the chimpanzee have a theory of mind? Behav. Brain Sci. **4**, 515–526 (1978)
2. Román, F., et al.: Argencog. Baremos del test de la mirada en español en adultos normales de Buenos Aires. Revistaneuropsicología latinoamericana **4**(3), 1–5 (2012)

3. Cho, I., Cohen, A.: Explaining age-related decline in theory of mind: evidence for intact competence but compromised executive function. PLoS ONE **14**(9), e0222890 (2019)
4. Wade, M., Prime, H., Jenkins, J.M., Yeates, K.O., Williams, T., Lee, K.: On the relation between theory of mind and executive functioning: a developmental cognitive neuroscience perspective. Psychon. Bull. Rev. **25**(6), 2119–2140 (2018). https://doi.org/10.3758/s13423-018-1459-0
5. Bradford, E., Jentzsch, I., Gomez, J.: From self to social cognition: theory of mind mechanisms and their relation to executive functioning. Cognition **138**, 21–34 (2015)
6. Baron-Cohen, S., Wheelwright, S., Hill, J., Raste, Y., Plumb, I.: The "Reading the Mind in the Eyes" test revised version: a study with normal adults and adults with asperger syndrome or high-functioning autism. J. Child Psychol. Psychiat. **42**(2), 241–251 (2001)
7. Boccadoro, S., et al.: Defining the neural correlates of spontaneus theory of mind (ToM): an fMRI multistudy investigation. NeuroImage **203**, 116193 (2019)
8. Björngrim, S., van den Hurk, W., Betancort, M., Machado, A., Lindau, M.: Comparing traditional and digitized cognitive tests used in standard clinical evaluation–a study of the digital application minnemera. Front. Psychol. **10**, 2327 (2019)
9. Bauer, R.M., Iverson, G.L., Cernich, A.N., Binder, L.M., Ruff, R.M., Naugle, R.I.: Computerized neuropsychological assessment devices: joint position paper of the American academy of clinical neuropsychology and the national academy of neuropsychology. Clin. Neuropsychol. **26**(2), 177–196 (2012)
10. Torres-Carrión, P.V., et al.: Improving cognitive visual-motor abilities in individuals with down syndrome. Sensors **19**(18), 3984 (2019)
11. Perner, J., Stummer, S., Sprung, M., Doherty, M.: Theory of mind its Piagetian perspective: why alternative naming comes with understanding beliefs. Cognit. Dev. **17**, 1451–1472 (2002)
12. Johnson, M., De Haan, M.: Perceiving and acting in the social world. In: Developmental Cognitive Neuroscience: An Introduction. 4th edition. Wiley-Blackwell, United Kingdom (2015)
13. Decety, J.: The neurodevelopment of empathy in humans. Dev. Neurosci. **32**, 257–267 (2010)
14. Wagner, G.P., Trentini, C.M.: Assessing executive functions in older adults: a comparison between the manual and the computer-based versions of the wisconsin card sorting test. Psychol. Neurosci. **2**(2), 195 (2009)
15. Schatz, P., Browndyke, J.: Applications of computer-based neuropsychological assessment. J. Head Trauma Rehabil. **17**(5), 395–410 (2002)
16. Al-Amri, S.: Computer-based vs. paper-based testing: does the test administration mode matter. In: Proccedings of the BAAL Conference (2007)
17. Choi, I., Sung, K., Boo, K.: Comparability of a paper-based language test and a computer-based language test. Lang. Test. **20**(3), 295–320 (2003)
18. Clionsky, M., Clionsky, E.: Psychometric equivalence of a paper-based and computerized (iPad) version of the memory orientation screening test (MOST). Clin. Neuropsychol. **28**(5), 747–755 (2014)
19. Vanderslice-Barr, J., Miele, A., Jardin, B., McCaffrey, R.: Comparison of computerized booklet versions of the TOMM. Appl. Neuropsychol. **18**(1), 34–36 (2011)
20. Carpenter, R., Alloway, T.: Computer versus paper-based testing: are they equivalent when it comes to working memory? J. Psychoeduc. Assess. **37**(3), 1–13 (2018)
21. Bowman, L., Dodell-Feder, D., Saxe, R., Sabbagh, M.: Continuity in the neural system supporting children's theory of mind development: longitudinal links between task independent EEG and task-dependent fMRI. Dev. Cognit. Neurosci. **40**, 100705 (2019)

22. Bauer, R., Iverson, G., Cernich, A., Binder, L., Ruff, R., Naugle, R.: Computerized neuropsychological assessment devices: joint position paper of American academy of clinical neuropsychology and the national academy of neuropsychology. Arch. Clin. Neuropsychol. **27**, 362–373 (2012)
23. Dede, E., Zalonis, I., Gatzonis, S., Sakas, D.: Integration of computers in cognitive assessment and level of comprehensiveness of frequently used computerized batteries. Neurol. Psychiatry Brain Res. **21**(3), 128–135 (2015)
24. Kane, R.L., Parsons, T.D.: The Role of Technology in Clinical Neuropsychology, 1st edn. Oxford University Press, New York (2017)
25. Kumazaki, H., et al.: Brief report: evaluating the utility of varied technological agents to elicit social attention from children with autism spectrum disorders. J. Autism Dev. Disord. **49**, 1700–1708 (2019)
26. Gillespie-Lynch, K., Kapp, S., Shane-Simpson, C., Shane, D., Huntman, T.: Intersections between the autism spectrum and the internet: perceived benefits and preferred functions of computer-mediated communication. Intelectual Dev. Disabil. **6**, 456–469 (2014)
27. Christy, J., et al.: Technology for children with brain injury and motor disability: executive summary from research summit IV. Pediatr. Phys. Therapy **28**, 483–489 (2016)
28. Heitzer, A., et al.: Computerized assessment of cognitive impairment among children undergoing radiation therapy for medulloblastoma. J. Neuro-oncol. **141**(2), 403–411 (2019)

Augmented Energy for Locomotion: How Do Users Perceive Energy Dynamics in Prototypical Mobility Scenarios?

Markus Gödker[✉] and Thomas Franke

Universität zu Lübeck, Ratzeburger Allee 160, 23562 Lübeck, Germany
goedker@imis.uni-luebeck.de

Abstract. Locomotion can be considered to represent a core need in human life and is—in its most natural form—achieved by using available muscular human energy to obtain relocation in space over time. In order to amplify the human capabilities of locomotion (e.g. expand the mobility space and pace) humans have developed mobility tools (i.e. vehicles) that augment the available human energy for motion. Yet, effectively controlling such mobility tools can increase the demands on cognitive human energy resources. The objective of the present research was to examine how people perceive and understand human vs. external energy consumptions in dynamic locomotion situations and how to strike the right path towards optimal energy-related human-machine symbioses. An online survey ($N = 108$) assessed which energy-related dynamics in locomotion situations are typically perceived by users in exemplary prototypical mobility scenarios and key variables to characterize user diversity in user interaction within such situations were assessed. Results are discussed from the perspective of potential integrated energy feedback systems for locomotion, with a focus on absent sensory information to understand and trace energy dynamics. This study might serve as a first basis for further research that aims at developing tools to optimally integrated human energy and external energy.

Keywords: Locomotion · Energy · Augmented energy · Energy awareness · Energy perception

1 Introduction

1.1 Locomotion and Energy

Human locomotion can be considered to represent a core need in human life that is achieved by transforming available muscular energy to kinetic and potential energy in order to obtain relocation in space over time. In its most natural form, humans use the bipedal plantigrade progression (walking), where coordinated body movements lead to conversions of potential and kinetic energy [1]. Interestingly, biologists study variances in energy costs for locomotion for different animal species to find influencing factors like for example muscle efficiency [2], body movements (gait kinematics) [3] or even the number of leg pairs [4]. Although humans tend towards reducing energy

D. Harris and W.-C. Li (Eds.): HCII 2020, LNAI 12186, pp. 150–160, 2020.
https://doi.org/10.1007/978-3-030-49044-7_14

expenditure, the human walking movements cannot be considered as optimally energy efficient in locomotion [1].

In order to amplify the human capabilities of locomotion, i.e. expand the space, reduce the time needed or reduce the invested muscular energy, humans have developed various mobility tools (i.e. vehicles). Some of these vehicles achieve higher mechanical efficiency compared to walking by using and controlling radial motion (e.g. bikes) other vehicles make use of "external" forms of energy, like wind, fuel, electricity in boats or cars. Effectively controlling such mobility tools can in turn increase the demands on cognitive human energy resources (e.g. increase in cognitive workload for safe locomotion induced by higher movement velocities, and therefore e.g. shorter reaction times, or by energy management demands). This is particularly the case in scenarios where external and/or human energy resources for locomotion are scarce or precious (e.g. the available range in battery electric vehicles, BEVs).

In the work context, a similar distinction of resources has been made [5], where "physical energy" might be considered as muscular and cognitive energy, and "other resources" include external energy resources. According to Quinn et al., human energy encompasses both, physical energy and so-called "energetic activation". Physical energy can be defined as the "capacity to do work" [5, p. 341], which means moving, thinking or doing something. This type of energy is basically glucose or adenosine triphosphate (ATP). Nevertheless, the presence or absence alone of these chemical structures does not automatically lead to an energization or exhaustion. In general, certain "energizers" or "de-energizers" [5, p. 340] moderate the engagement in a certain kind of action or task. De-energizers and energizers can be understood as certain energy-related events (e.g. sleep deprivation vs. well-restedness), interindividual differences (e.g. self-control vs. loss of thereof) or temporary (situation dependent) states (e.g. motivation, flow, positive affect). Yet, it remains unclear to what extent humans are aware of these dynamics and how they can be controlled.

In the context of locomotion, human energy is a limiting factor. Of course, in walking, where almost no external energy (e.g. as tailwind) is used, muscular human energy limits the locomotion ultimately and, hence, must be used efficiently and in a controlled manner. A bicycle is a vehicular tool that achieves higher mechanical efficiency. Therefore, the expended muscular energy and even the scarce external energy from tailwinds results in greater speeds and higher dynamics. This in turn means that cognitive demands increase and – besides the primary muscular energy – cognitive energy gains importance. With the use of greater amounts of external energy in a car, the connection between muscular energy and the covered distance in space dissolves almost completely and muscular energy becomes irrelevant. In a conventional combustion car, external energy can be refilled quickly and easily and is therefore not a limiting factor (rather the financial resources). In this case, cognitive energy is of greater relevance and might even pose a safety risk when depleted. Furthermore, the use of external energy in the field of electromobility is even more precious and complex. Here, an empty battery means that the trip must be interrupted to recharge the battery. This makes external energy a limiting factor and the scarcity and complexity of external energy raises the cognitive load and consequently demands cognitive energy.

In everyday mobility, not only the question about what ultimately limits locomotion is of relevance. It is obvious that people will not use their BEV until its external energy is depleted, and then just leave their BEV and complete the rest of the route on foot because they have so much muscular energy left. Everyday mobility demands further considerations regarding time, costs and comfort (among others), which in turn are related to different vehicular tools and energy expenditures [6, 7]. From a human-machine-system perspective, the question that arises is: How can we optimally achieve our mobility needs with the available vehicular tools and with efficient human and external energy use? To do so, the optimal human-machine symbiosis must be characterized by a high *controllability* of the limiting and relevant energy resources, which means that users are empowered and capable to control the energy flows according to their will. In other words, external energy should be as directly perceivable and understandable as one's own human energy and may be understood as an augmentation of the own human energy resources. We will call this *augmented energy* (for locomotion).

1.2   Energy Dynamics Awareness

For an optimal use of human and external energy, it is crucial for designers of technical systems to know how to support energy efficient behaviors and decisions. Regarding external energy, different approaches have already been examined. For example, DeWaters and Powers [8] introduced the term energy literacy as an individual predisposition that encompasses affective, cognitive and behavioral characteristics and empowers humans to make more efficient energy related decisions. Another approach is to examine system characteristics for example how transparent and perceivable energy dynamics are. Furthermore, additional supportive energy feedback human-machine interfaces (HMIs) [9] can then be designed and implemented. Energy feedback HMIs support energy related decisions by providing useful information (e.g. as live, aggregated or visualized feedback) that help form comprehension about the dynamic situation and serves as a requirement for rational and accurate energy decisions.

Regarding human energy, problems in accurately perceiving and understanding dynamic energy flows and expenditures arise from the gap between actual physical energy and energetic activation [5]. Nevertheless, in most of the human energy literature, no doubts about the accuracy of human energy estimates are raised. It is unclear what shortcomings might prevent an accurate perception and assessment of one's own human energy, and how an accurate assessment improves the management and regulation of human energy at all.

Situation awareness (SA) [10] as a human factor concept is described as a factor that facilitates appropriate behavior in dynamic situations. Situation awareness can be understood as the result from the situation assessment process [11] and arises from three levels: (1) perception, (2) comprehension and (3) projection. An accurate situation awareness guides the situation assessment process and leads to more accurate decisions and behavior. Originally, situation awareness literature refers to safety critical situations, but first attempts to transfer situation awareness to energy-related situations have already been introduced [12, 13]. Based on preliminary findings, energy dynamics awareness (EDA) [13] as an adaptation of situation awareness to the specific context of energy-relevant situations might facilitate energy-related decisions and behaviors.

Interindividual differences like energy literacy or system characteristics (including energy feedback HMIs) could be useful for the energy situation assessment process and help to build EDA. As an adaptation of the three levels of SA, two aspects are encompassed by EDA: (1) the *comprehension* of energy situations (refers to SA level 1 and 2) and (2) the capability to *control* energy flows and consumption (refers to SA level 3). While the first aspect means to accurately perceive and understand energy-related information and to form an appropriate comprehension of the situation, the second aspects arises from the short-termed projection of energy flows that empowers to plan and decide for energy efficient behaviors.

We postulate that energy dynamics awareness is a key factor in the optimal energy-related human-machine symbiosis and that energy dynamics awareness can help humans understand energy flows regarding human and external energy in locomotion, in order to optimally control limiting resources.

1.3   Present Research

The objective of the present research is to get a first insight into user's understanding of different energy flows in one form of locomotion. We aim at finding similarities and differences in the comprehension of those types of energies, how humans deal with energy resources and to what extent they are aware of different aspects of the temporal dynamics in human and external energy resource flows. Moreover, the present study is a very first step on the way to develop reasonable methods and assumptions to study human and external energy resource management. In the end, such knowledge is needed to achieve truly optimal human-machine integration and user-vehicle symbiosis, in order to obtain high levels of energy efficiency and sustainable resource utilization.

In this study, we want to examine whether a resource regulation perspective that considers both, human and external energy together, could be possible. To be precise, can we measure a concept of energy dynamics awareness in all types of energy?

As already mentioned, different types of energy are the limiting factor in human locomotion and therefore most relevant to regulating in different forms of locomotion. While walking and riding the bicycle, muscular energy is most relevant, external energy is the most limiting factor in an electric vehicle. Moreover, in an electric car as well as on a bicycle, cognitive energy plays an additional role. Whenever a type of energy is relevant, technical systems should be designed to optimally support an adequate energy dynamics awareness. In this study we aim at measuring energy dynamics awareness for different types of energies in different forms of locomotion and exploring differences and their link to the above-mentioned relevance. To be able to make comparisons between the above-mentioned forms of locomotion, we include two further forms of locomotion: taking an electric bus and riding an e-scooter. Taking a bus can be considered a form of locomotion where no energy resource become relevant, because it requires minimal human energy and external energy resources are already almost completely planned and managed. We decided for a battery electric bus, because it is reasonable to stick to one external energy resource (electricity) for comparative purposes. Secondly, e-scooters are a completely new form of locomotion in inner cities in Germany and demand for muscular (coordination), cognitive and external energy.

2 Method

2.1 Sample

After closing the online survey, 110 participants completed the survey. All participants whose time to complete was 1 min or less (1), or whose variance in all Likert scale answers was 0 (1) were excluded from the analysis. In the end, the final sample consisted of N = 108 participants (69% female, 30% male, 1% diverse) with an average age of M = 23.7 (SD = 3.5), 62% of the participants had a high school diploma and 25% an academic bachelor.

2.2 Procedure

To measure the ability to assess the energy dynamics in external, muscular and cognitive energy, we adapted the energy dynamics awareness scale [13] to a shorter version of 4 instead of 7 items and where the type of energy is specified in the item text. However, we kept the two aspects comprehension and control. Item 1 and 2 therefore refer to the aspect comprehension, item 3 and 4 to the aspect control. As a rating scheme, all items use a 6-point Likert scale (agreement rating) as follows: 1 = *completely disagree*, 2 = *largely disagree*, 3 = *slightly disagree*, 4 = *slightly agree*, 5 = *largely agree*, 6 = *completely agree*. All items are formulated positively and originally in German (Table 1).

Table 1. The adapted EDA scale for all three types of energy.

Item No.	Text (muscular/cognitive/external), *original German text in italics.*
1	In this form of locomotion, I get a very good overview of the temporal dynamics in the (muscular/cognitive/external) energy consumption of the (body/brain/vehicle). *In dieser Fortbewegungsart bekomme ich einen sehr guten Überblick über die zeitliche Dynamik im (muskulären/kognitiven/externen) Energieverbrauch des (Körpers/Gehirns/Fortbewegungsmittels).*
2	With this form of locomotion, it is possible to understand the (muscular/cognitive/external) energy consumption exactly. *Bei dieser Fortbewegungsart ist es möglich, den (muskulären/kognitiven/externen) Energieverbrauch genau zu verstehen.*
3	In this form of locomotion, it is easy for me to understand with which actions I can influence the (muscular/cognitive/external) energy consumption. *Es ist für mich in dieser Fortbewegungsart gut verständlich, durch welche Handlungen ich den (muskulären/kognitiven/externen) Energieverbrauch beeinflussen kann.*
4	In this form of locomotion, I feel well able to influence the (muscular/cognitive/external) energy consumption. *In dieser Fortbewegungsart fühle ich mich gut dazu imstande, den (muskulären/kognitiven/externen) Energieverbrauch zu beeinflussen.*

We conducted a scenario-based online survey to assess which energy-related dynamics in locomotion situations are typically perceived how by users in exemplary prototypical mobility scenarios, to test the three new EDA scales. We developed a scenario into which the study participants should put themselves:

"Please imagine that you are on the old town island of Lübeck and want to go to the university. The distance to the university is about 3.7 km and leads along a large inner-city road. [...] You have 5 possibilities to get to the university: (1) by public bus, (today a battery electric bus), (2) with the E-CarSharing car, which is a small battery electric vehicle from the local CarSharing provider, (3) by bicycle, (4) with the e-scooter and (5) by walking.

Please imagine for this survey, you are not under time pressure and you have unlimited access to all forms of transport, i.e. you are registered with StattAuto and VOI, have a bicycle and a bus ticket. The CarSharing vehicle is available and there are enough parking spaces at the university."

The route was additionally presented on a map. In order to avoid biases due to financial aspects, we added the information about unlimited access to all modes of transport. Since we recruited participants via the student e-mail list, we assumed that almost all participants were students who could easily put themselves in this situation. After the scenario description, all participants read an explanation and the definitions of the three types of energy.

After this introduction, each mode of mobility was presented, shortly described and an exemplary photo was shown in a randomized order. After each of the three modes that explicitly use external energy (bus, car and e-scooter), all three EDA scales were queried, and after bicycle and walking, the EDA scales for muscular and cognitive energy was queried. After all modes of mobility, the Affinity for Technology Interaction Scale (ATI) [14] was queried. As an incentive to participate in the study, we offered a raffle of three 20€ cash prizes as a bank transfer to all participants.

3   Results

3.1   Scale Analysis

We first examined the internal reliability of each EDA scale in all forms of locomotion (3×3 EDA scales and 2×2 EDA scales = 13 EDA scales), in order to summarize the four items into one scale. All 13 scales showed internal reliabilities between $\alpha_{cron} = .79$ and $\alpha_{cron} = .93$ and therefore, we summarized each four items into one EDA scale by calculating the unweighted mean. For clarity, the three EDA scales are subsequently called EDA-M (or muscular EDA), EDA-C (or cognitive EDA) and EDA-E (or external EDA) and the form of locomotion is indicated in the index (e.g. EDA-M$_{bicycle}$ = awareness for muscular energy dynamics while riding the bicycle).

The ATI scale showed an internal reliability of $\alpha_{cron} = .93$ and was also summarized by calculating the mean of all 9 items (items 3, 6 and 8 reversed). The descriptive statistics of the final scales are depicted in Table 2.

Table 2. Descriptive statistics of all scales.

Scale	N	M	SD	Min	Max
EDA-$M_{e\text{-}bus}$	108	2.94	1.36	1	6
EDA-$C_{e\text{-}bus}$	108	3.27	1.19	1	6
EDA-$E_{e\text{-}bus}$	108	2.98	1.16	1	6
EDA-$M_{e\text{-}car}$	108	2.70	1.33	1	6
EDA-$C_{e\text{-}car}$	108	3.58	1.12	1	6
EDA-$E_{e\text{-}car}$	108	4.27	1.20	1	6
EDA-$M_{e\text{-}scooter}$	108	3.08	1.18	1	5.75
EDA-$C_{e\text{-}scooter}$	108	3.13	0.97	1	5
EDA-$E_{e\text{-}scooter}$	108	3.43	1.23	1	6
EDA-$M_{bicycle}$	108	5.17	0.69	2	6
EDA-$C_{bicycle}$	108	3.88	1.02	1	6
EDA-$M_{walking}$	108	5.19	0.74	2.75	6
EDA-$C_{walking}$	108	4.16	1.14	1	6
ATI	108	3.52	1.09	1.33	6

3.2 Analysis of Variances

To test for main and interaction effects of the type of energy and the form of loco-motion on the mean EDA, two repeated measures analyses of variances (rmANOVA) were conducted. A visualization of the EDA score means is displayed in Fig. 1.

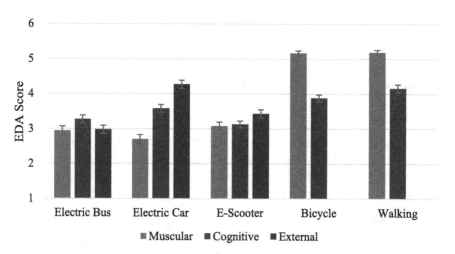

Fig. 1. Mean scores of all EDA scales for every form of locomotion. Error bars represent standard error. $N = 108$.

In the first rmANOVA (1st model), all five forms of locomotion but only regarding muscular and cognitive energy were analyzed, in the second rmANOVA (2nd model), only the three forms of locomotion with all three types of energy were analyzed. Both rmANOVAS are depicted in Table 3. Both rmANOVAS indicated strong effect sizes in the main effects and the interaction effect. To further explore mean differences, post-hoc tests with Bonferroni corrections were conducted.

Table 3. Repeated measures analyses of variances of EDA.

	Levels	df	F	p	η_p^2
1st Model (5 × 2)					
Forms of locomotion	5	4	118.7	<.001	0.53
Type of Energy	2	1	9.6	.002	0.08
Locomotion * Energy	10	4	81.3	<.001	0.43
2nd Model (3 × 3)					
Forms of locomotion	3	2	13.9	<.001	0.12
Type of Energy	3	2	17.4	<.001	0.14
Locomotion * Energy	9	4	28.6	<.001	0.21

In the first model, post-hoc comparisons of the forms of locomotion main effect revealed significant differences between walking/bicycle and e-bus/e-car/e-scooter (all $p_{bonferroni}$ < .001) but not between walking and bicycle or e-bus and e-car/e-scooter respectively. In the second model, the main effect of the forms of locomotion was due to significant differences between the electric car and both the electric bus and the e-scooter. This means that for all types of energy, in walking and riding a bicycle (and in electric cars) EDA scores were higher.

Regarding the main effect of the type of energy, in the first model, EDA scores for muscular energy were significantly higher than those for cognitive energy ($t(107) = 3.10, p_{bonferroni} = .002$) and in the second model, EDA scores for muscular energy were significantly lower than those for cognitive energy ($t(214) = -3.75, p_{bonferroni} < .001$) and for external energy ($t(107) = -5.83, p_{bonferroni} < .001$). Note that in the second model, the forms of locomotion walking and riding a bicycle were not included.

In both models an interaction effect between the forms of locomotion and the types of energy was observable. Significant mean differences were observable between the external EDA scores of the electric bus and the electric vehicle condition ($t(587) = -10.19, p_{bonferroni} < .001$). For the electric vehicle, all three EDA scores differed significantly from each other (EDA-M vs. EDA-K: $t(486) = -6.09, p_{bonferroni} < .001$; EDA-M vs. EDA-E: $t(486) = -10.86, p_{bonferroni} < .001$; EDA-K vs. EDA-E: $t(486) = -4.77, p_{bonferroni} < .001$). Interestingly, the cognitive EDA scores for walking and riding a bicycle did not show a significant difference ($t(756) = 2.12, p_{bonferroni} = $ n.s.).

We also tested for significant correlations of the ATI scale to any of the EDA scales. Only EDA-M$_{e-bus}$ ($r = .24, p = .013$) and EDA-M$_{e-car}$ ($r = .26, p = .007$) had a significant correlation with ATI.

4   Discussion

To sum up, there are indeed differences in the EDA scores for different forms of locomotion and types of energy. Most interestingly, we can observe that all EDA scores in the electric vehicle are significantly different with the highest scores in external energy, which might hint at the relevance of external energy resources in this vehicle. Additionally, the highest EDA scores can be found in the muscular EDA of walking and riding a bicycle, and neither the external EDA in electric cars/e-buses nor any EDA of the e-scooter are higher. This finding leads towards the assumption that whenever external resources are traded for muscular energy, the awareness of those energy flows is reduced. But at this point, it is not valid to draw strong conclusions about the consequences of this observation or the user-vehicle integration concerning the energy use due to some limitations of the study.

First, in this specific scenario the total distance was 3.7 km which is far from a trip that is perhaps limited by one of the types of energy. Moreover, to put oneself into a scenario might not be enough to provoke relevance of the types of energies, because time and comfort is not actually perceived. Secondly, in this study we did not ask about what the participants assumed how much of each type of energy is actually used. In fact, the lack of any objective estimation makes it impossible to say whether a high EDA score is in fact a high EDA score or just the participants' opinion about having a better EDA. Thirdly, we did not control for the actual use of any of these vehicles. Although we assumed that most students know all these forms of locomotion and can envision themselves using them, one must assume a difference in energy perception between actual users and those who never use these vehicles.

Nevertheless, it can be stated that the new constructed EDA scales (with 4 items) show enough reliability to be further examined concerning their internal and external validity.

This study is a very first attempt to integrate human and external energy into one concept of energy resource management. To our knowledge, this has never been documented in the scientific literature before. An integrative model of human and external energy might be useful for research that aims at understanding human-machine systems that uses external energy where humans are in place to make energy relevant decisions and where human and external energy resources are scarce (e.g. aviation, mobility).

5   Conclusion

The optimal human-machine symbiosis in mobility regarding energy management must be characterized by a high *controllability* of the limiting and relevant energy resources. Energy dynamics awareness could be a key factor and help humans to understand different energy flows in order to optimally control those limiting resources. Up until now, the approaches to minimize external energy consumption have been technically oriented. The underlying assumption is, that users must develop a correct mental model of the technical system to be in place to make appropriate decisions. To achieve this, the standard approach is to make technical relations transparent and comprehensive

[15–17]. We don't consider this approach to be wrong, but rather suggest an alternative: to understand external energy as augmented human energy. Provided that – along with our findings – human energy is indeed easier to assess, to close the gap between external and human energy management might be a chance for a better energy awareness. To do so, it is necessary to find out what the exact information is that humans perceive and use to build up an awareness (e.g. body heat/sweat, heart rate, respiratory frequency). This information might then serve as a basis for metaphors in the design of feedback systems for external energy.

References

1. Inman, V.T.: Human locomotion. Can. Med. Assoc. J. **94**(20), 1047 (1966)
2. Alexander, R.M.: Models and the scaling of energy costs for locomotion. J. Exp. Biol. **208** (9), 1645–1652 (2005)
3. Borghese, N.A., Bianchi, L., Lacquaniti, F.: Kinematic determinants of human locomotion. J. Physiol. **494**(3), 863–879 (1996)
4. Weihmann, T.: Leg force interference in polypedal locomotion. Sci. Adv. **4**(9), eaat3721 (2018)
5. Quinn, R.W., Spreitzer, G.M., Lam, C.F.: Building a sustainable model of human energy in organizations: exploring the critical role of resources. Acad. Manag. Ann. **6**(1), 337–396 (2012)
6. Haustein, S., Hunecke, M.: Reduced use of environmentally friendly modes of transportation caused by perceived mobility necessities: an extension of the theory of planned behavior. J. Appl. Soc. Psychol. **37**(8), 1856–1883 (2007)
7. Franke, T., Krems, J.F.: Interacting with limited mobility resources: psychological range levels in electric vehicle use. Transp. Res. Part A Policy Pract. **48**, 109–122 (2013)
8. De Waters, J.E., Powers, S.E.: Energy literacy of secondary students in New York State (USA): a measure of knowledge, affect, and behavior. Energy Policy **39**(3), 1699–1710 (2011)
9. Darby, S.: The effectiveness of feedback on energy consumption. A Review for DEFRA of the Literature on Metering, Billing and direct Displays. (2006)
10. Endsley, M.R.: Toward a theory of situation awareness in dynamic systems. Hum. Factors **37**(1), 32–64 (1995)
11. Endsley, M.R.: Situation awareness misconceptions and misunderstandings. J. Cogn. Eng. Decis. Making **9**(1), 4–32 (2015)
12. Rasmussen, H.B., Lützen, M., Jensen, S.: Energy efficiency at sea: Knowledge, communication, and situational awareness at offshore oil supply and wind turbine vessels. Energy Res. Soc. Sci. **44**, 50–60 (2018)
13. Gödker, M., Dresel, M., Franke, T.: EDA scale - assessing awareness for energy dynamics. In: Alt, F., Bulling, A., Döring, T. (eds.) Mensch und Computer 2019, MuC 2019, pp. 683–687. Association for Computing Machinery, New York (2019)
14. Franke, T., Attig, C., Wessel, D.: A personal resource for technology interaction: development and validation of the Affinity for Technology Interaction (ATI) scale. Int. J. Hum. Comput. Interact. **35**(6), 456–467 (2019)

15. Fors, C., Kircher, K., Ahlström, C.: Interface design of eco-driving support systems–truck drivers' preferences and behavioural compliance. Transp. Res. Part C Emerg. Technol. **58**, 706–720 (2015)
16. Strömberg, H., Andersson, P., Almgren, S., Ericsson, J., Karlsson, M., Nåbo, A.: Driver interfaces for electric vehicles. In: Kranz, M., Weinberg, G., Meschtscherjakov, A., Murer, M., Wilfinger, D. (eds.) International Conference on Automotive User Interfaces and Interactive Vehicular Applications, AutoUI, pp. 177–184. Association for Computing Machinery, New York (2011)
17. Gödker, M., Herrmann, D., Franke, T.: User perspective on eco-driving HMIs for electric buses in local transport. In: Dachselt, R., Weber, G. (eds.) Mensch und Computer 2018, MuC' 2018. Gesellschaft für Informatik e.V, Bonn (2018)

An Exploratory Study on the Perception of Optical Illusions in Real World and Virtual Environments

Sophie Giesa[1]([✉]), Manuel Heinzig[1], Robert Manthey[1], Christian Roschke[2],
Rico Thomanek[2], and Marc Ritter[1]

[1] Faculty Applied Computer Sciences and Biosciences,
University of Applied Sciences, Technikumplatz 17, 09648 Mittweida, Germany
{sophie.giesa,manuel.heinzig,robert.manthey,marc.ritter}@hs-mittweida.de
[2] Faculty Media Sciences, University of Applied Sciences,
Technikumplatz 17, 09648 Mittweida, Germany
{christian.roschke,rico.thomanek}@hs-mittweida.de

Abstract. Virtual environments offers the possibility to simulate the real world as well as the effects like optical illusions. They are images with special arranged content to pretend the human visual system to observe non-existent content for instance. In this study we use several well known optical illusions implemented in real world and in a virtual environment. They were present to subjects to explore their properties and their perception within the different environments.

Keywords: Optical illusions · Synthetic data · Virtual reality · Human perception study

1 Introduction

Immersion in virtual worlds offers the possibility to reflect experiences which simulate challenges of real life and generate completely new experiences. In recent years, virtual reality has opened up to a broad market and is accessible to the general public, enabling the possibility to evaluate their perceptual edges with respect to the field of optical illusions as illustrated in Fig. 1.

Optical illusions are known from the field of human visual perception consist of images or videos with content being arranged in special manner. This composition pretends the human visual system to observe properties of the content which are ambiguous or non-existent like the perception of a non-existent motion or the black dots at the non-focused intersections in Fig. 1. In real world the field of optical illusions is well known and scientifically investigated as in [3]. But the environment of virtual reality features its own properties and constraints as shown in the previous works [4,5]. Those differences limit as well as amplify the effect of the appearing optical illusions.

© Springer Nature Switzerland AG 2020
D. Harris and W.-C. Li (Eds.): HCII 2020, LNAI 12186, pp. 161–170, 2020.
https://doi.org/10.1007/978-3-030-49044-7_15

In our studies we use the virtual reality headset *HTC Vive*[1] with an OLED display. This device shows a good immersion to the user but still has perceivable individual pixels at this resolution because the distance between the display of the glasses and the eyes is only some centimeters in average. With the 3D game engine *Unity*[2] a virtual test environment was realized containing 3D models showing the optical illusions. To be able to determine differences between real world and virtual environment we used a real world room showing printed copies of the illusions from virtual reality. We form a data set by selecting some illusions like the Ponzo illusion which misled the viewer to defective length estimations, as described in [1]. The motion illusion produce the impression of movement inside a stationary image. Inside a scintillating grid a black dot occurs at every white line intersection which is not focused be the viewers eyes and afterimages temporarily produces shapes at homogeneous areas.

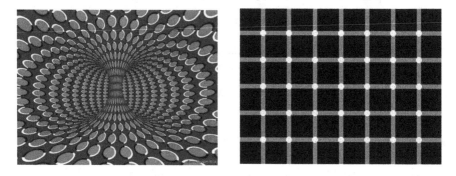

Fig. 1. Examples of the implemented optical illusions showing a motion illusion in the left and a scintillating grid illusion with non-existent black dots at the non-focused intersections.

In the following, experiments with test persons will be used to investigate whether illusions are perceived in the real world in the same way as in virtual reality. The basis for these experiments is a virtual gallery in which the participants can discover a selection of visual illusions. The spectrum of these illusions ranges from illusions of movement, size and depth illusions to afterimages, for instance. The remainder of this paper is organized as follows: Sect. 2 gives an overview about the system architecture and the design of our testing environment. Section 3 describes the system application and the workflow of the study followed by the results of in Sect. 4. A brief summary and an outlook into future work is given in Sect. 5.

---

[1] http://www.vive.com/de/.
[2] https://unity.com.

2   System Architecture and Design

We use the state-of-art and widely used 3D game engine *Unity* to create the virtual test environment as shown in Fig. 2. It comprise a building with some rooms of common size to be traversed by the test persons. Each room contain picture frames to be the place holders to present the optical illusions. Each test person would be able to move through the rooms of the building using the controls of the Vive regarding all the objects.

The virtual test environment is presented by the Vive displays which has a resolution of 1080×1200 pixels per eye, a color depth of 24 bit and a rate of 25 frames per second. The infrared sensors allow the user to move 360° within the range of movement. The field of view of the HTC Vive has a circumference of 110° and thus contributes a large part to immersion. 32 headset sensors allow 360° movement tracking of the head. The HTC controllers allow the participant to interact with selected objects in the virtual gallery. Sound is only transmitted via loudspeakers, not via the headphones, so that the user can make himself heard during the experiment and understand instructions from the experimenter.

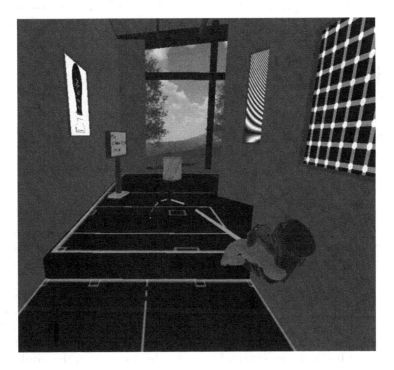

Fig. 2. Example of the virtual test environment simulating a room with different optical illusions at the walls.

A room of $5\,\mathrm{m} \times 5\,\mathrm{m} \times 3.5\,\mathrm{m}$ represents the real-world complement to the virtual test environment. As before, it contain picture frames to be the place holders to present the optical illusions. Each test person entering the room would be able to move through and to regard all the objects.

In both environments the dimensions of the rooms, of the objects and mainly of the optical illusions are similar of size to facilitate comparability. The real-world version of the illusions are printed and placed at a randomly selected picture frame of the room. Each illusion was created in three different sizes to recognize if size is a relevant feature.

Our architecture permits a random selection and combination of all test persons and optical illusions as shown in Fig. 3. This should prevent memorization of the test by the test persons or effects based on unfavorable sequence of illusions. In each environment another instance of the same type of illusion is chosen with the same intention.

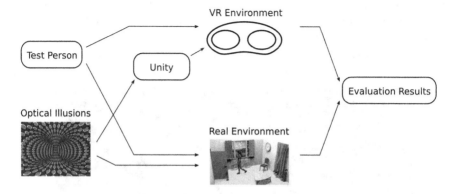

Fig. 3. Principle structure of our exploratory study. The test persons and the optical illusions are the inputs of the virtual environment and the real environment. The virtual environment is created with the Unity software. Within both environments the optical illusions were presented to the persons and the results are compared.

Some of the illusions be presented in the gallery should be explained in detail for better understanding by representative images. The first illusion shown in the gallery is called the illusion of movement like the left of Fig. 1. To exclude an error component, the image was implemented in four different sizes. The test persons should describe the picture without the hint that it is an illusion of movement, because this could influence the results. The second image showed a so-called Hermann grid like the right of Fig. 1. Since this is a very subtle kind of illusion, the experimenter took the liberty of asking for the color of the grid or the middle between the corners. It was avoided to mention the word dot, or color names.

Another illusion is an afterimage as described by [6] and exemplarily shown in Fig. 4a. Here the experimenter did not have to give any hints, because an information board was implemented, which tells the test persons how long they have to fix the picture. Then they should look at the empty wall next to it.

The next picture shows alternating white and black lines like in Fig. 4b. While the test person shakes his head, the outlines of a well-known work of art should appear. Of course the test persons are not told that it could be an art object. The following illusion shows a so called double image, as described by [2]. Here the test person should describe what is recognized first when looking at the picture. The picture shows a dancer as well as a skull. No double images were selected, which were generally treated during school education like in Fig. 4c.

The next two-dimensional illusion shows some circles, which are diagonally separated by red lines. The test persons are asked which course the red lines take. If they measure with a ruler, the lines are absolutely straight. However, the arrangement of the circles makes the lines appear wavy like in Fig. 4d.

Some afterimages shows the Batman logo in black and white, an afterimage should appear after close observation and subsequent change of viewpoint. In another image, the course of a red line is again asked for, but this time not interwoven with circles, but with overlapping rays.

3   Systems Application and Workflow

The introduced system architecture and the design of the study could be classified as a qualitative study. The knowledge goal is an independent basic scientific study and the study itself is empirical. The data basis is collected based on a primary analysis. Virtual reality technology is a comparatively new field, as it has only become accessible to a broad market in recent years. The field of optical illusions has not yet been further explored and therefore provides the basis for an exploratory study. The experimental group study took place without repetition under laboratory conditions. The results of this study serve as a basis for concept development and the formulation of new hypotheses.

The location for the virtual reality experiments is a room with a free movement area of at least $2\,m \times 1.5\,m$. The test person can move around in this area without any interference to real obstacles. The experimenter takes up an observational sitting position outside the free area. Via the connected PC he has the possibility to observe the field visible to the test person. If necessary, the experimenter may give hints for the use in virtual reality, should it concern the interaction possibilities. The test leader is thus able to follow the test person's reactions in real time without directly influencing the participant. The virtual environment is created with Unity and shown via an HTC Vive to the test persons.

In order to simplify the evaluation of the results, the statements made during the experience were recorded. To collect data, an interview was conducted based on a previously prepared questionnaire. The volunteers' verbal self-reports on selected aspects of their perception were recorded in writing during a personal interview. The questions were asked in such a way that, if possible, no simple yes or no answers could be given. In the perception of illusions, such a procedure would have led to one answer appearing more correct than the other.

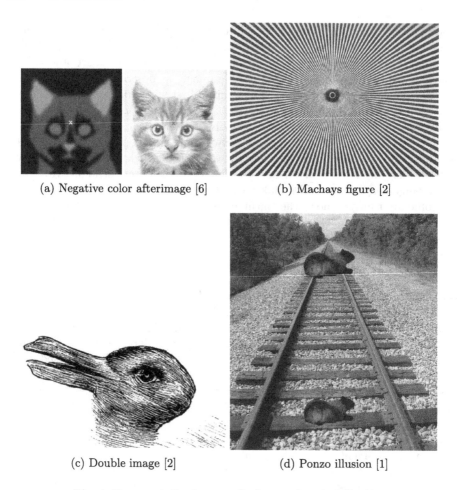

(a) Negative color afterimage [6]   (b) Machays figure [2]

(c) Double image [2]   (d) Ponzo illusion [1]

Fig. 4. Representative images of other used optical illusions

For instance:

Do you see the gray dots in the Hermann grid?

would already implies that points in the grid should be seen. Therefore we favor more unbiased questions like:

What do you see in the grid?

While this allows the respondent not to consciously perceive this subtle illusion, it rules out manipulation by the question (Table 1).

The selection of the test persons was done on a voluntary basis which results in a set of ten participants. They were between 18 and 24 years old and of different gender, with technical skills of different fields from institutions of higher education. They were divided into two groups by random selection to perform all the tests with virtual environment first and real world second, respectively in reverse order.

Table 1. Example of some questions of the experimenter to get the perceptions of the illusions from the test persons. Each question target to certain property of a certain illusion.

Question to the test person	Associated type of illusion
Does the illusion of movement work?	Motion illusion
What do you see in the grid?	Hermann grid
Look at the picture of the light bulb for at least 15 s. Now look at the empty wall next to it. What do you see?	Afterimage
The next picture shows alternating white and black lines. Now shake your head. What do you see?	Machays figure
What do you see first in the double image?	Double image
The spiral is pinning before your eyes. Look at the center for at least 20 s. What happens if you now look at another point?	Spiral illusion
Is the left or the right vertical line longer?	Ponzo illusion

Every test run was organized in the following steps according to the group. In the beginning, the respective volunteers of the first group were individually instructed in the use of virtual reality glasses, if they had never used them before. Afterwards the rough test procedure was explained to them. General information was given that this attempt was aimed at investigating illusions in virtual reality. In addition, they were instructed to describe their perceptions during the experience in verbal form. Details of the experiment were not revealed to prevent to influence the test person. We take care to ensure that the test persons were physically separated from the others so that the other test persons did not receive any information about the use of the test. Before the actual experiment began, a short interview was queried to get general background informations of the test persons. Afterwards the test person was informed about possible risks of virtual reality use and about legal disclaimers. The test persons were helped to find the correct focus of the glasses just before the experiment starts. After this part of the test in the virtual environment the second part in the real world was processed. Starting with a short introduction similar to the previous one and the recording of the perceptions during the experiment. The entire experiment lasted about 25 minutes per participant. The second group of test persons pass through a similar test setting but with changed order of virtual and real environment.

4   Evaluation and Results

The presented combination of system architecture, workflow and test design on the one hand and the set of ten voluntary test persons on the other hand enables an exploratory study of the recognition of optical illusions as well as their perception in real world and virtual environments presented with systems like the HTC Vive. It allows a comparison of the two environments and a determination of differences.

Some of the results will be presented in the following as well as Fig. 5. The evaluation of the questions targeting the illusions of movement shows that in virtual reality as well as in the real world do not always work for all individuals. Of all test subjects, 30% could not detect any movement illusion in the virtual environment. Those who could, did not all have this realization at the same size of the image. At the largest of the images, no subject could detect any movement. Others were able to detect the illusion of movement at several sizes, while 50% of the test subjects only noticed it at one size. This variance in the results suggests that the sensory impressions were processed differently by different individuals. While a subject is looking at the image, afterimages should appear in the eye, which overlap with each movement of the eye and compete with the previous afterimages. The following experiments will also show that this effect seems to be at least partially canceled out. Another factor could be the fly screen effect with the OLED displays from HTC Vive or its blurring as shown in [4]. This is hardly avoidable so far, because the eyes are too close to the displays resulting in the perception of individual pixels of the LEDs producing the image.

The evaluation of the Hermann grid experiment showed that 90% of the test persons could identify the non-focused points as gray. In virtual reality, the effect of lateral inhibition, the automatic increase of contrast perception, seems to work for surfaces of very different brightness. This effect only occurs when lines overlap at right angles.

The questions targeting the illusions of the visual afterimage illusions shown that the color receptors in the eyes of different people fatigue at different rates depending on the situation. 20% of the test persons did not recognize a negative afterimage in the virtual reality or in the real world. Only 50% of all test participants could perceive an afterimage in virtual as well as in the real world. Interestingly, 40% of them stated that the afterimage was less strong in virtual environment than the Batman logo in reality. Again, several factors may have played a role. First, the fly screen effect of the OLED displays, and second, the fatigue effect of the photoreceptors.

The used double image in the real world environment consisted of an old man's head and a rider on his horse. 80% of the test persons took the old man first. This result shows quite clearly that the human brain is designed to recognize faces faster than other structures. In the double image in virtual reality the face of the dancer is only very small, and the results of this test sequence are balanced. Another test using the spiral illusion could be perceived by all subjects of both test groups. However, it was shown that those who were presented with the illusion in virtual environment first noticed stronger effects. According to statements of the test subjects of the other group, who performed the presentation in the real world first and the subsequent test in virtual reality, the extension of the opposite circular movement in virtual reality was smaller. This corresponds with results that could already be seen in the afterimages. Here, too, the question arises as to the cause of the perceived phenomena.

Further illusions show that 90% of the test subjects were deceived by the course of the lines in virtual reality, whereas only 70% were deceived in the second group. A further test run, with further variations, would have to show whether the circular patterns with color gradient possibly had a stronger distracting effect on the test subjects than the distraction caused by jagged lines arranged in the pattern.

The Ponzo illusion causes an illusion of the senses through apparent differences in depth. One of the test persons said of herself that she had already seen a great many such illusions and therefore recognized at first glance that the lines were actually straight. Another test persons admitted that although they knew that the lines were straight, their perception tried to make them believe that the lines or tree trunks were of different lengths. This effect can be explained by the incorrectly applied size-distance scaling.

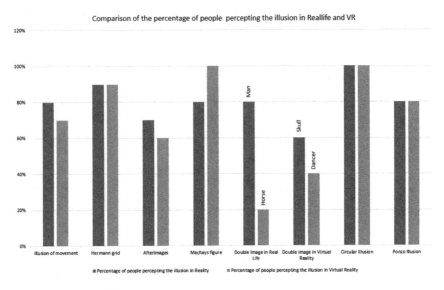

Fig. 5. The comparison of the some results from the study show no appreciable differences between real and virtual environment, like Hermann grid or the circular or Ponzo illusion. Others exhibit significant differences like the afterimage.

5   Summary and Future Work

In conclusion, we present an architecture and a workflow to be used to compare real world and virtual environment. With this it is possible to evaluate their properties as well as their constraints. We conduct an exploratory study on the perception of optical illusions presented to a group of test persons in both environments. The most notable differences were presented and could be subjects of further investigations. Some of detections seem to be the results of technical shortcomings of the used devices which may be equalized by technological advantages.

References

1. Goldstein, E.B.: Wahrnehmungspsychologie, 9th edn. Springer, Berlin (2015)
2. Gregory, R.L.: Auge und Gehirn: Psychologie des Sehens. Rororo Science, Hamburg (2001)
3. Idesawa, M.: A study on visual mechanism with optical illusions. J. Robot. Mechatron. **9**, 85–91 (1997)
4. Manthey, R., Ritter, M., Heinzig, M., Kowerko, D.: An exploratory comparison of the visual quality of virtual reality systems based on device-independent testsets. In: Lackey, S., Chen, J. (eds.) VAMR 2017. LNCS, vol. 10280, pp. 130–140. Springer, Cham (2017). https://doi.org/10.1007/978-3-319-57987-0_11
5. Ritter, M., et al.: Simplifying accessibility without data loss: an exploratory study on object preserving keyframe culling. In: Antona, M., Stephanidis, C. (eds.) UAHCI 2016. LNCS, vol. 9739, pp. 492–504. Springer, Cham (2016). https://doi.org/10.1007/978-3-319-40238-3_47
6. Sarcone, G.A., Waeber, M.J.: Optische Illusionen: Mit vielen Beispielen. Erklärungen und Experimenten. arsEdition, Munchen (2018)

Application of a Pictorial Scale of Human Energy in Ecological Momentary Assessment Research

Fabienne Lambusch[1]([⊠]) [iD], Oliver Weigelt[2]([⊠]) [iD],
Michael Fellmann[1] [iD], and Katja Siestrup[3] [iD]

[1] Business Information Systems, University of Rostock, Rostock, Germany
{fabienne.lambusch,michael.fellmann}@uni-rostock.de
[2] Organizational and Personnel Psychology, University of Rostock,
Rostock, Germany
oliver.weigelt@uni-rostock.de
[3] Work and Organizational Psychology, FernUniversität Hagen,
Hagen, Germany
katja.siestrup@fernuni-hagen.de

Abstract. Individual resource status plays a major role in the literature on employee strain and recovery. Many theoretical accounts draw on the analogy between humans and batteries to describe the ups and downs in individual resource status over time. Taking the battery-metaphor literally, we have developed a pictorial scale to capture momentary resource status in terms of levels of human energy (i.e., high level of subjective vitality and low levels of fatigue). A brief and face-valid single-item measure is particularly useful in ecological momentary assessment research, e.g. surveying employees multiple times over the course of a day. We present empirical results from a diary study across twelve days among 57 employees. The pictorial scale is highly correlated with momentary ratings of subjective vitality and fatigue at the intraindividual and the interindividual level (i.e., has convergent validity). Drawing on these results, we demonstrate how the pictorial scale can be applied in ecological momentary assessment research to track trajectories of energetic well-being over the course of the day. We discuss, how providing personalized feedback to end-users regarding their energetic peaks and troughs over the course of the day or over the course of the week, may be leveraged in technology-assisted management of human energy on the individual and organizational level.

Keywords: Human energy · Pictorial scale · Experience sampling · Energy management · Individual feedback

1 Introduction

1.1 Relevance of Tracking Resource Statuses

These days, many employees are confronted with very high work demands [1]. High workloads, complex and knowledge-intense tasks and increased expectations for

---

The first and the second author contributed equally to this paper.

© Springer Nature Switzerland AG 2020
D. Harris and W.-C. Li (Eds.): HCII 2020, LNAI 12186, pp. 171–189, 2020.
https://doi.org/10.1007/978-3-030-49044-7_16

flexibility intensify the work. Of concern in this context is that the boundaries between life domains are becoming increasingly blurred, while the intensification of work continues. Due to an extensive use of information and communication technology, multitasking and interruptions of the workflow have become daily business in many companies. All these circumstances can induce long-term stress, which can result in serious health problems. Worldwide, mental health problems are a major contributor to the overall burden of disease and are particularly concentrated in the working population, possibly leading to a loss of human capital [2].

In order to prevent such negative consequences and promote a sustainable management of personal resources, it is desirable that individuals reflect on their own behavior to discover strength as well as necessary changes. Observations of the own behavior constitute a basis for self-evaluation, which is in turn necessary for reinforcing desired behaviors [3]. While keeping a diary might be a good practice for gathering data, it might be time-consuming and impractical in many cases. Valid measures in combination with information technology (IT) could act as facilitators to keep track of the own resource status and desirable behaviors. IT can be supportive by collecting personally relevant information e.g. through mobile apps [4]. Such apps can incorporate self-assessments as well as mechanisms to automatically capture some data e.g. on the physical activity of the user. On the basis of data collected, visualizations can be generated that might help people in self-reflection [5]. Furthermore, tracking a person's state enables to create and trigger situational interventions. IT-based interventions were, for example, already used for promoting relaxation [6]. Thus, tracking of a person's state constitutes a fundament to provide support in reflection and development processes.

1.2   Dynamics in Individual Resource Status: Human Energy

The literature on occupational stress considers a broad range of aspects of individual well-being as reflected in states like fatigue [7], thriving [8], work-related rumination [9, 10] or engagement and burnout [11]. Studying these phenomena is important to promote and sustain health and quality of life [12] of individual workers and whole organizations in the long run. Given the major impact of resource theories [13] in industrial and organizational psychology [14], well-being is often conceptualized in terms of individual resource status [15]. In this sense, processes of strain and recovery from work are reflected in decreases and increases of individual resource status over time [16] – expending specific resources during work and returning to pre-stressor levels of functioning during breaks [17, 18]. Individual resource status is usually operationalized in terms of fatigue, (emotional) exhaustion, need for recovery, self-control capacity or vitality [19]. A common theme inherent in the aforementioned states is that they refer to different aspects of human energy [20]. Building on research in social and personality psychology [21], the literature on work breaks and energy management has focused on human energy in terms of high levels of subjective vitality and low levels of fatigue [22, 23]. According to Ryan and Frederick [24] subjective vitality encompasses a feeling of aliveness and energy. By contrast, the terms fatigue and exhaustion refer to states of low energy. A common theme in definitions of fatigue is extreme tiredness and reduced functional capacity [7]. Therefore, valid indicators of human energy should tap into the experience of subjective vitality and fatigue.

2   Development of a Pictorial Scale of Human Energy

2.1   The Value of Valid Measures in Experience Sampling Research

Experience sampling methodology has become a major approach to studying human affect, cognition, and behavior in various disciplines, such as psychology and management [25]. In a typical experience sampling study participants provide self-reports on focal variables multiple times over a period of several days. Recently, ecological momentary assessment has gained currency [26]. Ecological momentary assessment is a specific form of research within experience sampling methodology usually including hourly self-reports over the course of a day. Tracking individual experiences over time is highly relevant not only in psychology and behavioral sciences. Measuring the subjective experience is often required to facilitate the interpretation of objective indicators of individual health (e.g., blood pressure) or to put individual behavior (e.g., time spent on twitter during work) into context. Although, research in information technology has adopted experience sampling methodology alongside tracking objective user activity, the measures or scales applied are often taken for granted. It is quite common to shorten scales or adapt scales to the purposes of the study. Unfortunately, the adapted scales rarely undergo rigorous scrutiny regarding reliability and validity, although any changes to the original instruction and items might change the meaning of the construct. In other words, we cannot be certain that adapted scales meant to capture specific aspects of current affect (i.e., moods or emotions), such as exhaustion and vigor, actually reflect these subjective states as accurately as the original scales do. Accordingly, examination of validity and reliability should precede the application of adapted or new scales in experience sampling or ecological momentary assessment research. Of note, reliability and validity of measurement instruments in experience sampling research is crucial both from a theoretical and a practical perspective. For instance, if the scales applied lack reliability and validity, we may not only draw faulty conclusion regarding the antecedents of individual well-being, we may also risk giving wrong advice to end-users within recommender systems applying feedback from self-reports.

2.2   Developing a Brief and Valid Measure: A Pictorial Scale

Usually, experience sampling research precludes the application of comprehensive scales because of time constraints [27]. For one, comprehensive scales would put a high burden on participants increasing the risk of non-participation or dropout from the study. For the other, given that an experience sampling study consists of five, ten or more self-reports per participant, the length of each daily or momentary self-report might accumulate over time to produce reactivity of the study. In other words, a study originally planned to capture life unobtrusively may become a major disturbance in its own right. Accordingly, experience sampling researchers have to rely on shortened scales or brief measures to capture the variables of interest. The need for brief measures applies even more for ecological momentary assessment research which aims to capture life in situ by sampling individual experiences multiple times a day. Given the need for brevity particularly in applied field research, there is a long tradition in psychology applying pictorial scales to measure variables of interest. Probably the first pictorial scale in (applied) psychology was developed by Kunin to capture overall job satisfaction [28]. In the Kunin or faces-scale individuals choose one of five faces ranging from a (1) a frowning face to (5) a smiling face to describe their level of job satisfaction. There is meta-analytic evidence that the Kunin pictorial scale is a reliable and valid measure of job satisfaction as reflected in a correlation of $\rho = .67$ with verbal scales of overall job satisfaction [29].

Pictorial scales have also been applied in research on affect. For instance, the self-assessment manikin scales are among the most popular pictorial scales. Originally developed to assess pleasure arousal, and dominance of presented stimuli in lab experiments [30], the self-assessment manikin scale has been adapted to measure momentary affect. Recently, Weigelt et al. [31] have proposed a pictorial scale to capture human energy. These authors have drawn on the metaphor of batteries that is dominant in lay theories of well-being and the scientific discourse on aspects of human energy. For instance, the literature on recovery from job stress frequently refers to "charging the batteries" [18]. Drawing on the analogy between individuals and batteries, in the pictorial scale of human energy, participants may choose one of five symbols of a battery to describe their levels of energy. The battery scale ranges from (1) an almost empty battery to (5) a fully charged battery. In a cross-sectional survey study among 189 workers Weigelt et al. [31] found that the battery scale is reliable and valid as reflected in high correlations with several indicators of human energy. More specifically, the battery scale correlates highly with verbal scales of vigor ($r = .73$), fatigue ($r = -.72$), subjective vitality ($r = .69$), self-control capacity ($r = .68$), and emotional exhaustion ($r = -.66$). Retest-reliability across six weeks was rather low ($r = .40$). While low retest-reliabilities of trait measures are problematic, the rather high volatility of the battery scale suggests that this pictorial scale may be sensitive to situational factors, and hence, be particularly suited in settings where researcher strive for tracking and understanding within-person changes in subjective states or behaviors (e.g., experience sampling research). Although, these initial findings from survey research are encouraging, so far, the battery scale has not been examined in the context of experience sampling research. Given that the battery scale holds promise to be particularly valuable in ecological momentary assessment settings, we set out to scrutinize the reliability and validity of the battery scale in an experience sampling study.

3 Examination of the Pictorial Scale in the Context of Experience Sampling Research

3.1 Aims and Scope

Prior research on the battery scale has referred to general or "chronic" levels of human energy in the context of a self-report survey study. Hence, in this study we aim to adapt the battery scale from a chronic (or trait) version to a state version. Of note, it is neither certain nor trivial that the battery scale will fare equally well in experience sampling research. Accordingly, we explicitly examine the psychometric properties of the battery scale referring to momentary levels of human energy. Importantly, experience sampling data allow for analyzing associations among focal variables at different levels. In experience sampling research, researchers usually distinguish the within-person level of analysis from the between-person level of analysis [32]. The within-person level of analysis refers to fluctuation in human energy within an individual over time (high energy vs. low energy days). By contrast, the between-person level of analysis refers to differences between persons as reflected in different average levels of energy across all days (high energy persons vs. low energy persons).

In this study, we consider correlations of the battery scale with subjective vitality and fatigue as measured by established verbal scales. This is consistent with the description and operationalization of human energy in terms of high levels of vitality and low levels of fatigue [22, 23]. In line with the cross-sectional evidence reported above, we expect the battery scale to yield high positive correlations with subjective vitality and high negative correlations with fatigue at the within-person level and at the between-person level. We examine these associations at the two levels of analysis (within-person and between-person) applying multilevel structural equation modeling. Unlike classic structural equation modeling, multilevel (structural equation) modeling accounts for the nested data structure and allows separation of within-person variance from between-person variance [33, 34]. Hence, multilevel (structural equation) modeling allows for modeling associations among variables separately at different levels of analysis (within-person vs. between-person).

Furthermore, we examine the battery scale regarding the level of variability (or stability) across days by analyzing intra-class correlations coefficients ICC(1). We compare intra-class correlations as a measure of stability across the battery scale, the vitality-scale, and the fatigue-scale. We expect the battery scale to yield an ICC similar to those of vitality and fatigue. In other words, the battery scale should be equally sensitive to capture changes in human energy within person over time as the verbal scales do.

3.2   Methods

Procedure. This study is part of a larger project on occupational stress and recovery in leisure time. The study was pre-registered. We provide details through the open science framework[1]. We made sure that the study materials comply with the declaration of Helsinki. Our study fully complied with the ethical guidelines of the Department of Psychology at the University of Hagen, the university where this study was hosted. We obtained informed-consent from each participant before the study started. Participation was voluntary and participants were free to quit whenever they wanted to. We conducted an experience sampling study across 12 consecutive days beginning on a Monday and ending on a Friday. The self-reports covered a broad range of topics such as daily job demands, recovery activities, and recovery experiences. Participants provided self-reports of momentary energy level three times a day: In the morning upon

---

[1] Study details at https://osf.io/8rn42/.

getting up, in the afternoon upon leaving the workplace, and in the evening before going to bed. For each of the 36 surveys we sent participants an invitation via email containing a personalized link. Participants responded online to our electronic surveys through their web browser on their (mobile) electronic devices (computers, tablets, mobile phones).

Measures. We measured different aspects of energetic well-being to examine the convergent validity of the battery scale. First, we measured subjective vitality with three items of the subjective vitality scale developed by Ryan and Frederick [24]. The scale has been adapted to German and applied in an experience sampling methodology setting, recently [35]. A sample item is "Right now, I feel alive and vital." Second, we measured fatigue with three items of the profile of mood states (POMS) scales [36]. The POMS has been adapted to German [37]. We selected items from the fatigue-subscale and asked participants to describe how they felt right now. The selected items were "exhausted", "worn out", and "weary". The response format for subjective vitality and fatigue ranged from 1 (*strongly disagree*) to 5 (*strongly agree*). Third, we measured human energy applying the battery scale. We applied the following instruction: "People often describe how they feel right now referring to the metaphor of a battery ranging from exhausted to full of energy. Please indicate which of the following battery icons describes your current state best." The battery icons are displayed in Fig. 1.

Fig. 1. Battery icons applied to capture momentary level of human energy

Sample. Our sample consisted of 57 workers from diverse organizations, occupations, and industries. The majority was female (79%). Age ranged from 20 to 57 years ($M = 35.19$, $SD = 10.21$). Average tenure with the organization was 7 years ($M = 6.99$, $SD = 8.55$). On average, they worked 35 h per week ($M = 35.52$, $SD = 11.16$). Participants came mainly from healthcare (21%), industry (14%), public administration, (12%), commerce (11%), and the service sector (11%). Eighteen persons had a leadership position (31.6%). In total 599 complete self-reports in the morning, 535 self-reports in the afternoon, and 565 self-reports in the evening from 57 individuals were available for the focal analyses (1699 of the theoretically possible 2052 of self-reports $\sim$ 83% response rate).

Analytic Strategy. We analyzed data applying multilevel structural equation modeling in the "lavaan" package for R. In a first step, we examined the measurement model of the verbal scales. Accordingly, we specified a set of confirmatory factor analyses to examine the reliabilities of the verbal scales of energetic well-being (subjective vitality and fatigue). In a second step, we added a structural model by including regression paths from the battery scale to the subjective vitality-factor and the fatigue-factor. In a third step, we ran multilevel regression models in the "nlme" package for R. More specifically, we specified a null model to infer intra-class correlation coefficients from the variance components. The ICC reflects the degree of agreement between ratings from the same person across time. A value of 1 would suggest that the variable is totally invariant (i.e., perfectly constant) across the repeated measurements within person. A value of 0 would indicate that the variable of interest varies complete randomly within person. We estimated alpha reliabilities separately at the within-person and at the between-person level of analysis for subjective vitality and fatigue as proposed by Geldhof and colleagues [38]. We applied procedures proposed by Huang to estimate Alpha in R [39]. To avoid running three-level structural equation models (self-reports nested in days, days nested in persons), a feature currently supported by only few statistics packages, we ran analyses for the morning survey, the afternoon survey, and the evening survey separately. Besides pragmatic aspects, our approach provides the opportunity to examine whether the validity of the battery scale varies systematically over the course of the day.

3.3 Results

Preliminary Analyses. The correlations among the focal variables are presented Table 1. Correlations below the diagonal refer to correlations at the within-person level (uncentered or raw correlations). Correlations above the diagonal refer to the between-person level (Person-mean values across self-reports person). Table 1 also contains the means, standard deviations, and ICCs of all scales across surveys (morning, afternoon, evening). We also included coefficient alphas as a measure of reliability for the multi-item verbal scales. Of note, standard deviations and ICCs of the battery scale are very similar to the respective values of vitality and fatigue. The means of the battery scale and the vitality verbal scale were very similar. Applying confirmatory factor analysis, we compared a single-factor model in which vitality items and fatigue items loaded on a common factor with a two-factor model in which vitality items loaded on one factor and fatigue items loaded on the other factor. We specified homologous models [40]. That is, we assumed the same structure at the within-person level and at the between-person level. Although, the two-factor models fit better than the single-factor models, the absolute fit of these models was poor according to most fit indices (CFI < .90, TLI < .80, RMSEA > .15, SRMR $_{\text{Level 1}}$ > .08, SRMR $_{\text{Level 2}}$ > .13).

Modification of the Measurement Models. Inspection of the output and the modification indices suggested that the item "weary" yielded low loadings on the fatigue factor. Accordingly, we excluded this indicator from the fatigue-factor and retained it as a manifest indicator besides the latent factors (vitality and fatigue). Furthermore, we freely estimated the covariance of the third vitality item (alert and awake) and "weary" because the content of both items referred to the aspect of tiredness (the German item "muede" of the POMS actually corresponds better to the English word "tired"). This modified model fit considerably better and achieved excellent model fit. Accordingly, we treated vitality, fatigue and weariness as three distinct outcome variables.

Table 1. Correlations, means, standard deviations, reliabilities, and intra-class correlation coefficients among the focal variables

Variable	Sex	Morning survey			Afternoon survey			Evening survey		
		Battery	Vitality	Fatigue	Battery	Vitality	Fatigue	Battery	Vitality	Fatigue
Baseline survey (N = 57)										
Age	.02	.08	.16	−.26	.21	.18	**−.38**	.04	.01	−.24
Sex	–	−.17	−.18	.08	−.02	−.16	.10	−.21	**−.33**	.22
Morning survey (N = 599)										
Battery	–		**.83**	**−.81**	**.59**	**.36**	**−.44**	**.38**	**.29**	**−.29**
Vitality	–	**.77**		**−.82**	**.57**	**.54**	**−.52**	**.26**	**.35**	**−.31**
Fatigue	–	**−.72**	**−.74**		**−.53**	**−.41**	**.63**	**−.25**	−.22	**.41**
Afternoon survey (N = 535)										
Battery	–	**.46**	**.43**	**−.40**		**.75**	**−.63**	**.57**	**.54**	**−.54**
Vitality	–	**.37**	**.48**	**−.38**	**.78**		**−.77**	**.41**	**.61**	**−.56**
Fatigue	–	**−.37**	**−.45**	**.47**	**−.70**	**−.77**		**−.33**	**−.44**	**.66**
Evening survey (N = 565)										
Battery	–	**.26**	**.16**	**−.15**	**.37**	**.28**	**−.22**		**.86**	**−.78**
Vitality	–	**.18**	**.18**	**−.12**	**.31**	**.34**	**−.26**	**.79**		**−.81**
Fatigue	–	**−.23**	**−.22**	**.28**	**−.33**	**−.31**	**.39**	**−.73**	**−.76**	
M		3.30	3.18	2.23	3.20	3.34	2.24	2.76	2.93	2.61
SD Level 1		1.02	0.96	0.96	1.01	0.94	1.01	1.04	1.03	1.07
SD Level 2		0.76	0.69	0.62	0.63	0.63	0.64	0.72	0.72	0.74
Alpha Level 1		–	.86	.78	–	.89	.83	–	.88	.79
Alpha at Level 2		–	.96	.92	–	.98	.94	–	.97	.89
ICC(1)		.46	.45	.34	.31	.32	.31	.41	.40	.37

Note. Correlations below the diagonal refer to the within-person level (n = ranges from 535 to 599 occasions). Correlations above the diagonal refer to the between-person level (n = 57 individuals). Correlations in bold are significant at p = .05. ICC(1) = intra-class correlation coefficient.

Focal Multilevel Structural Equation Models. In the final models, vitality, fatigue and weariness were regressed on the battery scale. We ran separate models for the morning survey, afternoon survey, and evening survey. The three resulting models are depicted in Fig. 2, Fig. 3, and Fig. 4 respectively. All fit indices for all models are reported in Table 2. Factor loadings and standardized covariances among factors are displayed in Fig. 2 through Fig. 4. The focal models yielded excellent model fit according to common criteria for assessing model fit [41], such as:

- comparative fit index (CFI, values above .97 reflect good fit),
- Tucker Lewis index (TLI, values above .97 reflect good fit),
- root mean square error of approximation (RMSEA, values below .05 reflect good fit),
- standardized root mean square residual (SRMR, values below .05 reflect good fit).

Consistently across surveys (morning, afternoon, evening) all items yielded high loadings on their respective factors ($\lambda > .71$). The subjective vitality factor, the fatigue factor and the "weary"-item correlated moderately.

Table 2. Overview of Model Fit Statistics for the Confirmatory Factor Analyses and the Focal Structural Equation Models

Model	Survey	Chi-square	Df	CFI	TLI	RMSEA	SRMR Level 1	SRMR Level 2	AIC	BIC
Single-factor-model										
	Morning	365.4	18	.808	.679	.180	.079	.097	8023.3	8155.2
	Afternoon	315.3	18	.852	.753	.176	.063	.010	7021.6	7150.1
	Evening	362.2	18	.824	.707	.184	.074	.094	7912.6	8042.7
2-factor-model										
	Morning	216.6	16	.897	.807	.145	.079	.132	7878.5	8019.2
	Afternoon	187.7	16	.920	.849	.142	.066	.060	6898,0	7035.1
	Evening	238.1	16	.893	.800	.157	.081	.119	7792.4	7931.2
Modified 2-factor-model										
	Morning	14.1	12	.999	.998	.017	.011	.019	7684.0	7842.2
	Afternoon	18.3	12	.997	.993	.031	.010	.026	6736.7	6890.8
	Evening	15.6	12	.998	.996	.023	.010	.024	7577.9	7734.1
Final structural equation model (Figs. 2, 3 and 4)										
	Morning	31.4	18	.995	.988	.035	.015	.020	8643.3	8827.9
	Afternoon	35.9	18	.994	.985	.043	.014	.024	7598.9	7778.7
	Evening	34.7	18	.994	.986	.040	.014	.025	8516.2	8698.3

Note. CFI = comparative fit index, TLI = Tucker Lewis index, RMSEA = root mean square error of approximation, SRMR = standardized root mean square residual, AIC = Akaike information criterion, BIC = Bayesian information criterion. Level 1 refers the within-person level. Level 2 refers to the between-person level.

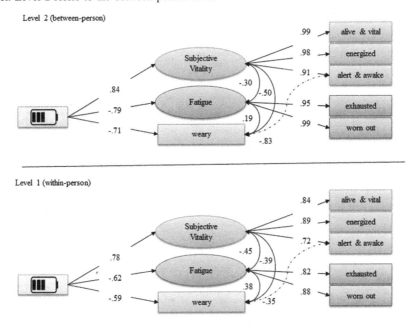

Fig. 2. Standardized loadings, covariances, and regression coefficients of the morning survey at the between-person level (top) and at the within-person level of analysis (bottom)

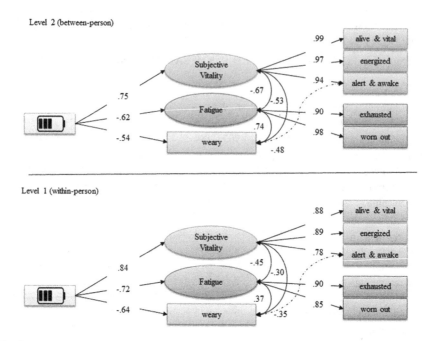

Fig. 3. Standardized loadings, covariances, and regression coefficients of the afternoon survey at the between-person level (top) and at the within-person level of analysis (bottom)

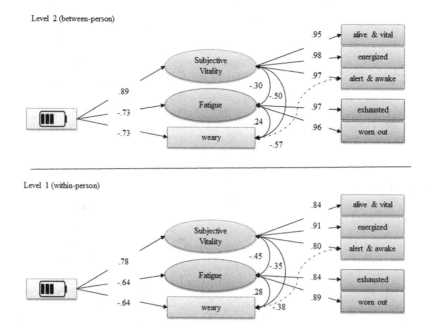

Fig. 4. Standardized loadings, covariances, and regression coefficients of the evening survey at the between-person level (top) and at the within-person level of analysis (bottom)

Associations of the Battery Scale with Vitality and Fatigue. Our focus in this study was examining the links between the battery scale and the prototypical indicators of human energy, namely subjective vitality and fatigue. Accordingly, the regression coefficients of the battery scale predicting vitality and fatigue were of major interest. In essence, the battery scale yielded high standardized regression coefficients predicting subjective vitality, fatigue, and tiredness as reflected in the weary-item at the within-person and the between-person level of analysis.

At the within-person level, regression coefficients ranged from .71 to .84 for subjective vitality and −.62 to −.72 for fatigue. The regression coefficients predicting tiredness as reflected in the weary-item ranged from −.59 to −.64. That is, momentary status as measured with the battery scale is strongly linked to momentary perceptions of vitality, fatigue, and tiredness as captured through established verbal scales of human energy. At the between-person level, the regression coefficients of the battery scale ranged from .84 to .89 predicting subjective vitality and ranged from −.62 to −.79 predicting fatigue. The battery scale was strongly associated with tiredness as reflected in the weary-item as evidenced in standardized regression coefficients ranging from −.54 to −.73. In other words, the average status within person across time or the person-mean of momentary status as captured with the battery scale corresponds highly to the person-mean in vitality, fatigue, and tiredness. Of note, the associations at the within-person level were a bit weaker than the associations at the between-person level. The strongest links across levels of analyses emerged between the battery scale and subjective vitality. In sum, although the battery scale provides a rather broad (or unspecific) assessment of momentary human energy status reflecting aspects of vitality, fatigue, and tiredness, the battery scale corresponds most closely to subjective vitality that is feeling energetic, alive, and alert.

3.4   Implications

In sum, our results suggest that the battery-scale is a reliable and valid measure of human energy when applied to capture momentary energetic state in experience sampling methodology studies. The links between ratings through the battery scale on the one hand and ratings through verbal scales of vitality and fatigue on the other hand are in line with prior research at the between-person level of analysis [31]. The strong associations of the battery scale with the other indicators of human energy seem to be homologous across levels. Hence, the battery scale is a useful and valid tool to track differences in energetic well-being between persons at a given point in time. At the same time, the battery scale is almost equally valid and sensitive to measure changes in energetic well-being within an individual over time, as reflected in the high correlations with vitality and fatigue and the ICCs, very similar to those of the verbal scales. Drawing on these findings, we discuss next how our pictorial scale can be applied for managing energy on an individual and organizational level.

4 Application on an Individual Level

The developed single-item measure is particularly useful in ecological momentary assessment research in order to survey people multiple times over the course of a day without a need for the people to spend too much time to it. Tools such as formr[2] not only offer the possibility of surveying participants online, but even to automatically analyze and visualize the data according to different aspects [42]. Thus, it is possible to generate and provide to the user individual feedback based on the collected data, such as individual energetic peaks and troughs. The pictorial scale offers the opportunity of collecting more data points by means of short, fast surveys. Due to the higher data density, trajectories relevant for the user can be examined not only in the long term, but even down to the course of a day. The pictorial scale thus provides the basis for observing the development of the individual energy level over time, gaining insights into individual patterns of human energy and supporting personal energy management through IT. In the following the motivation and approach for technology-assisted self-management in the area of human energy are presented.

4.1 Importance of Personal Energy Management

As human energy is experienced as a high level of subjective vitality and low level of fatigue, it is a limited and depletable, but also renewable resource. An effective energy management could enable people to reach their goals in life while maintaining their energy [43]. There are several phenomena connected to higher human energy that might be desirable for the individual, and often at the same time for the organization a person works for. As humans want to feel competent and effective functioning [24], it is worth noting that higher levels of energy are connected to more productivity and creativity [43]. Furthermore, research suggests that high levels of subjective energy are related to wellbeing, both in its psychological and physical part [24]. It is shown that there is a positive relationship between subjective vitality and mental health, as well as a negative relationship to ill-being [24]. In contrast, a lack of energy might be a problem to health. For example, Schippers and Hogenes associate burnout with a lack of energy [43]. Mental health problems contribute substantially to the overall disease burden worldwide and are especially concentrated in the working population, potentially leading also to a loss in human capital [2]. Thus, promoting sustained personal human energy might be an important aspect in future countermeasures, which are not only relevant to the individual, but in the larger context also to e.g. health insurances and employers.

However, handling one's own energy effectively is not easy, because there are several things to consider. For example, even if a person is experiencing high levels of energy, it is important that this energy is held stable and furthermore has a targeted direction rather than being depleted randomly [43]. In order to avoid problems resulting from a suboptimal use of personal resources, there are approaches to self-management that can also be utilized with regard to human energy. Kleinmann and König define

---

[2] https://formr.org/.

self-management as "all efforts of a person to influence the own behavior in a targeted way" [44]. Thus, self-management is necessary, whenever many alternatives of behavior are available. Given a set of behavioral alternatives with different consequences, self-management actions are usually maintained by desirable long-term consequences [3]. In line with these characteristics, we define personal energy management as all efforts of a person to influence behaviors associated with energy depletion and recovery in a targeted way that take into account desirable long-term consequences of behavioral alternatives on the own energy status.

4.2 Key Competences Self-reflection and Self-development

Self-management does not mean to change oneself, but to find and improve the own individual way to perform - and thus, it is essential to analyze and become aware of one's own behaviors [45]. For many people it is challenging to determine the activities, which help them to energize, and even more to reinforce them in order to change a situation actively [43]. However, reflecting on one's own behavior and utilizing the gained insights to implement changes represent two fundamental levels of self-management competence in which it is also possible to evolve [46]. Based on these both levels, we use the terms *self-reflection* and *self-development* for the corresponding competences, which can be developed by using different self-management procedures. Self-reflection can be promoted through the procedure of self-observation, which "involves systematic data gathering about one's own behavior" [3]. On the basis of observations and goals set, self-evaluation can take place [3], which we also consider a part of self-reflection. While achievements and strengths can be discovered through self-reflection, also potentials for improvement and necessary steps for change can be determined. When managing human energy, a person could e.g. reflect on the personal energy curve over the day in order to discover energizing and exhausting activities as well as influencing factors such as the time of day. This knowledge could then, for example, be used to increase the proportion of energizing activities, to utilize energy peaks for important tasks, or to establish recovery phases for energy troughs. Several self-management procedures are proposed that support implementing envisaged improvements for self-development. An example is cueing, which means altering the exposure to certain stimuli in a way that desirable behaviors are promoted while undesirable are limited [3]. One could e.g. place a picture or text in a prominent place as a reminder of regular recovery activities.

4.3 The Role of Technology

Since managing energy effectively can be challenging, information technology could act as a facilitator, both in self-reflection and self-development. The first step in self-reflection is to systematically gather data as a basis for later self-evaluation. Using validated instruments may help focusing the data collection on targeted aspects like the personal energy. As such, the developed battery scale provides the opportunity to regularly assess the levels of human energy without a need for the people to spend too much time to it. In order to minimize the effort and introduce a stimulus for self-assessment, a digital version of the scale e.g. as an application with a reminder could be used. As can be seen from the quantified self movement, IT can also be supportive by automatically collecting personally relevant information like physical activity, e.g. through mobile apps or smart devices [4]. This could constitute a good complement for personal energy management with context data or even reduce the need for self-assessments, where reliable measurements are possible.

As a next step, the data collected has to be prepared to be useful for self-evaluation. Choe et al. propose to provide rich visualizations to help people in self-reflection [5]. For energy management, a very simple approach is to visualize the personal energy curve over time. Similar to using a tool like formr in momentary assessment research, where it is possible to survey participants and automatically create plots based on the data [42], an application for personal purposes could collect data and generate visualizations. In order to create rich visualizations for personal energy management, further research is necessary to determine other relevant data in conjunction with the "battery status", which can be collected and integrated appropriately in order to reveal the contingencies between individual behaviors and energetic well-being to end-users.

In addition to supporting self-reflection, IT can also promote self-development, e.g. by cueing through alerts or recommendations. Such interventions should be created on the basis of the individual data collected, so that they are personalized. Alerts were already used, for example, in the form of changing light colors depending on mental workload of a person performing a task [47]. The use of recommendations is studied, among others, for promoting relaxation [6] or physical activity (see [48] for a review). Such IT-supported triggers might help people in implementing their targeted improvements. With regard to human energy further research is necessary to determine, which kind and combination of feedback procedures are appropriate to effectively support managing personal energy through IT.

5 Application on an Organizational Level

In the previous section, we emphasized the importance of managing human energy from the perspective of the individual. In a similar way, energy is also important for organizations as acknowledged by Bruch and Ghoshal [49]. The authors state that "without a high level of energy, a company cannot achieve radical productivity improvements, cannot grow fast and cannot create major innovations". Since organizations typically involve multiple persons, the construct of human energy has to be viewed on a collective level leading to a construct of organizational energy [50]. In the following approaches to the measurement, aggregation and mapping of organizational energy are discussed.

5.1 Measurements and Aggregation

Schiuma (2007) states that "the energy of employees is recognized as an important factor in their performance and in maximizing their overall contribution to the organization. Organizational energy is dynamic in nature; it is more than just the sum of the energy of its employees. It also includes the interaction and dynamics of teams and the organization as a whole" [51]. What all considerations of organizational energy have in common is that human energy forms an important component of organizational energy. Hence applying the concept of human energy on an organizational level implies to integrate it and melt it into organizational energy. Measuring energy within organizations has already been discussed previously e.g. in regard to energy assessments [51] or as part of research reviews on organizational energy [43]. Different kinds and models of energy may be relevant such as individual energy, relational energy, or productive energy, to name only a few [52]. The developed battery scale can be used to capture the individual energy level of a person ranging from exhausted to full of energy. As a brief measure, it could be integrated in information systems of organizations allowing to diagnose the employees' energy level at several moments at work, while requiring much less time than comprehensive scales. As mentioned above, organizational energy also includes aspects like team dynamics, so that assessments via the battery scale could be complemented with other scales like the productive energy scale by Cole's et al. [53]. While brief measures as the battery scale could then be used for regular assessments, more comprehensive scales could be used mainly at certain points in time, e.g. at times of organizational changes. Furthermore, self-assessments could be complemented by automated IT-based data analysis tools that try to infer the level of energy e.g. based on the duration of concentrated uninterrupted work on single content objects such as documents, presentations, or code. Next, according to Baker (2019), "individual-level measures can be aggregated to the group level as the sum, mean, or variability of individual members' energy" [52]. After measuring the organizational energy at various levels of aggregation, mapping it might be relevant in order to reflect on the current level of energy at various level of details (e.g. group level, department level, organization-wide) and to find ways of sustainably improving the energy level.

5.2   Energy Mapping and Analysis

An aggregated level of team energy could be mapped along the typical phases of a project showing the average level of energy of all involved team members (cf. Figure 5). This will give insights into the current state of the project and the energy increases or decreases. The latter might depend on the project or on factors outside the project which in either case must be analyzed when energy is lacking. In this direction, mappings and comparisons between different teams and departments might help team leaders and managers to identify needs for intervention and to help them reflect on their leadership style.

Furthermore, energy trajectories could be mapped to phases of organizational changes or interventions. Depending on the scope of changes such mappings could be created and analyzed on several aggregation levels of energy ranging from the team level to the departmental level up to the organizational level. In this way it would be possible to determine which managerial strategies are favorable or dysfunctional in the long term in order to unfold and promote human potential. On this basis, organizations can develop and achieve sustainable organizational energy.

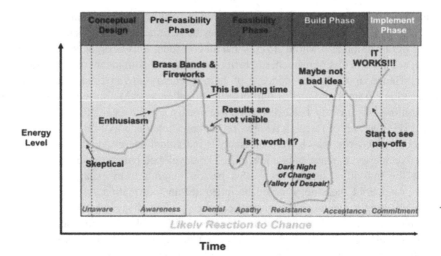

Fig. 5. Mapping of energy along typical phases of projects (source: [54], p. 154)

6 Conclusions

Human energy is an important construct that is connected to a person's wellbeing, health, creativity, and productivity. Thus, the energy of employees is an important factor for organizations regarding their innovations, overall productivity, and growth. Taking the battery-metaphor literally, we have developed a pictorial scale to capture momentary levels of human energy. We present empirical results demonstrating high correlations of the pictorial scale with momentary ratings of subjective vitality and fatigue at the intraindividual and the interindividual level. Hence, the battery scale is a useful and valid tool to track differences in energetic well-being between persons at a given point in time and is almost equally valid and sensitive to measure changes in energetic well-being within an individual over time. A brief and face-valid single-item measure is particularly useful in ecological momentary assessment research, e.g. surveying employees multiple times over the course of a day. By enabling higher data density, the battery scale provides the basis to examine the development of the energy level even down to the course of a day and to gain insights into individual patterns of human energy in greater detail. Furthermore, it could be applied for technology-assisted management of energy at the individual and organizational level. As the battery scale is a very brief measure, it could conveniently be integrated into personal applications of an individual or information systems of organizations in order to capture momentary energetic states over time and thus, provide the basis for reflection and development processes.

References

1. Parent-Thirion, A., et al.: 6th European Working Conditions Survey. Overview Report. Publications Office of the European Union, Luxembourg (2017)
2. James, S.L., et al.: Global, regional, and national incidence, prevalence, and years lived with disability for 354 diseases and injuries for 195 countries and territories, 1990–2017: a systematic analysis for the Global Burden of Disease Study 2017. Lancet **392**, 1789–1858 (2018)
3. Manz, C.C., Sims, H.P.: Self-management as a substitute for leadership: a social learning theory perspective. AMR **5**, 361–367 (1980)
4. Rapp, A., Cena, F.: Self-monitoring and technology: challenges and open issues in personal informatics. In: Stephanidis, C., Antona, M. (eds.) UAHCI 2014. LNCS, vol. 8516, pp. 613–622. Springer, Cham (2014). https://doi.org/10.1007/978-3-319-07509-9_58
5. Choe, E.K., Lee, B., Zhu, H., Riche, N.H., Baur, D.: Understanding self-reflection. How people reflect on personal data through visual data exploration. In: Oliver, N., Czerwinski, M., Czerwinski, M.P. (eds.) Proceedings of the 11th EAI International Conference on Pervasive Computing Technologies for Healthcare. PervasiveHealth 2017, 23–26 May 2017, Barcelona, Spain, pp. 173–182. ACM, New York (2017)
6. Fallon, M., Spohrer, K., Heinzl, A.: Wearable devices: a physiological and self-regulatory intervention for increasing attention in the workplace. In: Davis, F.D., Riedl, R., vom Brocke, J., Léger, P.-M., Randolph, A.B. (eds.) Information Systems and Neuroscience. LNISO, vol. 29, pp. 229–238. Springer, Cham (2019). https://doi.org/10.1007/978-3-030-01087-4_28
7. Frone, M.R., Tidwell, M.-C.O.: The meaning and measurement of work fatigue: development and evaluation of the Three-Dimensional Work Fatigue Inventory (3D-WFI). J. Occup. Health Psychol. **20**, 273–288 (2015)
8. Kleine, A.-K., Rudolph, C.W., Zacher, H.: Thriving at work: a meta-analysis. J. Organ. Behav. **40**, 973–999 (2019)
9. Weigelt, O., Gierer, P., Syrek, C.J.: My mind is working overtime—towards an integrative perspective of psychological detachment, work-related rumination, and work reflection. Int. J. Env. Res. Pub. Health **16**, 2987 (2019)
10. Wendsche, J., Lohmann-Haislah, A.: A meta-analysis on antecedents and outcomes of detachment from work. Front. Psychol. **7**, 2072 (2017)
11. Crawford, E.R., LePine, J.A., Rich, B.L.: Linking job demands and resources to employee engagement and burnout: a theoretical extension and meta-analytic test. J. Appl. Psychol. **95**, 834–848 (2010)
12. Reilly, N.P., Sirgy, M.J., Gorman, C.A. (eds.): Work and Quality of Life: Ethical Practices in Organizations. Springer, Dordrecht (2012). https://doi.org/10.1007/978-94-007-4059-4
13. Hobfoll, S.E.: Conservation of resources: a new attempt at conceptualizing stress. Am. Psychol. **44**, 513–524 (1989)
14. Halbesleben, J.R.B., Neveu, J.-P., Paustian-Underdahl, S.C., Westman, M.: Getting to the "COR": understanding the role of resources in conservation of resources theory. J. Manag. **40**, 1334–1364 (2014)
15. Ragsdale, J.M., Beehr, T.A.: A rigorous test of a model of employees' resource recovery mechanisms during a weekend. J. Organ. Behav. **37**, 911–932 (2016)
16. Zijlstra, F.R.H., Cropley, M., Rydstedt, L.W.: From recovery to regulation: an attempt to reconceptualize 'recovery from work'. Stress Health: J. Int. Soc. Invest. Stress **30**, 244–252 (2014)

17. Meijman, T.F., Mulder, G.: Psychological aspects of workload. In: Drenth, P.J.D., Thierry, H., de Wolff, C.J. (eds.) Handbook of Work and Organizational: Work Psychology, 2nd edn., vol. 2, pp. 5–33. Psychology Press/Taylor & Francis, Erlbaum/Hove (1998)

18. Zijlstra, F.R.H., Sonnentag, S.: After work is done: psychological perspectives on recovery from work. Eur. J. Work Organ. Psychol. **15**, 129–138 (2006)

19. Sonnentag, S., Venz, L., Casper, A.: Advances in recovery research: what have we learned? What should be done next? J. Occup. Health Psychol. **22**, 365–380 (2017)

20. Quinn, R.W., Spreitzer, G.M., Lam, C.F.: Building a sustainable model of human energy in organizations: exploring the critical role of resources. Acad. Manag. Ann. **6**, 337–396 (2012)

21. Thayer, R.E., Newman, J.R., McClain, T.M.: Self-regulation of mood: strategies for changing a bad mood, raising energy, and reducing tension. J. Pers. Soc. Psychol. **67**, 910–925 (1994)

22. Fritz, C., Lam, C.F., Spreitzer, G.M.: It's the little things that matter: an examination of knowledge workers' energy management. AMP **25**, 28–39 (2011)

23. Zacher, H., Brailsford, H.A., Parker, S.L.: Micro-breaks matter: a diary study on the effects of energy management strategies on occupational well-being. J. Vocat. Behav. **85**, 287–297 (2014)

24. Ryan, R.M., Frederick, C.: On energy, personality, and health: subjective vitality as a dynamic reflection of well-being. J. Pers. **65**, 529–565 (1997)

25. Ilies, R., Aw, S.S.Y., Lim, V.K.G.: A naturalistic multilevel framework for studying transient and chronic effects of psychosocial work stressors on employee health and well-being. Appl. Psychol. **65**, 223–258 (2016)

26. Syrek, C.J., Kühnel, J., Vahle-Hinz, T., Bloom, J.D.: Share, like, Twitter, and connect: ecological momentary assessment to examine the relationship between non-work social media use at work and work engagement. Work Stress **32**, 209–227 (2018)

27. Ohly, S., Sonnentag, S., Niessen, C., Zapf, D.: Diary studies in organizational research: an introduction and some practical recommendations. J. Pers. Psychol. **9**, 79–93 (2010)

28. Kunin, T.: The construction of a new type of attitude measure. Pers. Psychol. **8**, 65–77 (1955)

29. Wanous, J.P., Reichers, A.E., Hudy, M.J.: Overall job satisfaction: how good are single-item measures? J. Appl. Psychol. **82**, 247–252 (1997)

30. Bradley, M.M., Lang, P.J.: Measuring emotion: the self-assessment manikin and the semantic differential. J. Behav. Ther. Exp. Psychiatry **25**, 49–59 (1994)

31. Weigelt, O., Wyss, C., Siestrup, k., Fellmann, M., Lambusch, F.: Ein Bild sagt mehr als tausend Worte - Entwicklung und Überprüfung einer Piktogramm-Skala zu Human Energy (A picture is worth a thousand words - Development and validation of a single-item pictorial scale of human energy). In: Neue Formen der Arbeit in der digitalisierten Welt: Veränderungskompetenz stärken (2019). https://doi.org/10.13140/rg.2.2.33862.22082

32. Gabriel, A.S., et al.: Experience sampling methods: a discussion of critical trends and considerations for scholarly advancement. Organ. Res. Methods **22**, 969–1006 (2019)

33. McCormick, B.W., Reeves, C.J., Downes, P.E., Li, N., Ilies, R.: Scientific contributions of within-person research in management: making the juice worth the squeeze. J. Manag. **46**, 321–350 (2020)

34. West, S.G., Ryu, E., Kwok, O.-M., Cham, H.: Multilevel modeling: current and future applications in personality research. J. Pers. **79**, 2–50 (2011)

35. Schmitt, A., Belschak, F.D., Den Hartog, D.N.: Feeling vital after a good night's sleep: the interplay of energetic resources and self-efficacy for daily proactivity. J. Occup. Health Psychol. **22**, 443–454 (2017)

36. McNair, D.M., Lorr, M., Droppleman, L.F.: EDITS manual profile of mood states. Educational and Industrial Testing Service, San Diego, CA (1992)

37. Albani, C., et al.: Überprüfung der Gütekriterien der deutschen Kurzform des Fragebogens 'Profile of Mood States' (POMS) in einer repräsentativen Bevölkerungsstichprobe. = The German Short Version of 'Profile of Mood States' (POMS): Psychometric Evaluation in a Representative Sample. PPmP: Psychotherapie Psychosomatik Medizinische Psychologie **55**, 324–330 (2005)

38. Geldhof, G.J., Preacher, K.J., Zyphur, M.J.: Reliability estimation in a multilevel confirmatory factor analysis framework. Psychol. Methods **19**, 72–91 (2014)

39. Huang, F.L.: Conducting multilevel confirmatory factor analysis using R (2016). http://faculty.missouri.edu/huangf/data/mcfa/MCFAinRHUANG.pdf

40. Chen, G., Bliese, P.D., Mathieu, J.E.: Conceptual framework and statistical procedures for delineating and testing multilevel theories of homology. Organ. Res. Methods **8**, 375–409 (2005)

41. Schermelleh-Engel, K., Moosbrugger, H., Müller, H.: Evaluating the fit of structural equation models: tests of significance and descriptive goodness-of-fit measures. Methods Psychol. Res. **8**, 23–74 (2003)

42. Arslan, R.C., Walther, M.P., Tata, C.S.: formr: a study framework allowing for automated feedback generation and complex longitudinal experience-sampling studies using R. Behav. Res. Methods **52**, 376–387 (2019)

43. Schippers, M.C., Hogenes, R.: Energy management of people in organizations: a review and research agenda. J. Bus. Psychol. **26**, 193–203 (2011)

44. Kleinmann, M., König, C.J.: Selbst- und Zeitmanagement. Hogrefe, Göttingen (2018)

45. Drucker, P.F.: Managing oneself. Harv. Bus. Rev. **83**, 100–109 (2005)

46. Graf, A.: Selbstmanagement-Kompetenz in Unternehmen nachhaltig sichern. Leistung, Wohlbefinden und Balance als Herausforderung. Springer Gabler, Wiesbaden (2012)

47. Maior, H.A., Wilson, M.L., Sharples, S.: Workload alerts—using physiological measures of mental workload to provide feedback during tasks. ACM Trans. Comput.-Hum. Interact. **25**, 1–30 (2018)

48. Ghanvatkar, S., Kankanhalli, A., Rajan, V.: User models for personalized physical activity interventions: scoping review. JMIR mHealth and uHealth **7**, e11098 (2019)

49. Bruch, H., Ghoshal, S.: Unleashing organizational energy. MIT Sloan Manage. Rev. **45**, 45 (2003)

50. Hannah, S.T., Avolio, B.J., Cavarretta, F.L., Hennelly, M.J.: Conceptualizing organizational energy. Citeseer (2010)

51. Schiuma, G., Mason, S., Kennerley, M.: Assessing energy within organisations. Meas. Bus. Excellence **11**, 69–78 (2007)

52. Baker, W.E.: Emotional energy, relational energy, and organizational energy: toward a multilevel model. Ann. Rev. Organ. Psychol. Organ. Behav. **6**, 373–395 (2019)

53. Cole, M.S., Bruch, H., Vogel, B.: Energy at work: a measurement validation and linkage to unit effectiveness. J. Organ. Behav. **33**, 445–467 (2012)

54. Matthias, T.M.: A conceptual model of information system implementation within organisations (2009)

Using Guided Cognitive Illusions to Compensate for the Motion Limits of 4D Seats

Zhejun Liu[1] , Guodong Yu[1] , Jing Lin[2(✉)] , Tianrun Gu[1] ,
and Qin Guo[1]

[1] Tongji University, 1239 Siping Road, Shanghai 200092,
People's Republic of China
{wingeddreamer,1833496,1751344,1833497}@tongji.edu.cn
[2] Shanghai Academy of Spaceflight Technology, 3888 Yuanjiang Road,
Shanghai 201109, People's Republic of China
397446509@QQ.com

Abstract. Presently 4D seats are widely used as an approach of training and entertainment in various industries. However, most 4D seats can move and/or rotate within a pretty limited range, making them impossible to faithfully copy the motion of a real or simulated vehicle without any sacrifice. Considering the fact that vision provides most information to a human being for pose judgement, we proposed a scheme where eye-body inconsistency was deliberately created to guide a user's perception of body rotation so as to make more room for over-limit moves. We conducted two experiments to verify its effectiveness from the perspective of user experience and the findings were partially in accordance with our expectation.

Keywords: Usability methods and tools · Motion platform · 4D seat · Cognitive illusion

1 Introduction

With the development of science and technology, 4D seats are widely used as an approach of training [1–6] and entertainment [7, 8] in various venues, such as planetariums, museums and cinemas, so as to enhance the audience's experience and sense of immersion. They also play an important role in many fields such as the driving simulation of aircrafts [2, 3], ships [4, 5], cars [6] and other vehicles.

At present, however, most 4D seats have limited ranges of motion (move and rotation included) and the difference varies according to their types. A six-degree-of-freedom (abbr. 6-DOF) seat is the most common type for the simulations of land and overwater vehicles whose rotation angles usually stay within a small range unless overturned [9]. But they are not suitable for the simulation of underwater, airborne or spatial vehicles for which much larger ranges of rotation are needed. New types of motion platforms were invented to make up for this limitation. For example, our 2-DOF (two degree of freedom) seat can rotate from 0° to 40° on the X axis (in the direction of pitch) and virtually freely on the Z axis (in the direction of roll) as shown in Fig. 1.

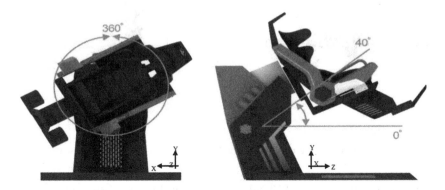

Fig. 1. 2-DOF 4D seat

Though a 2-DOF seat like this is a much better choice for aircraft or spaceship simulation, but the limited rotation on the X axis should not be overlooked. When used to simulate a jet flight, for example, the seat can roughly match the jet's pitch angle when it climbs up, but will fail to do so when it dives because the platform cannot lean forward. This problem makes a user sense the mismatch between vision and physical motion, deprive them of immersion or even cause simulator sickness [10]. In addition, it also places constraints on the design of an interactive application. Many simulators dodge this problem by simply reducing the possible operational range in the virtual environment.

Because vision is the most important sense and plays the dominant role in cognition [11], it can be used to creation illusions like the McGurk effect. [12, 13] this paper proposes a mechanism of intentionally making use of the deviation between vision and body rotation to create a guided cognitive illusion in order to solve the problems mentioned above.

2 Proposal and Hypotheses

2.1 Proposal

We assume that there is a maximal insensible threshold $\Delta\theta$ for the angle deviation between vision and body rotation and we can take advantage of it to compensate for the motion limits of a 4D seat.

Take the 4D seat in Fig. 1 as an example (free roll and 0° to 40° pitch), say $\Delta\theta$ is the maximal insensible angle deviation, we can purposefully set the pitch angle of the seat to $\Delta\theta$ when the simulated vehicle is maintaining a level flight. Then we define the pitch angle of the seat by packing the virtual vehicle's first-quadrant and second-

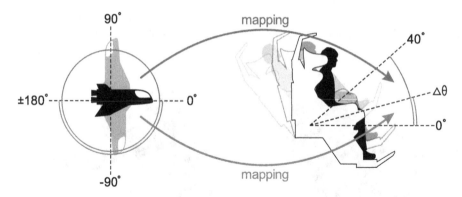

Fig. 2. Mapping the angle of a simulated aircraft to the 2-DOF 4D seat

quadrant pitch angles to range [Δθ, 40°], the third-quadrant and fourth-quadrant pitch angles to range [0°, Δθ] as shown in Fig. 2 using Eq. (1), a part of which is a sigmoid function.

$$
\alpha' = \begin{cases}
\Delta\theta\left(2 - \dfrac{2}{1+e^{\frac{\alpha}{18}}}\right) & -180 \leq \alpha < 0 \\
\Delta\theta & \alpha = 0 \\
\Delta\theta + (40 - \Delta\theta)\left(\dfrac{2}{1+e^{\frac{-\alpha}{18}}} - 1\right) & 0 < \alpha \leq 180
\end{cases}
\tag{1}
$$

2.2 Hypotheses

H1. The purposeful mismatch between vision and body rotation can create the cognitive illusion required. It is possible to make users believe they are in a state of horizontal motion when they are actually in a tilted state.

H2. The deviation of body rotation from vision has a limited range. If the difference goes beyond this limit, a user may sense this mismatch and the cognitive illusion is thus spoiled. We try to find out the boundary of the valid deviation acceptable to most users.

H3. If used properly, this guided cognitive illusion helps to enhance user experience and reduce discomfort such as simulator sickness in a 4D cinema environment.

3 Experiment

3.1 Design of the Experiment and Test Material

Two experiments were conducted to verify these hypotheses: Experiment A (abbr. Exp. A) for H1, H2 and Experiment B (abbr. Exp. B) for H3.

We developed two different programs for Exp. A and Exp. B with the Unity engine. Both were cruises in a modern city on a futuristic aircraft, as shown in Fig. 3.

Fig. 3. Example screenshots from the experimental programs

Experiment A. The procedure of Exp. A is as follows (Fig. 4).

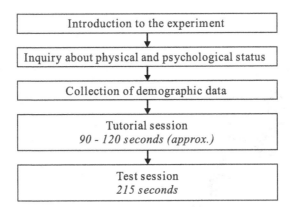

Fig. 4. Procedure of Experiment A

At the very beginning, subjects were introduced to this experiment and informed that its purpose was to test their sense of balance so they should do their best to get a final score as high as possible. It was not true but as an approach to keep subjects focused and attentive, avoiding possible bias caused by absent-mindedness.

After a subject was seated on the 4D seat and put on the HMD, he/she went through an interactive tutorial first. The aircraft flew flat for a while and then automatically dived. A voice told the subject to pull the stick backward to correct its pose and make the aircraft fly flat again. After a while, the aircraft automatically climbed and the subject was again guided to adjust the pose by pushing the stick forward. With these steps, the subject was taught how to use the stick for pose adjustment. Then the narrator told the subject that his/her task during the next few minutes was to do his/her best to keep the aircraft fly flat by pulling or pushing the stick. Despite that there might be a delay in the aircraft's response, he/she should keep doing the right thing until it flats. This tutorial session, depending on how quickly a subject finished all the tasks, usually lasted for 90 to 120 s approximately (Fig. 5).

Fig. 5. Aircraft adjustment by pulling or pushing the stick

When the tutorial session ended, the main cruise began. The route extended between buildings with ups and downs, all pre-animated and could not actually be controlled by the subject. There were carefully designed moments of sensory inconsistency: the

Fig. 6. Deliberately designed sensory inconsistency

HMD provided a perfectly levelled vision while the seat gradually tilted backward as shown in Fig. 6. To avoid cross effect and simplify the independent variables, roll was kept as 0 during the whole process.

Figure 7 shows the pitch curves of the virtual vehicle and the seat. The three parts with gray background were the most important windows to inspect. During these periods of time, the virtual cameras kept looking straight forward while the seat slowly tilted up. The expected discovery was whether and when the subjects would begin to feel the pitch and try to amend it with the stick controller.

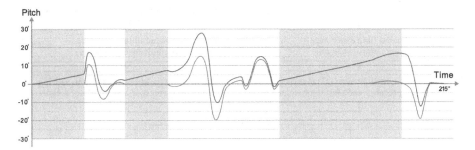

Fig. 7. Pitch curves of the virtual vehicle (green) and the seat (blue) (Color figure online)

We recorded the essential information in the following format at an interval of 40 ms as data files kept for later analysis.

[*execution time, virtual vehicle pitch angle, platform pitch angle, joystick input Y*]

Experiment B. The second experiment employed two program versions. They were both 215-s long cruises in the same modern city as the one used in Exp. A, but with different platform control schemes.

Version A used the default scheme. It offered a 1:1 copy when the pitch angle was between 0° and 40°. Angle values beyond this range were clamped. Version B used the scheme proposed by us where $\Delta\theta$ was set to 7° (See 2.1 & 5.3). The platform and the simulated aircraft had different pitch angles so that there was some freedom for the seat to nod when the aircraft dived with a <0° pitch. The automatic conversion, however, was not good enough so it was further finetuned manually by the experimenters based on the actual route and test users' feedback.

The experiment used within-subject design. Subjects were randomly assigned to group I or group II before the experiment started. The difference was that in group I, subjects experienced the cruise controlled with the default scheme first and then our scheme. Group II used a reversed order so that the carry over effect could possibly be minimized. Subjects were required to fill in 3 questionnaires, namely presence, immersion and simulator sickness questionnaires after each experience. They were also shortly interviewed at the end of the experiment. The procedure of Exp. B is shown in Fig. 8.

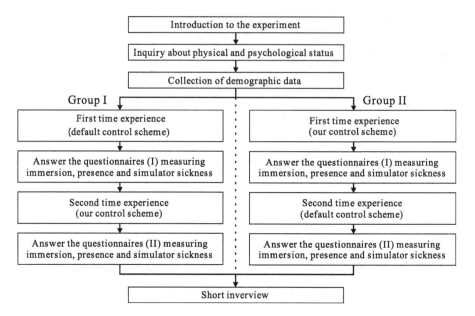

Fig. 8. Procedure of Experiment B

3.2 Measures

In Exp. A, no subjective report was required from the subjects. The only data recorded was how they manipulated the stick controller.

In Exp. B, presence, immersion and simulator sickness were measured using the presence questionnaire by Martin Usoh et al. [14], the immersion questionnaire by Jennett C et al. [15] and the simulator sickness questionnaire by Kennedy R S et al. [16]. They were all modified to suit the actual experimental material as shown in Table 1, Table 2 and Table 3.

Table 1. Presence questionnaire

	Questions & options
1	Did you have a sense of "being there" in the aircraft? (1: Not at all ~ 5: Very much)
2	To what extent were there times during the experience when the virtual space was the reality for you? (1:No time at all ~ 5: Almost all the time)
3	Do you think of the aircraft more as a miniature that you saw, or more as a real vehicle that you drove? (1:Miniature ~ 5: Real vehicle)
4	During the time of the experience, which was strongest on the whole, your sense of being in the aircraft, or of being elsewhere? (1: In the aircraft ~ 5: Elsewhere)
5	During the time of experience, did the motion of the chair feel realistic to you? (1: Very unrealistic ~ 5: Very realistic)
6	Consider your memory of being in the aircraft. How similar in terms of the structure of the memory is this to that of other transportation or entertaining vehicles you have used before? (1: Not similar at all ~ 2: Very Similar)

Table 2. Immersion questionnaire

	Questions & options
1	To what extent did you feel you were focused on the game? (1: Not at all $\sim$ 5. Very Focused)
2	To what extent did you lose track of time? (1: No time at all $\sim$ 5: Almost always)
3	To what extent did you feel as though you were separated from your real-world environment? (1: Not at all $\sim$ 5: Very much)
4	To what extent did you enjoy the graphics and the imagery? (1: Not at all $\sim$ 5: Very much)
5	To what extent did you enjoy the motion of the chair? (1: Not at all $\sim$ 5: Very much)
6	How much would you say you enjoyed the experience? (1: Dislike it very much $\sim$ 5: Enjoyed it very much)
7	Would you like to play the game again? (1: Not at all $\sim$ 5: Very eager to)

Table 3. Simulator sickness questionnaire

	Question: To what extent did you suffer from the following symptoms during the experience you just had?	Options
1	General Discomfort	1: Not at all
2	Fatigue	2: Slight
3	Headache	3: Moderate
4	Eyestrain	4: Severe
5	Difficulty focusing	
6	Salivation increasing	
7	Sweating	
8	Nausea	
9	Difficulty concentrating	
10	Fullness of the head	
11	Blurred vision	
12	Dizziness with eyes open	
13	Dizziness with eyes closed	
14	Vertigo	
15	Stomach awareness	
16	burping	

At the end of Exp. B, subjects were briefly interviewed and asked what they thought about the experience, how they liked it, why and when they felt uncomfortable or unnatural etc.

3.3   Subjects

17 subjects participated in Exp. A, another 17 subjects participated in Exp. B. Their demographic data is shown in Table 4. They were inquired about their health status

before the experiment to make sure that they had normal vision and no potential physical or psychological problem of experiencing VR on a 4D seat.

Table 4. Demographic data of the subjects participating in Exp. A and Exp. B

Experiment	Gender		Age			Profession	
	Male	Female	Average	Min	Max	Student	Other
Exp. A	9	8	23.0	19	34	17	0
Exp. B	6	11	22.9	19	25	16	1

3.4   Apparatus

The experiment used a high performance workstation running Windows 10 64bit version, as shown in Table 5. An HTC Vive Pro, featuring a 110° viewing angle and a resolution of 1440×1600 pixels each eye, was used as the HMD device in the experiment. This guaranteed that user experience would not be jeopardized by low frame rate or poor visual quality.

The controller used as the input device was an X56 from Logitech featuring a control stick and a throttle, but only the control stick was functional in this specific experiment.

Specifications of the 2-DOF platform used in the experiment can be found in the introduction part at the beginning of this paper.

Table 5. Hardware specification

CPU	Intel i7 9700	Graphic card	nVidia Geforce 2070
Memory	16 GB	Harddrive	512 GB SSD
Display	65-inch LCD TV	HMD	HTC Vive Pro
Controller	Logitech Saitek X56 controller for flight simulation		
4D Seat	2-DOF 4D seat produced by Topow Research Institute of VR Tech		

4   Results and Discussion

4.1   Results of Experiment A

The 17 subjects' operation of the stick controller was recorded at an interval of 40 ms as mentioned in Sect. 4.1. The input valued ranged between −1 and 1 while 0 was the neutral position (i.e. no push or pull). Using time as the X-axis, the input signals from a certain subject could be drawn into a curve as shown in Fig. 9.

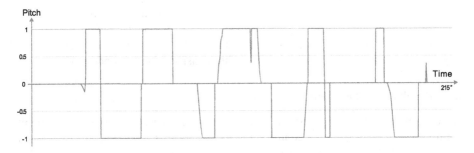

Fig. 9. An example curve of input signals

For this experiment, the purpose is to spot the moments when a subject perceived sensory inconsistency. Therefore, we marked these moments using red lines as shown in Fig. 10 based on the following criterium: the absolute value of the input must be greater than 0.5 and kept this way for more than 3 s, ignoring all the spikes and grooves shorter than 1 s.

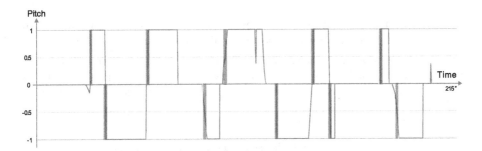

Fig. 10. Input signal curve with marked moments of inconsistency perception

Finally, all the red lines marked from the 17 subjects' data were piled together on top of the pitch curves of the virtual vehicle and the seat as shown in Fig. 11, where a concentration of lines means a strong tendency of inconsistency recognition.

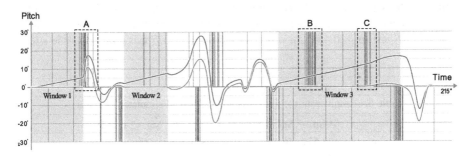

Fig. 11. the composed graph for Exp. A (Color figure online)

In the inspection window 1, 2 and 3 (with gray background, explained in 3.1) we may clearly see three concentrations in rectangle A, B and C, which reveals the attempts to tilt down the aircraft. Since concentration B and C are in the same window, only B, the first recognition, shall be taken into consideration in window 3.

4.2   Results of Experiment B

Presence was analyzed with SPSS v22. The data in each group (group I = default control scheme, group II = our control scheme) was normally distributed (1-sample KS test, $p_I = 0.549$, $p_{II} = 0.846$) and the variance was equal (Levene's test, $p = 0.412$) (Table 6).

Table 6. Result from the paired samples T-test of presence

	Mean	Std. deviation	t	df	Sig. (2-tailed)
Group I	19.290	5.850	−0.816	16	0.426
Group II	20.180	5.570			

A paired samples T-test was conducted to compare the presence of group I and group II, but no significance was found ($p > 0.05$). The result is as follows.

Immersion was analyzed with SPSS v22. The data in each group (group I = default control scheme, group II = our control scheme) was normally distributed (1-sample KS test, $p_I = 0.908$, $p_{II} = 0.664$) and the variance was equal (Levene's test, $p = 0.833$) (Table 7).

Table 7. Result from the paired samples T-test of immersion

	Mean	Std. deviation	t	df	Sig. (2-tailed)
Group I	24.290	6.517	−0.979	16	0.342
Group II	25.290	6.771			

A paired samples T-test was conducted to compare the immersion of group I and group II, but no significance was found ($p > 0.05$). The result is as follows.

Simulator Sickness was analyzed with SPSS v22. The data in each group (group I = default control scheme, group II = our control scheme) was normally distributed (1-sample KS test, $p_I = 0.451$, $p_{II} = 0.224$) and the variance was equal (Levene's test, $p = 0.347$).

Table 8. Result from the paired samples T-test of simulator sickness

	Mean	Std. deviation	t	df	Sig. (2-tailed)
Group I	123.577	36.121	12.861	16	0.000
Group II	34.320	42.049			

A paired samples T-test was conducted to compare the simulator sickness of group I and group II. Group II showed a much better result than Group I with significant difference ($p < 0.05$) as shown below (Table 8).

Short interviews were held after the subjects had experienced both programs to find out which they preferred and why. We categorized them into 4 types as shown in Table 9, where, for the sake of clarity and conciseness, the program using the default control scheme was abbreviated as "Default", and the other one as "Ours".

Table 9. Result from the interviews

	n	Pref.	Reason
Type A	4	*Default*	- *Default* is smoother, *ours* is more thrilling - Aircraft rides should be smooth, so *default* is more realistic
Type B	4	*Default*	- *Default* is dizzier - I feel dizzy on a roller coaster/plane in real life - *Default* is more realistic, because it's closer to real life
Type C	5	*Ours*	- *Default* is dizzier - I don't feel dizzy on a roller coaster/plane in real life - *Ours* is more realistic, because it's closer to real life
Type D	4	*Ours*	- *Default* is dizzier, *ours* is scarier - Both feel real - I feel scary but not dizzy on a roller coaster in real life - *Ours* is more realistic, because it's closer to real life

4.3 Discussion

Based on Fig. 11 from Exp. A, it's safe to conclude that our first hypothesis is proved. The majority of the subjects could hardly feel the tilt angle of the seat in window 1 and window 2. In window 3 most subjects sensed it only when the tilt angle was greater than approximately 5°. This proves that vision really affected the cognition of body rotation.

As for hypothesis 2, we may confidently conclude that there are boundaries where the illusion disappears, but it's difficult to define them numerically. In window 1 and 3 the thresholds were about 5°, but in window 2 most subjects did not sense the tilt even when it reached 7° approximately. Moreover, a closer inspection revealed that even a same subject might display different sensory threshold at different times. A reasonable guess is that multiple factors affected a subject's sense and the tilt angle of the seat did not have a dominant position.

Based on Exp. A, we chose the somewhat aggressive 7° as the $\Delta\theta$ value for Exp. B. The results from Exp. B did not conform with our hypothesis 3 completely. We expected our scheme to outperform the default one in presence, immersion and simulator sickness, but no significance was found in the former two. As for simulator sickness, however, our scheme showed a prominent advantage.

To understand the cause of it, we carefully analyzed what was obtained from the ending interviews and made the following deduction. Firstly, according to Table 9, our scheme truly did better in reducing simulator sickness, which was in accordance with the sensory conflict theory [17, 18]. Secondly, the reason why the subjects gave divers opinions on presence and immersion was largely that they tended to compare the virtual cruise with their real-life experience on a roller coaster or an airplane. Those felt more thrill or dizziness in a real-life experience tended to give higher scores to the virtual experience giving similar feelings when evaluating presence or immersion. But those felt the opposite way gave completely different answers. As a result, the evaluation of presence and immersion turned out to be pretty personal.

5   Conclusion

In this paper, we propose a motion control scheme for 2-DOF platforms with limited freedom of rotation on one or maybe multiple axes. By purposefully creating sensory inconsistency between vision and body rotation, our scheme is capable of responding better to the motion of a simulated vehicle. Via experimentation, we prove that vision plays an important role and can be used to guide cognition in a VR experience on a 4D seat. The motion control scheme proposed by us is also proven to be very valuable in reducing simulator sickness, though the effectiveness of enhancing presence and immersion still needs further investigation.

This paper shall be regarded as a pilot study on the use of guided cognitive illusions to compensate for the motion limits of 4D seats. We hope it can be used as a basic reference for future research in the exploration for better-defined theories and more quantitative solutions.

Acknowledgement. This work is part of a joint research project hosted by the College of Design and Innovation, Tongji University and Shanghai Academy of Spaceflight Technology, with the help from Topow (Jiangsu) Research Institute of Virtual Reality Technologies Co., Ltd.

References

1. Arioui, H., Hima, S., Nehaoua, L., et al.: From design to experiments of a 2-DOF vehicle driving simulator. IEEE Trans. Veh. Technol. **60**(2), 357–368 (2011)
2. Bürki-Cohen, J., Sparko, A.L., Jo, Y.J., et al.: Effects of visual, seat, and platform motion during flight simulator air transport pilot training and evaluation. In: 2009 International Symposium on Aviation Psychology, p. 373 (2009)
3. Dongsu, W., Hongbin, G.: Adaptive sliding control of Six-DOF flight simulator motion platform. Chin. J. Aeronaut. **20**(5), 425–433 (2007)
4. Likun, P.: Modeling and experiment study of a novel digital Hydraulic 2-DOF motion platform. J. Mech. Eng. (2011)
5. Murai, K., Okazaki, T., Hayashi, Y.: Basic study of body sway in artificial ship rolling and pitching by visual and motion platform. Toward efficient simulator-based training. IEEJ Trans. Electron. Inf. Syst. **130**(11), 2007–2012 (2010)
6. Kim, J.H., Lee, W.S., Park, I.K., et al.: A design and characteristic analysis of the motion base for vehicle driving simulator. In: IEEE International Workshop on Robot & Human Communication. IEEE (1997)

7. Zou, X., Chen, R., et al.: Design and analysis of a new 4D motion seat. Ind. Autom. **8**, 102–106 (2016)
8. Yoon, H.C., Lee, S.B., Park, J.Y., et al.: Development of racing game using motion seat. In: 2017 International Symposium on Ubiquitous Virtual Reality (ISUVR). IEEE, pp. 4–7 (2017)
9. Wentao, L.I., Hong, W.Z.: Motion controllability criterion of Steward platform. China Mech. Eng. (2001)
10. Groen, E.L., Bos, J.E.: Simulator sickness depends on frequency of the simulator motion mismatch: an observation. Presence: Teleoper. Virtual Environ. **17**(6), 584–593 (2008)
11. Ge, L.: Cognitive illusion of visual cheating. Popular Sci. Univ. **11**, 24 (2017)
12. Mcgurk, H., Macdonald, J.: Hearing lips and seeing voices. Nature **264**(5588), 746–748 (1976)
13. Macdonald, J.: Visual influences on speech perception processes. Percep. Psycho-phys. **24**, 253–257 (1978)
14. Usoh, M., Catena, E., Arman, S., et al.: Using presence questionnaires in reality. Presence: Tele-oper. Virtual Environ. **9**(5), 497–503 (2000)
15. Jennett, C., Cox, A.L., Cairns, P., et al.: Measuring and defining the experience of immersion in games. Int. J. Hum.-Comput. Stud. **66**(9), 641–661 (2008)
16. Kennedy, R.S., Lane, N.E., Berbaum, K.S., et al.: Simulator sickness questionnaire: an enhanced method for quantifying simulator sickness. Int. J. Aviat. Psychol. **3**(3), 203–220 (1993)
17. Oman, C.M.: Motion sickness: a synthesis and evaluation of the sensory conflict theory. Can. J. Physiol. Pharmacol. **68**(2), 294–303 (1990)
18. Lackner, J.R.: Simulator sickness. J. Acoust. Soc. Am. **92**(4), 2458 (1992)

Fostering Flow Experience in HCI to Enhance and Allocate Human Energy

Corinna Peifer[1]([⊠]) [iD], Annette Kluge[1] [iD], Nikol Rummel[2] [iD],
and Dorothea Kolossa[3] [iD]

[1] Faculty of Psychology, Ruhr University Bochum, Bochum, Germany
corinna.peifer@rub.de
[2] Faculty of Philosophy and Educational Sciences, Ruhr University Bochum,
Bochum, Germany
[3] Faculty of Electrical Engineering and Information Technology,
Ruhr University Bochum, Bochum, Germany

Abstract. Motivation explains the direction, intensity and persistence of human behavior and thus plays a crucial role in the mobilization and allocation of available energy. An experience that occurs during motivated action is *flow*. Flow is perceived as highly rewarding for its own sake and, thus, in flow all attention is directed towards the task at hand, leading to an experience of absorption. At the same time, attention is shielded from irrelevant stimuli and the activity feels easy and effortless. This suggests that flow is a highly efficient state in terms of energy expenditure. Studies addressing the physiology of flow support this assumption. Accordingly, for an optimal use of energy, it is of interest to promote flow in relevant work processes. In HCI, for example, in production work, flow promotion could be enabled by a real-time measure of the operator's flow state in combination with automated adjustments in the work system to achieve, sustain, or extend flow. Such a *real-time measure* should not interrupt a person, as traditional self-report measures do. A combination of physiological measures (e.g., heart rate variability, skin conductance, and blink rate) provides a promising starting point to find such a real-time measure. *Automated adjustments* first require the identification of design approaches that affect flow within the work system. Using the example of work in manufacturing, the concept of flow, its measurement, and potential design approaches for automated adaptation are presented, and their application in HCI processes is discussed.

Keywords: Human energy · Flow experience · Automated adaptation

1 HCI in Production Work – State of the Art

Based on the socio-technical system perspective, the human and the technical part of a production system need to be designed in an interwoven way [1]. Steghofer and colleagues [2] assume that the next generation of socio-technical systems will be underpinned by the most advanced and potentially most "intelligent" technology invented so far, while the "socio-" part—e.g. involving human behavior,

© Springer Nature Switzerland AG 2020
D. Harris and W.-C. Li (Eds.): HCII 2020, LNAI 12186, pp. 204–220, 2020.
https://doi.org/10.1007/978-3-030-49044-7_18

nondeterministic decision-making and interactions, complex social structures like organizations and institutions, culture, morality, ethics, and, above all values—will retain its validity.

Building on the socio-technical systems perspective, it has been acknowledged for some decades that the "technological imperative" has been overcome, and the awareness and the conviction have been disseminated that in order to optimize the whole production system, human and technical needs need to be designed in such a way that they optimally interact and maximize their potential. But for many years, the human aspect has been mainly limited to psycho-motor or cognitive demands, without having yet addressed the affective and emotional demands. Also, previous concepts of industrial psychology have primarily focused on cognitive processes of employees such as decision-making and fault finding, supported by mobile devices [3–5], and with performance-related targets, such as decreasing time and increasing quality [5–8]. And from the engineering side the main criteria for the design of devices in manufacturing and production were functionality and reliability [6, 9]. This means that, for a long time, the socio-technical systems approach referred to the psychomotor and cognitive aspects of system design, and meeting industry standards was the relevant criterion, thereby neglecting aesthetical aspects such as design and appearance, or affective states or motivation at the work place.

Then, with the increasing use and dissemination of smartphones since Apple's iPhone was introduced in 2007, and the increased user expectation concerning user experience and growing aesthetic needs, workers have started to become more demanding in expecting not only tools that are functional but also ones that are well and aesthetically designed.

Since then, the affective and motivational aspects of the execution of working tasks have increasingly become the focus of human-technology interaction and have also found their way into production. In that respect, the first field studies have shown that functionality and user experience are not mutually exclusive anymore [3, 10]. For example, the study by Borisov and colleagues [3] showed that the human/machine or human/computer interaction represents an important aspect in work related motivation and positive affect.

Aspects of user-centered design (ISO 9241-210:2019 Ergonomics of human-system interaction—Part 210: Human-centered design for interactive systems) have acknowledged the necessity to look at human needs more holistically by including not only the goals and required cognitive processes of a task to be fulfilled, but also requirements of sustained motivation and affect in order to provide interesting and challenging work places.

The development of smartphones and the apps that go with them was accompanied by the possibilities of increasing individualization of technical "helpers". While the classical socio-technical systems approach still represented a kind of solution for the entire working group, the new digital solutions allow an almost complete individualization of technical tools for individual operators. This can be illustrated using the examples of digital assistants for the future production context of industry 4.0 and the use of visualization systems and digital assistants for integrated learning and training in the workplace [11, 12]: Having the augmented and assisted operator in mind, examples are assistance systems that support employees individually in carrying out work steps based on their qualifications or competences already acquired. This enables, for example, job rotation through on-the-job training and execution support [12]. These kinds of assistance systems are known as situational employee qualification [11]. They support employees, even without many years of experience at a workplace, in special work activities or with problems that occur rather rarely [4], to carry out these activities within a defined standard, despite a lack of routine [11]. In the case of situational support, a systematic comparison of requirements and employee-related knowledge and skill takes place in the work process or in the process chain. These kinds of assistance systems are particularly relevant in heterogeneous and complex work, and they support the employees on-the-job, which can be done more flexibly with the assistance system [11]. But although this kind of adaptive assistant system has been developed with a focus on skill development, job enrichment, and enlargement, they have so far neglected the potential of assistant systems for assisting affective and motivational processes in production and manufacturing. That means that now, and *with scenarios of New Work, Industry 4.0, and the augmented operator in mind, we can go even further beyond user experience and affective design, and can build assistant systems to support positive affective and motivational states such as flow,* as they play a central role in directing human energy.

2   Flow Experience as a Concept to Allocate and Enhance Human Energy

Motivation explains the direction, intensity, and persistence of human behavior [13] and thus plays a crucial role in the mobilization and allocation of available energy. An intrinsically rewarding experience that occurs during motivated action is flow. Flow describes the pleasant experience of being fully absorbed with the task at hand [14]. In flow, all attention is directed towards the task while, at the same time, attention is shielded from irrelevant stimuli, leading to an experience of effortless action [15]. This suggests that flow is a highly efficient state in terms of energy allocation and expenditure. Already Csikszentmihalyi [16] suggested that "in flow, we are in control of our psychic energy" (p. 3). Studies addressing the physiology of flow support these assumptions [17, 18]. It was shown, for example, that flow is associated with moderate increases of cortisol [19]. Cortisol plays a role in the processing of relevant information, while inhibiting irrelevant information (for an overview see [20]). In addition, the secretion of cortisol enhances blood glucose levels and thus helps to provide additional and sustained energy resources to the acting person, and to direct them to the relevant task. Also, brain-imaging found indications of an association between focused attention and flow experience (for an overview see [21]): in flow, activation in the brain's multiple demand system—which is involved in task-relevant cognitive functions—was increased. This was combined with decreased activation in the default mode network [22, 23]—a network that is typically active when we are ruminating [24]. Other brain imaging studies found that flow was associated with a synchronization of brain regions involved in reward and cognitive control [25], again supporting the idea that flow promotes the sustained allocation of cognitive control towards the motivated action [26].

In line with these positive effects of flow on attention and energy control, we find positive short-term effects of flow on performance measures in the laboratory [27, 28] as well as in the field [29]. Furthermore, as operators in flow show increased sustained attention, they spend more time in concentrated action with immediate positive consequences on performance. Also, the rewarding character of flow leads to a higher motivation to repeat the respective activity, with the prolonged and repeated engagement in the activity resulting in increased practice, leading to increased skills and increased long-term performance.

According to the Self-Determination Theory [30] the experience of mastery as elicited through flow satisfies our need for competence, which is one essential precondition of well-being [30]. This might explain why we find not only positive effects of flow on performance, but also on well-being, as studies consistently show. Flow increased positive affect immediately after a flow activity [31]. When flow was experienced during an evening leisure activity, it had positive effects on affect the next morning [32]. Furthermore, flow at work was shown to be associated with increased energy afterwards [33]. On a larger scale, frequent flow experiences led to long-term increases in affective well-being and life satisfaction [34].

With respect to energy expenditure, however, there are several psychophysiological studies showing that flow is associated with increased physiological arousal. In more detail, arousal during flow was increased compared to a relaxed state [19, 35], but decreased compared to a stressful state [36], resulting in an inverted u-shaped relationship between flow and arousal [19]. Accordingly, despite its positive consequences on performance and well-being, it is still a state of increased arousal which needs to be counterbalanced by periods of relaxation. Supporting this assumption, Debus and colleagues [37] found that sufficient recovery in the evening is key to achieving the full potential to experience flow the following day. Reaching flow experience on a regular basis is thus only possible with sufficient recovery in between.

Taken together, flow is an experience that fosters both performance and wellbeing, while it is essential that enough recovery takes place between the flow episodes. Keeping that in mind, we may consider flow theory to be a promising approach for sustainable energy management, as it provides ideas for optimal use and enhancement of energy resources.

3 Physiological Measures of Flow Experience

In order to enhance flow experience in work processes during HCI, for example, in production work, we suggest identifying flow during the work process in real time in order to then be able to make automated adjustments in the work system. Traditional flow-measures are based on self-report questionnaires. However, such questionnaires interrupt the operators in their work flow, and any potential flow experience is likely to be gone afterwards [21, 38]. Thus, to identify flow in real time, we need to develop a new, interruption-free measure that is continuously assessed during the activity. This could potentially be realized using physiological measures; studies show that flow is associated with various physiological indicators [21, 38], including heart rate variability [19], electrodermal activity, respiration [39], blinking rate [40] and facial muscle activation [39, 41, 42]. Other indicators that were not particularly described for flow but rather for motivated performance include cardiovascular parameters such as heart rate, ventricular contractility, heart beat volume, and total peripheral resistance [43, 44]. In terms of brain physiology we also find associations with flow (compare [21] and Sect. 3); brain imaging is, however, difficult to apply during the work process.

Thus far, the existing research on physiological correlates of flow is still scarce; many inconsistencies exist, and many potential physiological measures have not yet been investigated in relation to flow. Also, there is no single physiological indicator that can reliably predict flow experience—and we consider it very unlikely, to find 'the one' physiological flow indicator in the future. Instead it is more likely to identify a pattern of physiological activation, consisting of a combination of different physiological indicators, that predicts flow.

Accordingly, the next step towards a real-time flow measure is to identify a flow-typical physiological pattern out of many available and feasible physiological indicators.

4 Using Machine Learning to Identify a Real-Time Flow Measure

Identifying such a pattern requires machine learning methods that can provide statistically optimal fusion where multimodal sensor data have to be merged on several time scales and with different, time-dependent reliability and validity. A challenge lies in the diversity of the sensor data under consideration: video data, which provide information on blinking, pupil width and movement, and activation of the facial muscles; and ECG signals as well as skin conductance measurements developed on their own time scales.

Previous related work on learning from multi-modal time series has often focused on statistical modeling, e.g. through Bayesian inference in graphical models. Graphical models provide a powerful mathematical language to describe dependency structures, as depicted in Fig. 1 below, for the specific case at hand, in a very much simplified form.

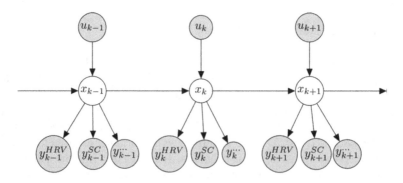

Fig. 1. Example of a graphical model, describing flow state x over discrete time k

Here, observable variables are shown in grey shaded circles, while hidden (or so-called *latent*) variables are denoted by open circles. With this notation, Fig. 1 describes the case where the likelihood of the user being in flow state at time point k is described by the latent (unobservable) variable that is denoted as x_k and is assumed to evolve over time. It depends both on the state of flow at the previous time instant, x_{k-1}, and on the current operator (or user) activity, which for reasons of simplicity is just described by a single variable u_k that subsumes all inputs to and activity performed by the user at time k, but which can also be split into a set of variables as necessary, to separately describe conditions at the workplace, stimuli, and activity. Finally, the observations (here, the heart rate variability y^{HRV}, skin conductance y^{SC}, and any other measurements that are just hinted at by ellipses) are dependent on this state of flow, and any other dependencies of these variables can again be described as additional nodes and arrows in the graph.

Based on this dependency structure, so-called *inference algorithms* can be selected and implemented which achieve an optimal estimate of all latent variables of interest. In our case, these could be affective and motivational variables, or, as shown in the graphic, the single flow state of the user, x_k.

This type of inference has a long history (see [45] and references therein), and it has a number of helpful characteristics for our application: the clear model structure allows for easy and modular incorporation of new observed modalities, as well as of additional, influential but unobserved latent variables. As an additional benefit, inference in the model yields probabilistic estimates, so that not only is the estimated flow state available, but it is accompanied by an estimated variance (or, more generally, an uncertainty) at all points in time. This uncertainty information is valuable to design, e.g., *active learning* systems, that ask the user to provide ground-truth information when the system becomes too unsure in its estimate.

In addition, statistical graphical models have recently been updated in a valuable manner: while their observation models have typically been constrained to statistical models with only a few parameters (e.g. considering all observations as Gaussian-distributed in the form

$$p(y_k|x_k) = \mathcal{N}(y_k; \mu(x_k), \sigma(x_k)) \tag{1}$$

that cannot easily accommodate the complex nature and dependency relationships that are present in biophysical measurements) recent work (by, e.g., [46]) has developed new inference algorithms that circumvent this limitation. Unlike statistical models such as the one in Eq. (1), they allow the observation models to become much more flexible and data-driven, allowing $p(y_k|x_k)$ to be computed by a learnable neural network model, which can be adapted to the training data through backpropagation and can then be used during test time to infer the most likely motivational or flow state from the observations y. Similarly, hybrid (neuronal/statistical) topologies are also developed there [47], for the purpose of multi-modal (gaze/pointing/speech-based) reference resolution.

We propose here to follow the same route to learn a model of physiological data depending on flow: we model the dependency relationships between the observable physiological data and the unobservable (or only rarely observable) latent state of the user as hybrid neuronal/statistical models, composed of a graphical model as shown above (Fig. 1) to describe all relevant dependency relationships, and of a neural-network based observation model. This model is learned in a training phase, estimating all model parameters based on collected, annotated data. After this initial training, it can be fine-tuned via active learning, where it will only ask the user to provide information when it becomes too uncertain in its classification. Finally, after these two learning phases, the system can be employed, allowing each observation modality to be optimally integrated in the probabilistic estimation of affective and motivational targets, or of a compound variable that describes the likelihood of flow.

5 Task Characteristics that Foster Flow Experience

In HCI, the identified flow-level of the operator can then be used to do automated adjustments in task design to reach or maintain a flow state. In another preparatory step, this requires the identification of appropriate task characteristics and related design approaches that affect flow within the work system.

There are a large number of studies and findings on situational conditions in the work context, in which flow-promoting factors have been identified (for an overview see [48, 49]. For the context of production work, task difficulty, autonomy, feedback, and pauses [49–51] are examples for useful design approaches.

Task difficulty has received much attention in flow research from its very beginnings. Already Csikszentmihalyi [14] predicted that flow occurs when a person perceives a balance between the demands (i.e. difficulty) of the task and the person's own skills (see Fig. 2). Many studies have confirmed the role of perceived demand-skill balance in the emergence of flow [28, 52], and experimental studies have used an automated adaptation of difficulty to successfully manipulate flow (e.g., [52, 53]). Accordingly, when the operator is bored during the work process, an increase of difficulty might help to bring the operator (back) into a flow state. Vice versa, when the operator is stressed, difficulty should be decreased. In practice, an adaptation of difficulty could be achieved using more or less support of a digital assistant (see Fig. 2).

Feedback refers to the knowledge of one's own progress and results. Also feedback was already proposed by Csikszentmihalyi as a flow-promoting factor. Only if a person knows about his or her results can he or she realistically perceive if a balance of demands and skills is achieved, and thus, it is (together with task difficulty) a prerequisite of a perceived demand-skill balance. Studies confirm that feedback in general [54, 55], and particularly positive feedback, has positive effects on flow [56–58]. In practice, feedback may be provided on the progress of task fulfillment (e.g. pieces produced) and/or the remaining time for the current task. Also, positive and motivating feedback could be applied by providing messages to the operator (such as "well done!" or "almost done – keep going!").

Autonomy refers to a person being responsible and in control of his/her actions [59]. Studies find positive effects of autonomy on flow experience [51, 60, 61], which qualifies it as a target condition to foster flow. Having degrees of freedom regarding how and when to solve a task gives a person the opportunity to act according to his/her skills and resources, which again provides better chances to reach a demand-skill

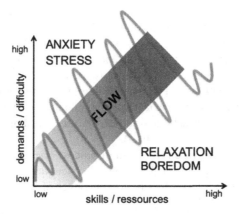

Fig. 2. Demand-skill balance as a predictor of flow with dynamic adaptations to achieve the flow state

balance. However, depending on their expertise as well as on their personality, people differ in how much they profit from different degrees of autonomy [62]. In practice, autonomy may be varied using more or less concrete and mandatory work instructions.

Last but not least, a balance between challenge and recovery is an important condition of flow both during and after the work process. Baumann and colleagues [63] found that a dynamic demand-skill balance led to higher flow values than a stable balance. A dynamic balance means alternating periods of challenge and recovery, that is periods during which demands are slightly higher than skills are followed by periods of demands being lower than skills. On a larger scale and as already noted, Debus and colleagues [37] could show the importance of being rested in the morning on the pattern of flow during the workday. Accordingly, an interplay between more and less demanding working periods and sufficient *pauses* during and after work are important for sustainable flow promotion. In practice, this implies a variation of higher and lower demands and the integration of pauses in between (compare Fig. 2).

6   Translating Flow-Promoting Task Characteristics into the Design of Adaptive Support

After having identified the operator's flow level (Sect. 4) and factors that promote flow (Sect. 5), the final step is now to bring it all together in order to dynamically adapt the work process according to the operator's flow level. Here we can build upon research on adaptive support systems in educational settings using affective-emotional aspects (e.g., [64, 65]). Research in this area is increasingly concerned with the question of how information on the affective state of a learner can be used to adapt support for enriching individual learning experiences and achieving higher learning gains. Such adaptive support systems collect relevant data of the learner in real-time, compare it automatically to a model and finally respond with personalized assistance [65]. Existing adaptive learning systems use, for example, speech, facial expressions, gaze data, physiological parameters, keystrokes, mouse movements, interactions with the learning environment, self-report, or a combination of these various sorts of information to identify a user's affective state (c.f. [64]). On the basis of the comparison with the model, an intelligent learning system is then able to provide specific support tailored to the user's current affective state. When flow is used as a target state, flow-specific parameters (Sect. 4) should instead be entered into the learning system.

Three broad approaches to an adaptive support that take up the described task characteristics difficulty, feedback, and autonomy can be distinguished based on the framework of Rummel [66] (see also [67]):

Mirroring as the first approach is a form of process-oriented feedback. It refers to the real-time reflection of the collected data by the system in a visually processed form. For example, the system can mirror the current level of task completion, or other parameters, which informs the operator about his or her work progress. Support systems based on this approach are characterized by a low level of directivity, and offer high levels of autonomy and responsibility for the operator since he or she decides independently how to deal with the presented information. Thus, mirroring systems build on the idea of self-regulation. They help the operator to know about his or her results and whether the operator was so far able to cope with the demands in the sense of a demand-skill balance. At the same time, mirroring systems do not reduce difficulty, as the decision regarding further action relies solely on the operator.

Alerting systems, which represent the second approach to adaptive support, are also forms of process-oriented feedback. They inform the operator about critical moments in task completion, for example by visually and/or acoustically announcing the elapse of available time. Although individuals can also decide on their own how to react to this information, their overall autonomy is somewhat restricted in this situation, because the evaluation of critical moments is predefined and not individually adaptable. Thereby task difficulty is decreased to some extent, as less attentional resources are needed to detect critical moments.

Guiding, the third design approach to adaptive support, aims at providing specific action-guiding prompts generated on the basis of the collected real-time data [67]. In contrast to mirroring or alerting systems, guiding support does not provide feedback about the operator's results, but it instead delivers concrete work instructions which the operator should or must follow. By implementing guiding, individual autonomy is revoked to ensure a certain further course of action. At the same time, difficulty is reduced, as the system decides how to proceed, thus preserving cognitive resources of the operator.

In line with these three approaches we would like to introduce adaptive support interventions based on data reflecting the individual's current level of *arousal* (compare [21, 38]). These interventions refer to the finding that flow occurs at a level of moderately increased arousal [19] (compare Fig. 2) and that sufficient recovery is needed to sustainably experience flow [37]. Thus, they are based on the degree of fatigue or stress of the user. Arousal management can be accomplished using each of the three approaches described above. When using mirroring, the operator receives feedback about his/her arousal level and the operator decides by him/herself how to deal with this information, e.g. if the production speed is modified, if the task is changed to a more or less activating one, or if a break is taken. Within the alarming approach, the system would, for example, remind the user to take a break at a certain point in the working process which is both early enough for efficient recovery and reasonable in terms of work re-uptake after the break. When arousal management is done with a guiding approach, breaks could, for example, be forced when the operator is tired: in this case a machine could switch off after the ongoing production step has been finished, and the machine could only be reactivated after the arousal level of the operator has been recovered. The automatic regulation of breaks can be helpful as, in a flow state, operators do not notice time passing and might thus not notice when recovery becomes necessary. During the break, resources can be recharged and, thus, coping with task demands will be easier thereafter, leading to decreased perceived task difficulty.

Similar kinds of interventions have already been applied by studies originating in the learning sciences. They have introduced and evaluated several adaptive support interventions based on data that reflect the individual's level of arousal and, beyond that, the individual's *affective state*. For example, Woolf and colleagues [68] used a pedagogical agent for delivering empathic messages aligned with the learner's current state of arousal and emotion detected by hardware sensors recognizing indicators such as facial movements and postures. If the sensors detected, for instance, signs of fatigue in the learner, the agent captured this low level of arousal (e.g. "Is this getting tiring?") and suggested switching to an activating task (e.g. "Shall we switch to something more fun?"). Beyond the state of arousal, the system also captured the learner's emotional state and adjusted the feedback messages accordingly. For example, if the system recognized that the learner is bored, it generated a corresponding message reflecting this affective state (e.g. "Maybe this is boring?") and suggested moving to a more challenging task (e.g. "Would you like to move to something more challenging?"). Connecting it to the three broad approaches to adaptive support outlined above, this intervention thus combines mirroring and guiding. The intervention applies individually tailored motivational feedback by providing suggestions, but no strict instructions on what to do next. Thus, in terms of promoting flow, autonomy remains high, and also difficulty is not affected.

Research investigating the effectiveness of adaptive support interventions also reveals that there are forms of support that are more beneficial when users are in a positive affective state (e.g. when they are already in flow), while other forms of support have been found to be more gainful when users are in a negative emotional state (e.g. when they are bored or confused). The first mentioned forms of support primarily aim at maintaining or even increasing the positive affective state of the user, whereas the latter forms of support aim at transforming the negative affective state of the user into a positive one. Thus, in the present context, both forms of support appear to be relevant to either sustain and extend the level of flow, or to overcome negative emotional states and achieve a flow experience in the first place. Grawemeyer and colleagues [64, 69], for instance, analyzed speech data to detect the affective state of users of an interactive learning platform and found that high-interruptive feedback during the learning activity (in form of a pop-up window that has to be dismissed before the user can proceed) was more beneficial for learners in a state of boredom than for learners who were in flow. However, learners who were already in flow preferred low-interruptive feedback in form of a glowing bulb indicating available feedback.

Moreover, the results of their research indicate that adapting and altering feedback presentation (high-interruptive feedback or low-interruptive feedback) and feedback type (instructional feedback or reflection prompts) in relation to the operators' affective states can reduce boredom, minimize off-task times, help users to move from a negative into positive emotional state, and consequently enhance task engagement [64]. In addition, Shen and colleagues [70] demonstrated that providing solution examples for a given problem was especially beneficial when learners experienced confusion, while delivering a video or music was effective to overcome boredom.

To sum up, research indicates that taking the current level of affective state into account when deciding on implementing particular adaptive support interventions appears to be a promising approach to providing tailored assistance, which produces, enhances, or maintains flow experience. Depending on whether individuals are in a positive or negative state, whether they already have experience in completing a task or are performing a task for the first time, or to what extent their work progress has already been advanced, it may be appropriate to offer more or less directive support. For example, it seems reasonable to grant experienced operators more autonomy and responsibility for their work process than inexperienced operators. Thus, high-experienced operators may already benefit from an indirect support raising a certain amount of awareness of task completion through a mirroring approach. In contrast, low-experienced individuals may initially need guided support in favor of practicing appropriate procedures.

7 Next Steps and Practical Implications

With the presented approach—using the flow state of the operator as the basis of dynamic adaptation of the work process—we aim to stimulate a more individualized and human-centered HCI for efficient and sustainable energy expenditure.

As the next steps, this approach should be tested in a controlled experimental setting, ideally, in a simulated work environment under laboratory conditions, e.g. in a so called special purpose setting [71] such as a learning factory. A special-purpose setting might be a laboratory setting that is equipped like a production setting or shop floor, or an industrial site that is used for experimental studies. Special-purpose settings cease to exist when research has been completed, and are designed for intentional manipulation of the independent variable. For instance, a learning factory is a special-purpose setting with high physical and psychological fidelity.

In such a simulated production process, operators' sensor data should be collected and the presented design approaches should be adapted according to the assessed flow level. Mirroring, alerting, and guiding could be realized using mixed-reality applications (e.g. Hololens) to provide visual feedback.

At first, the design approaches should be tested one by one, later also in combination.

Several challenges will need to be overcome in future implementation of our approach: it might well be that a specific combination of sensor data will provide indication for specific design approaches to be adapted in order to promote flow. Also, the flow-promoting physiological pattern may vary between activities—depending on the activation required for the activity; different tasks require different 'optimal' activation, for example, for operating a machine, a different activation is required than for programming or for gardening work. In addition, optimal activation certainly differs somewhat from machine to machine and from person to person. *Active learning* (see, e.g. [72]) can be used to ask the user for input when the system becomes too uncertain in its estimation of flow, which is indicated by a high variance of the estimated flow in the graphical model, cf. Sect. 4.

Furthermore, feasibility of the measurement needs to be taken into account. While video and sensor data are easy to assess at a workplace, brain imaging is rather not an option for field-research—although techniques like near-infrared spectroscopy may have potential here. However, even within video and sensor data, a reasonable choice based on a cost-benefit analysis should be done. And, importantly, ethical data use is an important issue to be thought of when implementing such adaptive support systems based on affective and motivational data in the workplace.

In the future, further flow-promoting task characteristics and related design approaches could be addressed. Examples are increasing the perceived fit of one's own skills to the task demands through job crafting. This could be realized using situational employee qualification (cf. Sect. 1) to adapt the task to a person's strengths.

As an outlook, e.g. task difficulty and cognitive demanding tasks can be varied according to the amount of problem solving activities involved. A working station can be assumed which requires problem solving non-routine activities, e.g. due to special customer requirements in addition to routine activities. In that respect, problem solving can be guided by problem solving assistance, by providing heuristic rules and guidelines for exploring the problem space and the solution space.

Furthermore, flow can be enhanced through additional social interaction options via mixed realities applications, e.g. in routine and for monotonous activities, in which social communicative exchange between workers or their avatars facilitates performance (instead of distracting from focused attention and concentration). Mixed reality applications could also include appropriate forms of entertainment, e.g. music, especially when cognitive demands and mental workload for attention allocation are low at particular times during the working day.

In summary, technological advancements available nowadays allow for more human centered evidence-based imagined scenarios regarding how work settings can be altered and changed towards more affectively attractive and motivationally flow-supporting settings, in which the concept of human energy is constructively used to strive for the development of work places which serve both production and efficiency requirements, and the human needs for meaningful and fulfilling activities—with a contribution to society and personal well-being.

References

1. Trist, E.: The evolution of socio-technical systems. Occasional Paper, no. 2 (1981)
2. Steghofer, J.-P., Diaconescu, A., Marsh, S., Pitt, J.: The next generation of socio-technical systems: realizing the potential, protecting the value [introduction]. IEEE Technol. Soc. Mag. **36**(3), 46–47 (2017)

3. Borisov, N., Weyers, B., Kluge, A.: Designing a human machine interface for quality assurance in car manufacturing: an attempt to address the "Functionality versus User Experience Contradiction" in professional production environments. In: Advances in Human-Computer Interaction (2018)

4. Frank, B., Kluge, A.: Can cued recall by means of gaze guiding replace refresher training? An experimental study addressing complex cognitive skill retrieval. Int. J. Ind. Ergon. **67**, 123–134 (2018). https://doi.org/10.1016/j.ergon.2018.05.007

5. Kluge, A., Termer, A.: Human-centered design (HCD) of a fault-finding application for mobile devices and its impact on the reduction of time in fault diagnosis in the manufacturing industry. Appl. Ergon. **59**, 170–181 (2017). https://doi.org/10.1016/j.apergo.2016.08.030

6. Grant, A.M., Fried, Y., Juillerat, T.: Work matters: job design in classic and contemporary perspectives. In: APA Handbook of Industrial and Organizational Psychology, Vol. 1: Building and Developing the Organization. APA Handbooks in Psychology®, pp. 417–453. American Psychological Association, Washington, DC (2011)

7. Thomaschewski, L., Herrmann, T., Kluge, A. (eds.): Unterstützung von Teamwork-Prozessen durch Augmented Reality (AR): Entwurf einer arbeitspsychologisch fundierten Taxonomie. [Supporting teamwork-processes through augmented reality (AR): Proposal of a work psychological taxonomy]. Arbeit interdisziplinär analysieren – bewerten – gestalten [Interdisziplinary Analyzing, Evaluating and Designing Work] (2019)

8. Weyers, B., Frank, B., Kluge, A.: A formal modeling framework for the implementation of gaze guiding as an adaptive computer-based job aid for the control of complex technical systems. Int. J. Hum.–Comput. Interact. 1–29 (2019). https://doi.org/10.1080/10447318.2019.1687234

9. Backman, J., Helaakoski, H. (eds.): Mobile technology to support maintenance efficiency—Mobile maintenance in heavy industry. In: 2011 9th IEEE International Conference on Industrial Informatics, 26–29 July 2011 (2011)

10. Thomaschewski, L., Weyers, B., Kluge, A.: A user-centered design approach to develop an augmented reality-based assistance system to support the temporal coordination of spatially dispersed teams with ambient awareness. In: 1st IEEE International Conference on Human-Machine Systems, ICHMS 2020 - Companion Technology Regular Track, Rome, Italy (2020)

11. Kreggenfeld, N., Kuhlenkötter, B.: Situational enabling for employees in "Industrie 4.0". ZWF Zeitschrift für wirtschaftlichen Fabrikbetrieb **111**(10), 658–661 (2016). https://doi.org/10.3139/104.111607

12. Prinz, C., Kreimeier, D., Kuhlenkötter, B.: Implementation of a learning environment for an Industrie 4.0 assistance system to improve the overall equipment effectiveness. Proc. Manuf. **9**, 159–166 (2017). https://doi.org/10.1016/j.promfg.2017.04.004

13. Kanfer, R., Frese, M., Johnson, R.E.: Motivation related to work: a century of progress. J. Appl. Psychol. **102**(3), 338 (2017). https://doi.org/10.1037/apl0000133

14. Csikszentmihalyi, M.: Beyond Boredom and Anxiety. Jossey-Bass Publishers, San Francisco (1975)

15. Bruya, B.: Introduction: toward a theory of attention that includes effortless attention and action. In: Bruya, B. (ed.) Effortless Attention: A New Perspective in the Cognitive Science of Attention and Action, pp. 1–28. MIT Press, Cambridge (2010)

16. Csikszentmihalyi, M.: Flow: The Psychology of Optimal Experience. Harper & Row, New York (1990)

17. de Manzano, Ö., Theorell, T., Harmat, L., Ullén, F.: The psychophysiology of flow during piano playing. Emotion **10**, 301–311 (2010). https://doi.org/10.1037/a0018432

18. Harris, D.J., Vine, S.J., Wilson, M.R.: Is flow really effortless? The complex role of effortful attention. Sport Exerc. Perform. Psychol. **6**, 103–114 (2017). https://doi.org/10.1037/spy0000083

19. Peifer, C., Schulz, A., Schächinger, H., Baumann, N., Antoni, C.H.: The relation of flow-experience and physiological arousal under stress - can u shape it? J. Exp. Soc. Psychol. **53**, 62–69 (2014). https://doi.org/10.1016/j.jesp.2014.01.009

20. Oitzl, M.S., Champagne, D.L., van der Veen, R., de Kloet, E.R.: Brain development under stress: hypotheses of glucocorticoid actions revisited. Neurosci. Biobehav. Rev. **34**(6), 853–866 (2010). https://doi.org/10.1016/j.neubiorev.2009.07.006

21. Peifer, C., Tan, J.: Psychophysiology of flow experience. In: Peifer, C., Engeser S., (eds.) Advances in Flow Research, 2nd edn. Springer, New York (in press)

22. Ulrich, M., Keller, J., Hoenig, K., Waller, C., Grön, G.: Neural correlates of experimentally induced flow experiences. NeuroImage **86**, 194–202 (2014). https://doi.org/10.1016/j.neuroimage.2013.08.019

23. Ulrich, M., Keller, J., Grön, G.: Neural signatures of experimentally induced flow experiences identified in a typical fMRI block design with BOLD imaging. Soc. Cogn. Affect. Neurosci. **11**, 496–507 (2016). https://doi.org/10.1093/scan/nsv133

24. Zhou, H.-X., et al.: Rumination and the default mode network: meta-analysis of brain imaging studies and implications for depression. NeuroImage **206**, 116287 (2020). https://doi.org/10.1016/j.neuroimage.2019.116287

25. Huskey, R., Wilcox, S., Weber, R.: Network neuroscience reveals distinct neuromarkers of flow during media use. J. Commun. **68**, 872–895 (2018). https://doi.org/10.1093/joc/jqy043

26. Huskey, R., Craighead, B., Miller, M.B., Weber, R.: Does intrinsic reward motivate cognitive control? A naturalistic-fMRI study based on the synchronization theory of flow. Cogn. Affect. Behav. Neurosci. **18**, 902–924 (2018). https://doi.org/10.3758/s13415-018-0612-6

27. Christandl, F., Mierke, K., Peifer, C.: Time flows: manipulations of subjective time progression affect recalled flow and performance in a subsequent task. J. Exp. Soc. Psychol. **74**, 246–256 (2018). https://doi.org/10.1016/j.jesp.2017.09.015

28. Engeser, S., Rheinberg, F.: Flow, performance and moderators of challenge-skill balance. Motiv. Emot. **32**(3), 158–172 (2008). https://doi.org/10.1007/s11031-008-9102-4

29. Peifer, C., Zipp, G.: All at once? The effects of multitasking behavior on flow and subjective performance. Eur. J. Work Organiz. Psychol. **28**(5), 682–690 (2019). https://doi.org/10.1080/1359432X.2019.1647168

30. Ryan, R.M., Deci, E.L.: Self-determination theory and the facilitation of intrinsic motivation, social development, and well-being. Am. Psychol. **55**(1), 68 (2000). https://doi.org/10.1037/0003-066X.55.1.68

31. Fullagar, C.J., Kelloway, E.K.: Flow at work: an experience sampling approach. J. Occup. Organiz. Psychol. **82**(3), 595–615 (2009). https://doi.org/10.1348/096317908X357903

32. Peifer, C., Syrek, C., Ostwald, V., Schuh, E., Antoni, C.H.: Thieves of flow: how unfinished tasks at work are related to flow experience and wellbeing. J. Happiness Stud. (2019). https://doi.org/10.1007/s10902-019-00149-z

33. Demerouti, E., Bakker, A.B., Sonnentag, S., Fullagar, C.J.: Work-related flow and energy at work and at home: a study on the role of daily recovery. J. Organiz. Behav. **33**(2), 276–295 (2012). https://doi.org/10.1002/job.760

34. Bassi, M., Steca, P., Monzani, D., Greco, A., Fave, A.D.: Personality and optimal experience in adolescence: implications for well-being and development. J. Happiness Stud. **15**(4), 829–843 (2013). https://doi.org/10.1007/s10902-013-9451-x

35. Keller, J., Bless, H., Blomann, F., Kleinbohl, D.: Physiological aspects of flow experiences: skills-demand-compatibility effects on heart rate variability and salivary cortisol. J. Exp. Soc. Psychol. (2011). https://doi.org/10.1016/j.jesp.2011.02.004

36. Peifer, C., Schächinger, H., Engeser, S., Antoni, C.H.: Cortisol effects on flow-experience. Psychopharmacology **232**, 1165–1173 (2015). https://doi.org/10.1007/s00213-014-3753-5

37. Debus, M.E., Sonnentag, S., Deutsch, W., Nussbeck, F.W.: Making flow happen: the effects of being recovered on work-related flow between and within days. J. Appl. Psychol. **99**(4), 713 (2014). https://doi.org/10.1037/a0035881

38. Peifer, C.: Psychophysiological correlates of flow-experience. In: Engeser, S. (ed.) Advances in Flow Research, pp. 139–165. Springer, New York (2012). https://doi.org/10.1007/978-1-4614-2359-1_8

39. De Manzano, O., Theorell, T., Harmat, L., Ullén, F.: The psychophysiology of flow during piano playing. Emotion **10**(3), 301–311 (2010). https://doi.org/10.1037/a0018432

40. Peifer, C., Butalova, N., Antoni, C.H.: Dopaminergic activity or visual attention? Spontaneous eye blink rate as an indirect measure of flow experience. In: 6th World Congress on Positive Psychology, Melbourne, Australia (2019)

41. Kivikangas, J.M.: Psychophysiology of flow experience: an explorative study [Master Thesis]. University of Helsinki, Helsinki, Finland (2006)

42. Nacke, L.E., Lindley, C.A.: Affective ludology, flow and immersion in a first-person shooter: measurement of player experience. Loading **3**(5) (2009). https://doi.org/10.1016/j.neulet.2008.02.009

43. Blascovich, J., Tomaka, J.: The biopsychosocial model of arousal regulation. Adv. Exp. Soc. Psychol. **28**, 1–51 (1996)

44. Tozman, T., Peifer, C.: Experimental paradigms to investigate flow-experience and its psychophysiology: inspired from stress theory and research. In: Harmat, L., Ørsted Andersen, F., Ullén, F., Wright, J., Sadlo, G. (eds.) Flow Experience, pp. 329–350. Springer, Cham (2016). https://doi.org/10.1007/978-3-319-28634-1_20

45. Bilmes, J.A., Bartels, C.: Graphical model architectures for speech recognition. IEEE Sig. Process. Mag. **22**(5), 89–100 (2005). https://doi.org/10.1109/MSP.2005.1511827

46. Johnson, M.J., Duvenaud, D.K., Wiltschko, A., Adams, R.P., Datta, S.R. (eds.): Composing graphical models with neural networks for structured representations and fast inference. In: 30th Conference on Neural Information Processing Systems, NIPS 2016, Barcelona, Spain (2016)

47. Kleingarn, D., Nabizadeh, N., Heckmann, M., Kolossa, D. (eds.): Speaker-adapted neural-network-based fusion for multimodal reference resolution. In: Proceedings of the 20th Annual SIGdial Meeting on Discourse and Dialogue (2019)

48. Peifer, C., Wolters, G.: Bei der Arbeit im Fluss sein. Konsequenzen und Voraussetzungen von Flow-Erleben am Arbeitsplatz. [Being in flow at work. Consequences and predictors of flow experience in the workplace]. Wirtschaftspsychologie **19**(3), 6–22 (2017)

49. Peifer, C., Wolters, G.: Flow experience in the context of work. In: Peifer, C., Engeser, S. (eds.) Advances in Flow Research, 2nd edn. Springer, New York (in press)

50. Hackman, J.R., Oldham, G.R.: Development of the job diagnostic survey. J. Appl. Psychol. **60**(2), 159 (1975)

51. Tausch, A., Peifer, C.: Auswirkungen von Autonomie auf Flow, Motivation und Leistung: Eine Studie im Schaltanlagenbau. Wirtschaftspsychologie **2019**(4), 83–100 (2019)

52. Keller, J., Bless, H.: Flow and regulatory compatibility: an experimental approach to the flow model of intrinsic motivation. Pers. Soc. Psychol. Bull. **34**(2), 196–209 (2008). https://doi.org/10.1177/0146167207310026

53. Harmat, L., de Manzano, Ö., Theorell, T., Högman, L., Fischer, H., Ullén, F.: Physiological correlates of the flow experience during computer game playing. Int. J. Psychophysiol. **97**, 1–7 (2015). https://doi.org/10.1016/j.ijpsycho.2015.05.001
54. Maeran, R., Cangiano, F.: Flow experience and job characteristics: analyzing the role of flow in job satisfaction. TPM-Test. Psychometr. Methodol. Appl. Psychol. **20**(1), 13–26 (2013)
55. Rau, R., Riedel, S.: Besteht ein Zusammenhang zwischen dem Auftreten von positivem Arbeitserleben unter Flow-Bedingungen und Merkmalen der Arbeitstätigkeit? Zeitschrift für Arbeits- und Organisations psychologie A&O **48**(2), 55–66 (2004). https://doi.org/10.1026/0932-4089.48.2.55
56. Jackson, S.A.: Factors influencing the occurrence of flow state in elite athletes. J. Appl. Sport Psychol. **7**(2), 138–166 (1995). https://doi.org/10.1080/10413209508406962
57. Peifer, C., Schönfeld, P., Wolters, G., Aust, F., Margraf, M.: Well done! effects of positive feedback on perceived self-efficacy, flow and performance. Front. Psychol. (accepted). https://doi.org/10.3389/fpsyg.2020.01008
58. Swann, C., Crust, L., Keegan, R., Piggott, D., Hemmings, B.: An inductive exploration into the flow experiences of European Tour golfers. Qual. Res. Sport Exerc. Health (2015). https://doi.org/10.1080/2159676x.2014.926969
59. Karasek, R.A.: Job demands, job decision latitude, and mental strain: implications for job redesign. Adm. Sci. Q. **24**(2), 285–308 (1979)
60. Emanuel, F., Zito, M., Colombo, L.: Flow at work in Italian journalists: differences between permanent and freelance journalists. Psicologia Della Salute (2016)
61. Fagerlind, A.-C., Gustavsson, M., Johansson, G., Ekberg, K.: Experience of work-related flow: does high decision latitude enhance benefits gained from job resources? J. Vocat. Behav. **83**(2), 161–170 (2013). https://doi.org/10.1016/j.jvb.2013.03.010
62. Schüler, J., Sheldon, K.M., Prentice, M., Halusic, M.: Do some people need autonomy more than others? Implicit dispositions toward autonomy moderate the effects of felt autonomy on well-being. J. Pers. **84**(1), 5–20 (2016). https://doi.org/10.1111/jopy.12133
63. Baumann, N., Lürig, C., Engeser, S.: Flow and enjoyment beyond skill-demand balance: the role of game pacing curves and personality. Motiv. Emot. **40**(4), 507–519 (2016). https://doi.org/10.1007/s11031-016-9549-7
64. Grawemeyer, B., Mavrikis, M., Holmes, W., Gutiérrez-Santos, S., Wiedmann, M., Rummel, N.: Affective learning: improving engagement and enhancing learning with affect-aware feedback. User Model. User-Adap. Inter. **27**(1), 119–158 (2017). https://doi.org/10.1007/s11257-017-9188-z
65. Walker, E., Rummel, N., Koedinger, K.R.: CTRL: a research framework for providing adaptive collaborative learning support. User Model. User-Adap. Inter. **19**(5), 387 (2009). https://doi.org/10.1007/s11257-009-9069-1
66. Rummel, N.: One framework to rule them all? Carrying forward the conversation started by Wise and Schwarz. Int. J. Comput.-Supp. Collab. Learn. **13**(1), 123–129 (2018). https://doi.org/10.1007/s11412-018-9273-2
67. van Leeuwen, A., Rummel, N., van Gog, T.: What information should CSCL teacher dashboards provide to help teachers interpret CSCL situations? Int. J. Comput.-Supp. Collab. Learn. **14**(3), 261–289 (2019). https://doi.org/10.1007/s11412-019-09299-x
68. Woolf, B., Burleson, W., Arroyo, I., Dragon, T., Cooper, D., Picard, R.: Affect-aware tutors: recognising and responding to student affect. Int. J. Learn. Technol. **4**(3–4), 129–164 (2009). https://doi.org/10.1504/IJLT.2009.028804
69. Grawemeyer, B., Holmes, W., Gutiérrez-Santos, S., Hansen, A., Loibl, K., Mavrikis, M. (eds.): Light-bulb moment? Towards adaptive presentation of feedback based on students' affective state. In: Proceedings of the 20th International Conference on Intelligent User Interface. ACM, New York (2015)

70. Shen, L., Wang, M., Shen, R.: Affective e-learning: using "Emotional" data to improve learning in pervasive learning environment. J. Educ. Technol. Soc. **12**(2), 176–189 (2009)
71. Stone-Romero, E.F.: Research strategies in industrial and organizational psychology: nonexperimental, quasi-experimental, and randomized experimental research in special purpose and nonspecial purpose settings. In: APA Handbook of Industrial and Organizational Psychology, Vol. 1: Building and Developing the Organization. APA Handbooks in Psychology®, pp. 37–72. American Psychological Association, Washington, DC (2011)
72. Lughofer, E.: Hybrid active learning for reducing the annotation effort of operators in classification systems. Pattern Recogn. **45**(2), 884–896 (2012). https://doi.org/10.1016/j.patcog.2011.08.009

New Production Development and Research Based on Interactive Evolution Design and Emotional Need

Tianxiong Wang[✉] and Meiyu Zhou

School of Art Design and Media, East China University of Science
and Technology, NO. 130, Meilong Road,
Xuhui District, Shanghai 200237, China
wangtx_2018@163.com, zhoutc_2003@163.com

Abstract. Due to the continuous release of new products, manufacturers must design products constantly to meet the diversified and differentiated needs of customers. In order to avoid the displacement by market competitors, enterprises and manufacturers must study the multiple combinations of product shapes to design products for meeting user's needs. At the same time, users' consumption levels and aesthetic concepts are constantly improved, consumer demand has become more personalized and diversified, and then the development of manufacturing and information industries has made people experience material results while paying more attention to their emotional needs. As an evolutionary optimization algorithm, the individual fitness values of the interactive genetic algorithm are directly obtained from the user's own preferences and the user could give a higher fitness degree to his favorite design individual, or directly selects his satisfied individuals as the next generation of individuals in the evolutionary process. Hence, the IGA method is used to effectively design innovative productions based on users' emotional demand. However, the inaccurate judgment and identify of the user's demand for product image style will increase the complexity of the design, and resulting in increased user fatigue. To respond to the challenge, this study propose a combination method of interactive genetic algorithm and fuzzy kano model (FKM) research methods, in which FKM is used to more accurately excavate the product image style that satisfies the user's perceptual needs, thus guiding the direction of product modeling evolution, and achieving user demand-driven production evolution design. Finally, through the application of the electric bicycle case to prove the practicability and effectiveness of the method. In addition, the proposed method has increased the user satisfaction in the NPD. This method is also applicable to the styling aesthetics study for other industrial products.

Keywords: Interactive genetic algorithm · Customer demand · User fatigue · User emotional preference · Production design

© Springer Nature Switzerland AG 2020
D. Harris and W.-C. Li (Eds.): HCII 2020, LNAI 12186, pp. 221–237, 2020.
https://doi.org/10.1007/978-3-030-49044-7_19

1 Introduction

Given the fierce competition on today's market, and with increased industrial intensification, product differentiation has gradually become difficult (Ding and Bai 2019). Meanwhile, product design has shifted from the past product orientation to market orientation, and eventually towards customer orientation. Whether consumers buy a certain product or not is determined not only by the technical ability and applicability of the product but also by the emotional reaction triggered by the product appearance. Therefore, a product's appearance plays a pivotal role in consumers' preference for a product. The matching quality is mainly judged by the match or mismatch between users' mental model and product design which is actually carried out based on designers' mental model (Yadav et al. 2013). At the same time, users' consumption level and aesthetic concepts are being constantly improved and their demands are more personalized and diversified. The development of the manufacturing industry and the information industry makes people pay more attention to their emotional needs while experiencing material achievements. Therefore, they expect products to demonstrate their personality, preference and values. Furthermore, manufacturers must focus on consumers' demands in the product design and production process, thus turning from large-scale production to large-scale customization (Chuang et al. 2001; Wang et al. 2016). Accordingly, product research, development and manufacturing enterprises tend to develop products by means of enhancing user satisfaction, quickly respond to users' diverse demands and dynamic market changes so as to strengthen their core competitiveness (Ji et al. 2014). As a result, it is crucial to accurately identify and acquire information on consumers' demands (Yoo 2016).

The main challenge for NPD is to assess the user's image perception and need accurately and subsequently to design products that match these needs. In recent years, some scholars have adopted the approach of Kansei engineering (Nagamachi 1995) to explore users' feelings. Kansei engineering (KE) is defined as "the translating technology of a user's feeling and image for a product into design elements." The designer can accurately obtain user's image perception and effectively convert it into a product design form that can meet user's emotional needs (Ding and Bai 2019). However, the general statistical methods and techniques adopted in Kansei engineering are insufficient to describe the effectiveness of overall product modeling that involves a number of design elements, and are unable to address the problem of dependency between these attributes (Tsuchiya et al. 1996). Yet, creative problem-solving through evolution is most exciting. In design, evolution can help investigate the space for design novelty and innovation (Lee and Chang 2010). With the introduction of computer technology, computer-aided design (CAD) has been extensively applied. Scientists have proposed multiple heuristic algorithms by simulating varied evolution phenomena in the biological sphere, among which interactive genetic algorithm (IGA) is an algorithm that underlines synergistic interaction and user participation, being widely used in the product concept design system (Dou et al. 2018; Kowaliw et al. 2012). In this method, consumers directly determine the individual adaptability of the population in the design solution. Users provide an adaptive value to replace the fixed fitness function. In this way, interaction between IGA and users is achieved. Consumers can give a higher adaptive value to the individual they like or directly choose an individual as the parent for the next generation in the population iteration process. Hence, IGA allows user demand to fully influence the trend of population evolution, caters to the specific demands of consumers in a more pertinent manner, and meanwhile incorporates individual diversity. All that's required is to pick out the solution which users are most satisfied (Dou et al. 2016).

Meanwhile, the diverse design style for product modeling leads to a multitude of emotional image vocabulary, which inevitably limits the efficiency in the interaction process of IGA. A workable solution is to effectively identify the image style of product modeling itself to increase evolution efficiency and thus reduce user fatigue. The Fuzzy Kano model is used to analyze the product image styles, classify them and single out attributes of excitement type help users make decisions and conduct evaluation (Lee et al. 2011), thus improving the evolution efficiency of IGA and eventually alleviate user fatigue. On this basis, this study adopts the Fuzzy Kano model to further analyze and extract key emotional image words obtained in the user survey, single out attributes with which users are highly satisfied, integrate image results into the evolution system process, and then conduct iteration and optimization on product modeling via IGA. Hence, the purpose of this study is to accurately investigate users' emotional images about the product, combine and optimize product modeling elements with IGA system so that product modeling can meet users' emotional needs, and help clients or developers quickly work out products with a high degree of satisfaction.

2 Literature Review

2.1 GA

Genetic algorithm (GA) was proposed for the first time by Professor John Holland from University of Michigan in the 1970s. It is a self-adaption technique of probability optimization based on species heredity and evolution mechanism. It is applicable to the optimization calculation of complicated systems (Holland 1973). GA introduces biological evolution and natural section mechanisms into the optimization calculation of complicated systems. It uses chromosome form to simulate the genetic inheritance and evolution mechanism, and represent the solutions of specific problems. With the fitness function as the optimization basis and the coding group as the evolution basis, it offers a selection and inheritance mechanism for individual bit strings in a group. Furthermore, an iteration process is established with step-by-step optimization. The individual adaptive value is improved through a series of genetic operations (select, cross and mutate). Eventually, satisfying solutions for the problem will be found out (Dou et al. 2016). Randomly re-structuring and encoding important genes in the bit strings make the new generation of bit strings superior to the old generation, and gradually approach the optimal solution through constant evolution. Finally, the problem will be solved. In this process, chromosome and genes of excellent individuals will be inherited generation by generation while those individuals that cannot adapt to environmental changes will be eliminated. This pattern happens to reflect the operation mode of the product market mechanism. When a product is well received and sold among the public, enterprises will launch mass production. In other words, varied modules of the product will be retained generation by generation. In fact, GA is suitable for tackling complicated and non-linear optimization problems which are hard to handle by traditional methods. It's because it is a new-type evolution algorithm that can search problems and optimize problem-solving with a limited cost. Moreover, usually it is not restrained by the background of to-be-solved problems (Renner and Ekárt 2003). Compared to other methods, GA is more effective in searching overall space (Hsiao et al. 2010). In recent years, it has been applied by the majority of scholars to industrial product design (Beale 2007; Diego-Mas and Alcaide-Marzal 2016; Hsiao et al. 2010). The only information which GA uses during calculation is the fitness function value of population individuals. Generally, GA allocates fitness value through individuals and evaluates the predefined fitness function. Therefore, GA has multiple advantages such as global perspective, universality, randomness, convenience and strong robustness. GA has been widely applied in combinatorial optimization, machine learning and many other fields.

2.2   IGA

Factors preferred by users are often ambiguous and differentiated, so it is extremely difficult to work out preference factors for the fitness function regarding users' optimization solutions (Lee and Chang 2010). As a result, people began to think about combining human intelligent evaluation with the optimization ability of GA, thus introducing interactive genetic algorithm based on traditional GA. Interactive genetic algorithm (IGA), proposed by Dawkins in 1986, can effectively conduct index optimization for implicit ambiguous problems (Dawkins 1986). The individual adaptive value of IGA is directly retrieved from users' own preferences. Users provide the adaptive value rather than fitness function for each individual, and confer greater fitness to the design solution they like or directly choose their favorite individual as the parent for the next generation in the group evolution process. In this process, customers' participation and subjectivity in product design are emphasized (Franke et al. 2010).

So far, experts and scholars have actively conducted studies on the application of IGA in the engineering fields. These studies involve image processing, product modeling optimization design, apparel design, color collocation, book cover design, industrial design and design decision making. For instance, in their study on mobile phone design, Ji-Hyun Lee and Ming-Luen Chang integrated IGA into the mechanism which generates interactive creativity, and updated it every day according to users' evaluation data in order to improve the design process (Lee and Chang 2010). P.Y. Mok and Jie Xu et al. came up with a custom-made user-friendly fashion design system so that non-professional users (ordinary customers) could create their favorite fashion designs (Mok et al. 2013). KIM H S and CHO S B discussed the apparel design model based on IGA in their study (Kim and Cho 2000). Through their study, Alexandra Melike Brintrup and Jeremy Ramsden et al. stressed the necessity of officially combining the qualitative and quantitative standards of ergonomic design, and meanwhile provided an IGA-aided design framework to enhance the solution for chair back design (Brintrup et al. 2008). Zahra Sheikhi Darani and Marjan Kaedi developed a kind of IGA in which candidate elimination algorithm is used to gradually learn user preference. Besides, they assessed the book cover design (Sheikhi Darani and Kaedi 2017).

Poirson et al. guide users to perceive product shapes based on IGA and verified it through the case of automobile dash board design (Poirson et al. 2013).

The optimization performance of IGA should be enhanced first in the IGA research process. Therefore, the main problem that needs to be solved is user fatigue arising in the user evaluation process. On this basis, while analyzing the characteristics of product modeling, this paper applies the satisfaction identification method of the Fuzzy Kano model to explore image style emotional words that influence user satisfaction. Besides, it converts such words into product evaluation carriers in IGA, so as to relieve users' psychological burden and fatigue in the evaluation process, and provide effective support for the user-oriented decision making process.

2.3 Fuzzy Kano Theory

Kano et al. put forward a two-dimensional quality model (Fig. 1) (Kano et al. 1984). The horizontal coordinate is used to show the degree of a product quality possessed. The right half indicates the possession of such quality element, and the closer to the right, the higher the degree of possession. The left half means the lack of quality elements, and the closer to the left, the higher the degree of deficiency. The vertical coordinate is used to express customer satisfaction. The upper half represents the degree of satisfaction and the higher the altitude, the higher the degree of satisfaction. The lower half refers to the degree of dissatisfaction, and the lower the altitude, the higher the degree of dissatisfaction. Based on the correlations between the two coordinates, quality elements can be divided into five categories. Furthermore, the Kano model is adopted to analyze user demands and formulate the questionnaire to obtain the effectiveness of user demands in positive and negative aspects. In other words, will the users be satisfied if such demand is provided or will the users be dissatisfied if such demand is not provided? The user demand is classified by analyzing whether there's a linear relationship between the two of them. The results are classified as follows: attractive quality (A), one-dimensional quality (O), indifference quality (I), reverse quality (R) and must-be quality (M). The classification of Kano's model explains the relationship between user satisfaction and product or service performance, as shown in Table 1 (Matzler and Hinterhuber 1998).

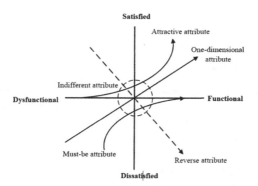

Fig. 1. Kano 模型

Table 1. Kano's evaluation table

Customer needs		Dysfunctional				
		Like	Must-be	Neutral	Live-with	Dislike
Functional	Like	Q	A	A	A	O
	Must-be	R	I	I	I	M
	Neutral	R	I	I	I	M
	Live-with	R	I	I	I	M
	Dislike	R	R	R	R	Q

However, the results obtained by the traditional Kano model are not accurate, this method may ignore the uncertain answer factors caused by customers' complex and changeable psychology. To handle this problem, L.A. Zadeh, a professor of automation at the University of California, Berkeley, first proposed fuzzy sets in 1965 (Zadeh 1965). He used mathematical methods to express uncertainty factors in real life. The main contribution of fuzzy set theory is to represent fuzzy data. More specifically, fuzzy sets is a class of elements with membership in continuity or dispersion and the feature of such set is that each element is allocated between 0 and 1. Therefore, the fuzzy Kano model (Yadav et al. 2013) has become a more valuable research topic, which can solve the problem of fuzzy emotional characteristics of consumer demand for products, and then the decimal point between the interval [0, 1] is used to express the user's awareness so as to make the result more accurately. Thus, the users' key needs are excavate accurately. This theory has been used in some industrial design studies (Chen and Ko 2009; Lee and Huang 2009; Chai et al. 2015; He et al. 2017; Shahin et al. 2017). Accordingly, it can be seen that there is the sufficient theoretical basis to meet the needs of users through the application of the fuzzy kano model, but few studies focus on the application of fuzzy Kano model tools as the guide direction for the evolutionary design of product modeling styles. Therefore, this study directly integrates the classification results of FKM perceptual image attributes into the IGA step process, and then to use this as the starting point to explore the evolution and innovation issues in product modeling, and optimizes the product form generation process. Furthermore, the user's fatigue problem in the IGA is alleviated significantly.

3   Method

3.1   Implementation Strategy

All the practice and research of product design must be carried out around the user center, and the focus of satisfying human needs has gradually shifted from physiology to psychology. Therefore, in order to better study the product development paradigm from the perspective of psychology and emotion, this study proposes the production form design decision-making method driven by user's kansei image through the combination method of EGM and IGA. Specifically, this paper analyzes the reasons of users' attractive factors for products, the upper perceptual image and the specific factors of product form characteristics based on the expert interview method of EGM, so as to build a database of user image and specific form characteristics of products. In order to accurately obtain the satisfaction of users for the perceptual needs of products, the kansei image attributes of users are analyzed by using fuzzy Kano model, and the perceptual image factors with high satisfaction of users are extracted so as to set as the carrier of user interaction evaluation in IGA, which makes the scope of product modeling evaluation index items required by users in the evaluation process narrow. Then, the system will automatically generate the adaptive value of evolutionary individuals after weighting according to the value of user evaluation. Accordingly, this method could not only reduce the burden of user decision-making in the process of product evolution, but also greatly improves the accuracy of user evaluation. Furthermore, the users' favorite products form proposal are explored in the production evolution process.

3.2 Exploring the Attractive Factors of Production Form

Miryoku Engineering originated from the category of psychology, and it was adapted from the book "*The Psychology of Personal Constructs*", published by Kelly (1955) in 1985, and was created by Junichiro Sanui and Masao Inui. The main concept of Miryoku Engineering is to catch the idea of personal cognition and arrange it into grid char. As an expert evaluation tool of Miryoku Engineering (Miryoku Engineering Forum 1992), which is named evaluation gird method (EGM) and this method was developed by Sanui in 1996 (Sanui 1996) on the basis of repertory grid method proposed by Kelly (1955). In fact, the goal of this method is to extract consumer language to understand the evaluation project and the various factors that make up the grid. Through in-depth investigation of individual attractive factors, especially for the structure of semantic visualized hierarchy structure, from the abstract to the concrete, the structure of consumers' value by the structure of originate evaluation, upper-level and down-level items can be visualized and concreted (Shen et al. 2015; Sanui 1996). Therefore, it has been widely used in product design and development (Ho and Hou 2015; Shen et al. 2009; Shen 2013).

3.3 Assessing the User Demand Attribute

In this stage, we use the FKQ and FKM model to capture users' emotional needs. Through the fuzzy Kano model, the image semantic level explored in the EGM interview is introduced into the fuzzy Kano model for further quality classification. Through to utilize the positive and negative questionnaires, the user's emotional response is investigated based on each image semantic performance is sufficient or insufficient. Then, according to the research results, the quality attributes of product image semantics are rationally determined. Hence, the relationship between the attractive factors and customer satisfaction of electric bicycle are further analyzed. Specifically, according to the research results of the fuzzy Kano model, the demand attributes of evaluation factors are divided into five categories: namely one dimensional quality (O), attractive quality (A), indifferent quality (I), must-have quality (M), and reverse quality (R). Thus, the two-dimensional concept in the Kano model enables describing the relationships of customer satisfaction with product quality. Kano's model, which studies the nature of customer needs, provides a way for a better classification of customer needs.

The FKQ questionnaire could enable interviewees to more comprehensively present the ideas and solutions that they usually face and match them with the human thinking model more precisely. Therefore, Kano model and quality attribute classification could be more real (Lee and Huang 2009). Accordingly, the FKQ questionnaire is assigned the corresponding proportion according to the fuzzy emotional thinking of users, and then the setting of FKQ is shown in Table 2.

Table 2. Fuzzy Kano questionnaire

Product Function	Like	Must-be	Indifferent	Live-with	Dislike
Functional	0.1	0.8	0.1		
Dysfunctional		0.2	0.8		

The basic steps of Fuzzy kano algorithm has three steps: Firstly, assume that a function can include the functional and dysfunctional, can realize the matrix fun = [0.1 0.8 0.1 0 0], cannot achieve the matrix dys = [0 0.2 0.8 0 0], by $(fun)^t \times (dys)$ interaction matrix of fuzzy relations generated Z is defined in Formula (1).

$$Z = \begin{bmatrix} 0 & 0.02 & 0.08 & 0 & 0 \\ 0 & 0.16 & 0.64 & 0 & 0 \\ 0 & 0.02 & 0.08 & 0 & 0 \\ 0 & 0 & 0 & 0 & 0 \\ 0 & 0 & 0 & 0 & 0 \end{bmatrix} \qquad (1)$$

Secondly, after get Z matrix, according to the experience of the previous literature, two-dimensional attribute classification will be obtained based on Matzler and Hinterhuber (1998) model as shown in Table 1, then get demand category of the membership degree vector T, and is defined in Formula (2).

$$T = \left\{ \frac{0}{M}, \frac{0}{O}, \frac{0.1}{A}, \frac{0.9}{I}, \frac{0}{R}, \frac{0}{Q} \right\} \qquad (2)$$

Finally, in order to research more satisfactory and identification, the consensus standard a-cut is introduced to get $\{T_h\}\alpha$. The threshold value of $a \geq 0.4$ as an example (Meng and He 2013), If an element in the membership vector is large than or over a, "1" will be represented; Otherwise it's 0. Therefore, the customer demand attribute vector is converted to T = [0, 0, 0, 1, 0, 0], which shows that the quality attribute is indifferent. Consequently, interview participants' different feelings about the degree of attributes sufficiency, the largest identification frequency of quality attribute fuzzy Kano model is the result agreed by majority.

3.4   Product Evolutionary Design Mechanism

3.4.1   The Evolution Process of IGA

In fact, the IGA is an improved algorithm of GA, which could fully emphasize the "interaction" behavior with users. In this algorithm, the user could directly evaluate the score to individual instead of the fitness function (Wu et al. 2018). In the process of user evaluation, the user evaluates the individual performance presented in the human-computer interaction interface based on the personal cognitive preference, and directly to score or evolve from multiple evolutionary individuals for selecting the best one so that the individual fitness value is given to replace the automatic calculation process of the fitness function of the fitness value in GA. Thus, in the development of population, the chromosomes and genes of excellent individuals will be inherited generation by generation, and those individuals who could not adapt to the changes of environment will be eliminated. Moreover, IGA combines GA with human subjective evaluation, which is essentially the result of the combination of intelligent technology and human intelligence. Through the interaction between the decision-maker and the computer, the intervention and guidance of the evolution process are realized to solve the implicit goal decision-making problem that the traditional GA and other intelligent optimization algorithm may not solve it, so that the user preferences can directly affect the direct direction of population evolution in the production design.

3.4.2   Crossover and Mutation

As a global optimization random search algorithm, the crossover operation plays a key role in the interactive genetic algorithm, that is, according to some rules, to adjust some genes of two paired parents, so as to generate two new offspring individuals. This method is closely related to the actual optimization problem, which can not destroy the dominant genes in the parents too much. At the same time, it can effectively generate offspring with high fitness value. The simulation of crossover operation also simulates the process of biological gene crossover in nature to make individuals show some characteristics. The individual crossover in binary system is specifically reflected in the alternation of corresponding strings, while the mutation operation will randomly

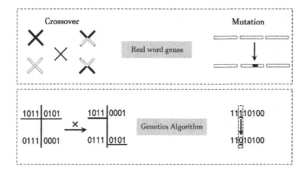

Fig. 2. Comparison of crossover and mutation operations in the real world and genetic algorithms

change some corresponding values in the binary coding of individual chromosomal system with a low probability. The specific way of crossover and mutation is described in Fig. 2.

3.4.3   Product Form Element Coding

The 0-1 string encoding method proposed by Professor Holland is used to represent the individual's chromosome values, and each chromosome consists of a binary coded bit string. In binary encoding process, the encoding unit formed by the combination of 0 and 1 is also a gene. A product may consist of several product modeling project modules which are paralleled. Each project module possesses the unique modeling feature unit. The position and the front and back sequence of form characteristics of each product in the chromosome are fixed.

3.4.4   Population Size and Initialized Population

Considering the user experience and usability of user to adopt IGA, the population size N of IGA is generally smaller than GA (Dou et al. 2016). Refer to previous experimental experience, we take N = 4 − 9 (Mok et al. 2013). In addition, in order to avoid excessive fatigue during the evaluation process, the maximum number of evolutionary iterations of interactive genetic algorithms cannot generally exceed 20 (Takagi 2001).

3.4.5   Obtaining Individual Fitness

The index weight value of kansei image vocabulary of product modeling style is calculated, and then the individual fitness value F_i is obtained by weighted calculation combining with the score value of user evaluation image semantics. Then, the formula for F_i is as follows:

$$F_i = \sum_{i=1}^{n} C_i \cdot W_i \tag{3}$$

The n is the statistically specific number of perceptual images, F_i is the comprehensive evaluation score of the product, C_i is the individual score of the image semantics, W_i is the weight value of the image semantics, and the sum of the weight values of the i items is equal to 1. The individual fitness value generally reflects the user interaction evaluation of the individual population, and conveys the user's implicit preference image emotion.

4   Case Study

4.1   Analyzing the Attractive Factors of Electric Bicycle

In this section, we choose electric bicycles as the research object. First, we need to collect pictures of popular electric bicycle products sold in the market to study the attractive factors of the products. 21 popular electric bicycle products are used as test samples based on user favorite which are searched from online shopping platforms, magazines and related books, and then the samples are processed to A4 paper size by the computer plane processing software PS, and the resolution of the picture is set to 150 dpi. At the same time, the picture is decolorized and the background of the picture is removed, which makes the picture sample unified. At the beginning of the experiment, a group of 12 experienced designers and users with five years of electric bicycle riding experience need to participate in the EGM survey of product attractiveness factors, which includes 6 men and 6 women. The age of the survey population is from 23 to 40. The original evaluation items of the samples' specific attractiveness (middle-level) and the specific characteristics of the attractiveness (lower-level) and its brought about emotional factors (upper-level) are obtained based on the EGM method, then the results can include 34 middle-level evaluation items, 113 lower-level evaluation items, and 205 upper-level evaluation items. Furthermore, the KJ method is adopted to merge similar vocabulary and specific features of shape so as to result in 4 middle-level evaluation items, 19 lower-level evaluation items and 12 upper-level evaluation items. Therefore, the hierarchical relationship diagram of attractive factors for electric bikes is constructed (Fig. 3).

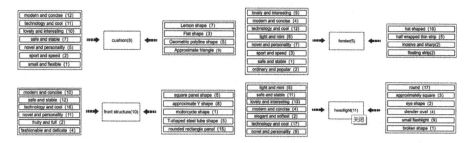

Fig. 3. The hierarchical diagram of electric bicycle production

4.2 Discussing the Classification of Emotional Needs

Through the EGM expert interview qualitative analysis for the relationship between the products' morphological attractiveness characteristics and the users' kansei images, and then to utilize the KJ simplification method to further simplify the obtained kansei image words. At the same time, according to the user's mentioning times in the EGM interview process, there are 12 adjectives, which are the ordinary and popular, light and mini, modern and concise, technology and cool, lovely and interesting, safe and stable, novel and personality, sport and speed, small and flexible, fruity and full, fashionable and delicate, elegant and softest. In fact, these kansei image words are only emotional factors of users' attractive factors to product modeling, so it may lack of the satisfaction factors of users' demand. Therefore, it is necessary to introduce the fuzzy kano model to discuss and classify the types of customer satisfaction based on users' demands, and then to define the demand as five categories: namely one dimensional quality (O), attractive quality (A), indifferent quality (I), must-have quality (M), and reverse quality (R). In this study, eight key evaluation items are selected according to the mentioning times of user's kansei image in descending order, and the user satisfaction is discussed based on the fuzzy Kano model. Then, 120 graduate students and undergraduates which majoring in industrial design are selected for the experimental research. Finally, the questionnaire is collected. Then, according to the formula (1)–(2), the statistical calculation results is shown in Table 3. More specifically, modern and concise, technology and cool, small and flexible, lovely and interesting, fashionable and delicate are attractive demand. When these five kansei image factors are outstanding, these factors can greatly affect and improve the satisfaction level of users. In order to improve the user satisfaction of the production, these five kansei words which could affect the user satisfaction are selected from the results of the fuzzy Kano model analysis to use in the subsequent IGA research.

Table 3. Fuzzy Kano model statistical results

Factors	M	A	I	O	R	Q	Category
Modern and concise	5	47	16	31	1	0	A
Safe and stable	6	23	20	49	2	0	O
Technology and cool	21	35	28	15	1	0	A
Novel and personality	23	23	34	16	4	0	I
Small and flexible	20	37	33	9	1	0	A
Fashionable and delicate	18	38	19	21	4	0	A
Sport and speed	14	18	31	36	1	0	O
Lovely and interesting	9	48	26	16	1	0	A

4.3   Construction of Interactive Genetic Algorithm System

4.3.1   Individual Coding Mode

In this study, the median level words extracted by the EGM to interview experts (the reason why the user likes the product) were taken as the component of the production. Therefore, the design system of the electric bicycle product in this paper is divided into four parts: the front structure, cushions, headlights and fenders. There are 5 models in the front structure, which are represented by 3-bit binary code. There are 4 models in the cushion part, which are represented by 2-bit binary code. There are 6 models in the headlight section, which are represented by 3-bit binary code. There are 4 types of fender parts, which are represented by 2-bit binary code. Accordingly, the entire individual genotype of the electric bicycle product consists of 10-bit binary code.

4.3.2   Parameter Settings

The population size is the key parameter attribute value to be considered in interactive genetic algorithms. In this study, the population size N is set to 8 in order to better balance the evaluation time with the population quality. The specific parameters are set as: the crossover probability is 0.8 and the mutation probability is 0.15. In addition, there are two cases for setting the termination condition, the iterations of the termination evolution is 20, and then it will stop automatically; If the user finds a satisfactory individual before terminating the evolutionary process, the user can manually control the evolution process to terminate. The specific parameters of the evolutionary system are shown in Table 4.

Table 4. The genetic evolution parameter

Population size	Encoding way	Crossover	Mutation	Terminate iteration
8	Binary	0.8	0.15	20

4.3.3 Contained Information of Interactive Interface

An important component of interactive genetic algorithms is the Graphical User Interface (GUI). As the most basic part of human-computer interaction, the GUI mainly has two functional characteristics: one of the main functions is to convert the individual's genotype in the genetic algorithm to a visual phenotype and pass it to the user for evaluation; the other function is receiving user evaluation values as individual fitness values so as to guide the genetic optimization operation. Accordingly, the GUI could handle two kinds of information, one is from algorithm to user, and the another is from user to algorithm (Jaksa and Takagi 2004).

The human-computer interaction interface of IGA also displays the other four parts of the user evaluation. The first part is the individual's phenotype and the fitness index of the kansei word. The user can evaluate the kansei word for different product designs according to their own preferences. Then, the interactive genetic system calculates the comprehensive score of the product based on the users' scoring value combined with the weight of the kansei word, which not only avoids the directionality and fuzziness of the user in the evaluation process, but also makes the evaluation process more in line with the characteristics of people's cognitive thinking, thus making the product modeling optimization result more effective. The second part is the setting of the weight of the image index of the product, that is, the weight value of the kansei image word is displayed. The third part is the command button that controls the evolution of the product population, such as "update population", "save and exit". The fourth part includes the information prompt of population evolution, specifically including display of the current best individuals and the change chart of maximum fitness value of individuals in each generation.

4.3.4 Results and Analysis

This research randomly invited eight experts in the field of industrial design. These experts were divided into two groups to participate in the research on the optimization of electric bicycle products forms. Then, the optimal evolutionary individuals are investigated to meet the users' preferences, and to count the maximum adaptive value of each individual population iteration for which the user finds the most satisfied individual, and the average time for them to evaluate the individual. Hence, the evolution changes and trends of the corresponding individual fitness values and evaluation hesitation time during the evolution of the two groups of experts are shown in Fig. 4 and 5.

According to Fig. 4 and 5, the product evolution process in this study is convergent. The data shows that as the evolution iteration increases, the comprehensive adaptive value obtained after weights are added to users' evaluation value for emotional words takes on an ascending tendency. The optimization quality of each generation of products is constantly enhanced and individual adaptive value gradually

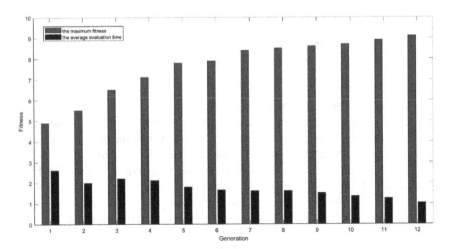

Fig. 4. The first group of experts' individual evaluation value

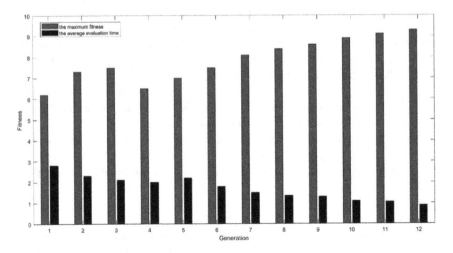

Fig. 5. The second group of experts' individual evaluation value

increases. This implies that the algorithm demonstrates desirable global convergence. Through the system, users can quickly locate product individuals they are satisfied with. Meanwhile, the logic behind the fluctuation in the maximum evaluation value of the optimal individual in each iteration is that the subjects' evaluation of new individuals is affected by experimental subjects subjectivity and the degree of fatigue in the experiment process. As a result, it's not a direct continuous increase. However, the comprehensive evaluation adaptive value demonstrates a rising trend on the whole, which indicates that the method adopted in this study can better reflect users' cognition model about individuals, and then to help users clarify their own needs and thus enhance user satisfaction about the product.

5 Conclusions

This study offers an approach for product form evolution design that's driven by user needs, and seeks out the problem-solving procedures which combine the FKM and IGA evolution designs. First of all, this study applies expert interview method to obtain attractive emotional image and modeling elements, and could merge and simplify them with KJ method and screens out attractive forms and emotional images. Then, we further explores user satisfaction with FKM, and digs out attractive qualities, including modern and concise, technology and cool, small and flexible, lovely and interesting, fashionable and delicate. When priority is given to these five users' kansei image factors, user satisfaction will be greatly augmented. Meanwhile, take these five key demands as the essential carrier to direct the direction of evolution solving of IGA and promotes the new generation production after the genetic operation. The experimental results show that our method could effectively enhance the product design process and increase the convergence efficiency of product evolution, and thus improve design innovative quality.

References

A Kelly, G.: The Psychology of Personal Constructs New York. Norton, New York (1995)

Beale, R.: Supporting serendipity: using ambient intelligence to augment user exploration for data mining and web browsing. Int. J. Hum Comput Stud. **65**, 421–433 (2007). https://doi.org/10.1016/j.ijhcs.2006.11.012

Brintrup, A.M., Ramsden, J., Takagi, H., Tiwari, A.: Ergonomic chair design by fusing qualitative and quantitative criteria using interactive genetic algorithms. IEEE Trans. Evol. Comput. **12**, 343–354 (2008). https://doi.org/10.1109/tevc.2007.904343

Chai, C., Bao, D., Sun, L., Cao, Y.: The relative effects of different dimensions of traditional cultural elements on customer product satisfaction. Int. J. Ind. Ergon. **48**, 77–88 (2015). https://doi.org/10.1016/j.ergon.2015.04.001

Chen, L.-H., Ko, W.-C.: Fuzzy approaches to quality function deployment for new product design. Fuzzy Sets Syst. **160**, 2620–2639 (2009). https://doi.org/10.1016/j.fss.2008.12.003

Chuang, M.C., Chang, C.C., Hsu, S.H.: Perceptual factors underlying user preferences toward product form of mobile phones. Int. J. Ind. Ergon. **27**, 247–258 (2001). https://doi.org/10.1016/s0169-8141(00)00054-8

Dawkins, R.: The blind watchmaker. J. Anim. Ecol. **16**, 423–424 (1986)

Diego-Mas, J.A., Alcaide-Marzal, J.: Single users' affective responses models for product form design. Int. J. Ind. Ergon. **53**, 102–114 (2016). https://doi.org/10.1016/j.ergon.2015.11.005

Ding, M., Bai, Z.: Product color emotional design adaptive to product shape feature variation. Color Res. Appl. **44**, 811–823 (2019). https://doi.org/10.1002/col.22402

Dou, R., Lin, D., Nan, G., Lei, S.: A method for product personalized design based on prospect theory improved with interval reference. Comput. Ind. Eng. **125**, 708–719 (2018). https://doi.org/10.1016/j.cie.2018.04.056

Dou, R., Zhang, Y., Nan, G.: Application of combined Kano model and interactive genetic algorithm for product customization. J. Intell. Manuf. (2016). https://doi.org/10.1007/s10845-016-1280-4

Franke, N., Schreier, M., Kaiser, U.: The "I designed it myself" effect in mass customization. Manage. Sci. **56**, 125–140 (2010). https://doi.org/10.1287/mnsc.1090.1077

He, L., Ming, X., Li, M., Zheng, M., Xu, Z.: Understanding customer requirements through quantitative analysis of an improved fuzzy Kano's model. Proc. Inst. Mech. Eng. Part B J. Eng. Manuf. **231**, 699–712 (2017). https://doi.org/10.1177/0954405415598894

Ho, C.H., Hou, K.C.: Exploring the attractive factors of app icons. KSII Trans. Internet Inf. Syst. **9**, 2251–2270 (2015). https://doi.org/10.3837/tiis.2015.06.016

Holland, J.H.: Genetic algorithms and the optimal allocation of trials. SIAM J. Comput. **2**, 88–105 (1973). https://doi.org/10.1137/0202009

Hsiao, S.-W., Chiu, F.-Y., Lu, S.-H.: Product-form design model based on genetic algorithms. Int. J. Ind. Ergon. **40**, 237–246 (2010). https://doi.org/10.1016/j.ergon.2010.01.009

Jaksa, R., Takagi, H.: Tuning of image parameters by interactive evolutionary computation, pp. 492–497 (2004). https://doi.org/10.1109/icsmc.2003.1243863

Ji, P., Jin, J., Wang, T., Chen, Y.: Quantification and integration of Kano's model into QFD for optimising product design. Int. J. Prod. Res. **52**, 6335–6348 (2014). https://doi.org/10.1080/00207543.2014.939777

Kano, N., Seraku, N., Takahashi, F., Tsuji, S.: Attractive quality and must-be quality. J. Jpn. Soc. Qual. Control **14**, 39–44 (1984)

Kim, H.S., Cho, S.B.: Application of interactive genetic algorithm to fashion design. Eng. Appl. Artif. Intell. **13**, 635–644 (2000)

Kowaliw, T., Dorin, A., Mccormack, J.: Promoting creative design in interactive evolutionary computation. IEEE Trans. Evol. Comput. **16**, 523–536 (2012)

Lee, J.-H., Chang, M.-L.: Stimulating designers' creativity based on a creative evolutionary system and collective intelligence in product design. Int. J. Ind. Ergon. **40**, 295–305 (2010). https://doi.org/10.1016/j.ergon.2009.11.001

Lee, Y.-C., Huang, S.-Y.: A new fuzzy concept approach for Kano's model. Expert Syst. Appl. **36**, 4479–4484 (2009). https://doi.org/10.1016/j.eswa.2008.05.034

Lee, Y., Witell, L., Lin, S., Wang, Y.: A new Kano's evaluation sheet. TQM J. **23**, 179–195 (2011). https://doi.org/10.1108/17542731111110230

Matzler, K., Hinterhuber, H.H.: How to make product development projects more successful by integrating Kano's model of customer satisfaction into quality function deployment. Technovation **18**, 25–38 (1998). https://doi.org/10.1016/s0166-4972(97)00072-2

Meng, Q., He, L.: Fuzzy-KANO-based classification method and its application to quality attributes. Ind. Eng. J. **16**, 121–125 (2013). https://doi.org/10.3969/j.issn.1007-7375.2013.03.020

Miryoku Engineering Forum: Miryoku Eng. Eng. Forum (1992)

Mok, P.Y., Xu, J., Wang, X.X., Fan, J.T., Kwok, Y.L., Xin, J.H.: An IGA-based design support system for realistic and practical fashion designs. Comput. Des. **45**, 1442–1458 (2013). https://doi.org/10.1016/j.cad.2013.06.014

Nagamachi, M.: Kansei engineering: a new ergonomic consumer-oriented technology for product development. Int. J. Ind. Ergon. **15**, 3–11 (1995). https://doi.org/10.1016/0169-8141(94)00052-5

Poirson, E., Petiot, J.-F., Boivin, L., Blumenthal, D.: Eliciting user perceptions using assessment tests based on an interactive genetic algorithm. J. Mech. Des. **135** (2013). https://doi.org/10.1115/1.4023282

Renner, G., Ekárt, A.: Genetic algorithms in computer aided design. Comput. Des. **35**, 709–726 (2003). https://doi.org/10.1016/s0010-4485(03)00003-4

Sanui, J.: Visualization of users' requirements: introduction of the evaluation grid method. In: The 3rd Design and Decision Support Systems in Architecture and Urban Planning Conference, pp. 365–374 (1996)

Shahin, A., Barati, A., Geramian, A.: Determining the critical factors of radical innovation using an integrated model of fuzzy analytic hierarchy process-fuzzy kano with a case study in

Mobarakeh steel company. EMJ - Eng. Manag. J. **29**, 74–86 (2017). https://doi.org/10.1080/10429247.2017.1298182

Sheikhi Darani, Z., Kaedi, M.: Improving the interactive genetic algorithm for customer-centric product design by automatically scoring the unfavorable designs. Hum.-Centr. Comput. Inf. Sci. **7** (2017). https://doi.org/10.1186/s13673-017-0119-0

Shen, K.-S., Chang-yu, P., Lu, Y., Liu, Z., Chuang, C., Ma, M.: A study on the attractiveness of heavy duty motorcycle. World Acad. Sci. Eng. Technol. **30**, 1116–1120 (2009)

Shen, K.S.: Measuring the sociocultural appeal of SNS games in Taiwan. Internet Res. **23**, 372–392 (2013). https://doi.org/10.1108/10662241311331781

Shen, K.S., Chen, K.H., Liang, C.C., Pu, W.P., Ma, M.Y.: Measuring the functional and usable appeal of crossover B-car interiors. Hum. Fac. Ergon. Manuf. **25**, 106–122 (2015). https://doi.org/10.1002/hfm.20525

Takagi, H.: Interactive evolutionary computation: fusion of the capabilities of EC optimization and human evaluation. Proc. IEEE **89**, 1275–1296 (2001). https://doi.org/10.1109/5.949485

Tsuchiya, T., Maeda, T., Matsubara, Y., Nagamachi, M.: A fuzzy rule induction method using genetic algorithm. Int. J. Ind. Ergon. **18**, 135–145 (1996). https://doi.org/10.1016/0169-8141(95)00076-3

Wang, Z., Zhang, M., Sun, H., Zhu, G.: Effects of standardization and innovation on mass customization: an empirical investigation. Technovation **48–49**, 79–86 (2016). https://doi.org/10.1016/j.technovation.2016.01.003

Wu, Z., Lin, T., Li, M.: A computer-aided coloring method for virtual agents based on personality impression, color harmony, and designer preference. Int. J. Ind. Ergon. **68**, 327–336 (2018). https://doi.org/10.1016/j.ergon.2018.09.003

Yadav, H.C., Jain, R., Shukla, S., Avikal, S., Mishra, P.K.: Prioritization of aesthetic attributes of car profile. Int. J. Ind. Ergon. **43**, 296–303 (2013). https://doi.org/10.1016/j.ergon.2013.04.008

Yoo, J.W.: A mathematical formulation for interface-based modular product design with geometric and weight constraints. Eng. Optim. **48**, 985–998 (2016)

Zadeh, L.A.: Fuzzy sets. Inf. Control **8**, 338–353 (1965). https://doi.org/10.1016/s0019-9958(65)90241-x

Shared Mental Model Processing in Visualization Technologies: A Review of Fundamental Concepts and a Guide to Future Research in Human-Computer Interaction

Nor'ain Mohd Yusoff[1]([✉]) and Siti Salwah Salim[2]([✉])

[1] Faculty of Computing and Informatics, Multimedia University,
Persiaran Multimedia, 63100 Cyberjaya, Selangor, Malaysia
norain.yusoff@mmu.edu.my
[2] Faculty of Computer Science and Information Technology,
University of Malaya, 50603 Kuala Lumpur, Malaysia
salwa@um.edu.my

Abstract. This paper presents a review of fundamental concepts behind the shared mental model and its processes. Shared mental model has two properties which are similarity and accuracy where both of these properties have different emphasis. Types of shared mental model refers to four (4) different ways on what type of cognitive process is being shared in team, which includes task-specific knowledge, task-related knowledge, knowledge of teammates and attitude or beliefs. These four (4) types of shared mental model are grouped into two (2) major content domains, i.e. task-work and team-work. It further describe the aspects of shared mental model on how cognitive process is being shared, includes shared vs. overlapping, similar vs. identical, compatible vs. complimentary and distributed. Shared mental model has to be evaluated using three (3) aspects of characteristics in order to show its operationalization: elicitation method, structure representation, and representation of emergence. A specific evaluation technique called cognitive task analysis that focuses on the analysis of difficulties in cognitive structures is introduced in evaluating shared mental model. This paper also discusses Collaborative Visualisation as technological approach in shared mental model, which includes the benefit of user participation in the aspects of joining or leaving, floor control, privacy and global view. Specific areas on big data, visual analytics, multimedia interface, mobility, disability, awareness and learning analytics that can benefit from the shared mental model approach is discussed.

Keywords: Shared mental model · Cognitive processing · Shared visualization

© Springer Nature Switzerland AG 2020
D. Harris and W.-C. Li (Eds.): HCII 2020, LNAI 12186, pp. 238–256, 2020.
https://doi.org/10.1007/978-3-030-49044-7_20

1   Introduction

One (1) type of cognitive architecture in domain of team work and collaboration that has received substantial research attention in Human-computer Interaction is Shared Mental Model (SMM). SMM is derived from the root of mental model construct from the discipline of cognitive psychology (DeChurch and Mesmer-Magnus 2010). Shared mental model (SMM) refers to "knowledge structure held by members of a team that enables them to form accurate explanations and expectations for the task, and in turn, to coordinate their actions and adapt their behavior to demands of the task and other team members" (Cannon-Bowers et al. 1993, p. 228). According to Payne (2003), even though the research interest in SMM has become overwhelming in HCI, more emphasis on theoretical, empirical and conceptual work pertaining to the team cognition and computer interaction are still required.

Earlier in 1993, mental model has been discussed by Staggers and Norcio (1993), concerning the mental model formation, characteristics and models. Other reviews have been conducted by Isenberg et al. (2011), which it provides a detailed review on five (5) scenarios on how SMM being applied in collaborative visualisation tools. Other study conducted by Grimstead et al. (2005) provide a review of how SMM are being used in forty two (42) collaborative visualisation systems across four application areas: collaborative problem-solving environments, virtual reality environments, multi-player online games and multi-user enabling of single user applications. Most recently study is a review and analysis of shared visual representation for building a shared mental model (Nor'ain and Siti Salwah 2015). These studies only described how important is the SMM application in particular interactive systems, however, none of them provide a thorough understanding on the theoretical concepts behind the SMM processing such as properties, aspects and types of SMM, as well as techniques being used in evaluating SMM in visualization technologies.

The aim of this paper is to provide a theoretical review about this concept and guide to future research. The search for this review studies was performed through popular electronic databases i.e. Science Direct, ACM Digital Library, IEEE Explore Digital Library, ISI Web of Knowledge, Scopus Online, Taylor and Francis Online and Springer Link. The search strategy was limited to English articles only. The initial keywords search include: ('shared mental model' OR 'shared cognition' OR 'visualisation technology'). The final step was to narrow down the search to the most relevant researched area in Human-Computer Interaction studies.

2   Shared Mental Model in Human-Computer Interaction

In HCI, team cognitive research is characterized as the study of a team as an information-processing unit (Salas et al. 2008). In order to understand the cognitive processes of the team that they want to study, HCI researchers learn from cognitive models that describe the fundamental concept.

Yusoff and Salim (2014) reported two (2) types of approaches in studying team cognition; socially shared cognitive approach (SSC) and shared situation awareness approach (SSA). SSC is a shared cognitive approach views "how dyads, groups and larger collectives create and utilize interpersonal understanding" (Thompson and Fine 1999, p. 3). On the other hand, SSA refers to "degree that team members possess the same awareness of shared situation awareness requirements, within a volume of time and space, as well as the comprehension of their meaning and projection of their status in the near future" (Endsley et al. 2003, p. 13).

SMM is a type of information-processing model which is developed underlying the SSC approach (Yusoff and Salim 2014). This model views that group members have a separate and independent memory structures. It suggests that group member who is able to access to other member's memory stores can effectively expand their storage and retrieval, thus leading to development of group interaction. Conversely, SMM has also seen to support awareness situation. For example, works by Haig et al. (2006) shows a SMM development to support situation awareness among clinicians as well as work by Entin and Entin (2000) who found team situation awareness in SMM using simulated military missions. Hence, the SMM can be used to support both SSC as well as SSA.

Using an underlying input-process-output (IPO) framework, the greatest focus of concern on team cognitive research in HCI has been on enhancing communication and collaboration among team members as well as optimizing the performance of the team as a whole. The theoretical concept of SMM is discussed further in next section.

3   Theoretical Concept of Shared Mental Model

SMM refers to "knowledge structure held by members of a team that enables them to form accurate explanations and expectations for the task, and in turn, to coordinate their actions and adapt their behavior to demands of the task and other team members" (Cannon-Bowers et al. 1993, p. 228). SMM develops when team members interact and that converge the individual team member's mental model, resulting in similar to, or sharing with, that of their team member's mental model. The terminology SMM has also been introduced in many ways, for example, team mental models and compatible mental model (McComb 2008). Throughout this paper, the two (2) terms SMM and shared cognition are being used interchangeably to describe essentially the same concept.

3.1   Properties of Shared Mental Model

SMM consists of two (2) properties; similarity and accuracy (Mohammed et al. 2010). Similarity in SMM refers to "sharedness" or the "degree to which members' mental models are consistent or converge with one another and does not signify identical mental models" (Mohammed et al. 2010, p. 880). Examples of SMM studies focusing on similarity are the study on similarity of knowledge structures between two (2) members (Mathieu et al. 2000), the effect of cross-training for the similarity of teammates' team-interaction model (Marks et al. 2002) and the effect of similar mental models for high-performance team (Zou and Lee 2010). On the other hand, accuracy in SMM refers to the "true score" or "similarity of knowledge ratings about other members and one's own corresponding self-ratings" (Espinosa 2001, p. 2103).

The studies in sharedness or similarity have given more emphasis than the accuracy in the literature even though some studies attempted to study both properties are also found (Mohammed et al. 2010). For examples; Burtscher et al. (2011) investigate how the similarity and accuracy and two (2) forms of monitoring behavior i.e. team versus systems interacted to predict team performance in anesthesia, and Resick et al. (2010) examined the relationships between team cognitive ability and personality composition in relation to the similarity and accuracy of team task-focused mental models.

3.2 Importance of Shared Mental Model

According to Cannon-Bowers and Salas (2001), constructing SMM is important due to following three (3) reasons:

- Firstly, SMM provides an explanatory mechanism that helps to understand team performance. It explains the effectiveness of teams' interaction with one another without the need to communicate.
- Secondly, SMM construction can be valuable to predict variable in teams such as identifying potential performance problems and providing insight into how the problems can be fixed.
- Thirdly, SMM can diagnose problems such as identifying poor communication that may derive from lack shared of knowledge.

Due to the importance of SMM, the application of SMM can lead to three (3) outcomes (Cannon-Bowers and Salas 2001) as below:

- First, SMM could lead to better task performance, such as in terms of the accuracy, efficiency, quality of output, volume, timeliness. This outcome is defined as task-specific.
- Second, SMM leads to better team processes, which in turn lead to better task performance such as more efficient communication, more accurate expectations and predictions, consensus, similar interpretations, and better coordination. This outcome is defined as task-related.
- Another expected outcome from SMM is referred to motivational outcomes. This includes cohesion, trust, morale, collective efficacy and satisfaction with the team. However, (Cannon-Bowers and Salas 2001) stresses that the motivational outcomes have a looser association with task performance than the previous two.

Cannon-Bowers and Salas (2001) argued that there is a need to clarify which kind of outcomes that is expected from the SMM so that the types and aspects of shared cognition can be determined. The types and aspects of shared cognition are explained in the next sections.

3.3 Types of Shared Mental Model

Types of SMM refer to what cognitive processes are shared. There are four (4) types of cognitive categories on what is shared in team (Cannon-Bowers and Salas 2001):

1. *Task-specific Knowledge* - This type of shared cognition allows the team members to coordinate without the need to communicate overtly and act on knowledge without discussion. The nature of knowledge being shared is highly task-specific, which involves specific procedures, sequences, actions and strategies to perform a task.
2. *Task-related Knowledge* - This type of shared cognition allows team members to have common knowledge about task-related processes such as what it is, how it operates and its importance, which contribute to the team's ability to accomplish the task. In contrast to task-specific knowledge, it is not task-specific, but it can hold variety of similar tasks.
3. *Knowledge of Teammates* - This type of shared cognition allows team members to understand each other in terms of their preferences, strengths, weaknesses, and tendencies in order to maximize performance. It views that team learns the distribution of expertise within the team over time. It is also a task-related knowledge but not necessarily task-specific.
4. *Attitude or Beliefs* - This type of shared cognition allows team members to have similar attitudes and beliefs that lead to effective decisions. It involves the notions of shared beliefs and cognitive consensus. This shared cognition type covers a broad category of knowledge, where it does not related to task-specific or task-related.

These four (4) types of SMM are categorized into two (2) major content domains; task-work, and team-work (Mathieu et al. 2000). Task work domain refers to the work goals and performance requirements, while the team work domain refers to the interpersonal interaction requirements and skills of other team members (Mohammed et al. 2010). The integration between the two (2) major domains and four (4) types of SMMs are presented in Table 1.

Table 1. Mathieu et al. (2000)'s major domains and Cannon-Bowers et al. (1993)'s types of SMMs

Major domain	Types of model	Knowledge content	Description	Stability of model content
Task-work (what needs to be accomplished)	Technology/equipment	Equipment functioning Operating procedures System limitations Likely failures	Likely to be the most stable model in terms of content. Probably requires less to be shared across team members	High
	Job/task	Task procedures Likely contingencies Likely scenarios Task strategies or techniques	In highly proceduralized tasks, members will have shared task models. When tasks	Moderate

(*continued*)

Table 1. (*continued*)

Major domain	Types of model	Knowledge content	Description	Stability of model content
		Environmental constraints Task component relationships	are more unpredictable, the value of shared task knowledge becomes more crucial	
Team-work (how work needs to be accomplished)	Team interaction	Roles/responsibilities Information resources Interaction patterns Communication channels Role interdependencies Information flow	Shared knowledge about team interactions drives how team members behave by creating expectations. Adaptable teams are those who understand well and can predict the nature of team interaction	Moderate
	Team	Teammates' knowledge Teammates' skills Teammates' abilities Teammates' preferences Teammates' tendencies	Team-specific knowledge of teammates helps members to better tailor their behavior to what they expect from teammates	Low

3.4   Aspects of Shared Mental Model

Aspects of SMM refer to how cognitive processes are shared. There are four (4) different categories of ways of how cognition is shared in team (Cannon-Bowers and Salas 2001):

1. *Shared vs. Overlapping* - This refers to situations where two (2) or more team members need to have some common knowledge but should not be redundant.
2. *Similar vs.* Identical - This refers to the need to hold similar or identical knowledge. Team members need to hold similar attitudes and beliefs in order to draw common interpretations that can drive towards effective performance. For example, surgeon and nurse working together in an operation theater are not expected to have identical knowledge, but portions of their knowledge bases are needed to be shared (Undre et al. 2006). This category of shared cognition is associated with the task that must be common among members.

3. *Compatible vs. Complimentary* - This refers to team which possess specialized roles and knowledge that is crucial to task performance. A multidisciplinary team where each member possesses specialized expertise to solve a problem may have dissimilar knowledge, however still can lead them to complementary behavior.
4. *Distributed* - This refers to the knowledge that is distributed across members. This aspect of shared cognition is applied in many high performance teams, such as military combat teams, where the systems and tasks are complex and difficult. Therefore, if the team members' knowledge is specialized and distributed, team members need to coordinate their knowledge effectively in order to achieve SMM.

4   Evaluation for Shared Mental Model

This section describes in detail approaches of evaluations in shared mental model.

4.1   Ways to Measure Shared Mental Model

Shared knowledge can be measured in two (2) ways (Cannon-Bowers and Salas 2001). Firstly is by assessing the structure of team member knowledge and secondly is to measure the content of team member knowledge. Mohammed et al. (2010, p. 884) refers the structure as "how concepts are organized in the minds of participants" whereas content is the "knowledge that comprises cognition". It is stated that assessing the team knowledge structure is more straightforward; however in practical ways it is rather very difficult. On the other hand, measuring contents has been seen as more possible to conduct.

Steps to Measure Shared Mental Model
DeChurch and Mesmer-Magnus (2010) further have described steps to measure SMM, which involves three (3) aspects of characteristics: elicitation method, structure representation, and representation of emergence. These three (3) characteristics are needed as they can show the operationalization of SMM. They are as described as follows:

a. *Elicitation Method*
   It refers to the technique used to determine the components or content of a mental model. Techniques include:

- Similarity ratings - Participants are presented with a grid and they will be requested to consider each pair of task nodes and report their perceptions of the relation between the two (2) nodes.
- Concept maps – Participants are asked to elicit contents and place the actions into a meaningful organizational scheme.
- Rating scales – Participants are asked to elicit the content of the model and respond to questions about the task on fixed-response formats such as strongly agree to strongly disagree.
- Card sorting tasks – Participants are asked to sort numbers of cards and categorize or list them based on their understanding of the structure and relationship.

- Interactively elicited cause mapping - Participants are asked to provide data through questionnaires and/or interviews using interactive ways.
- Text-based cause mapping – Participants are asked to provide post hoc analyses of data such as systematic coding of documents or transcripts.

b. *Structure Representation*

It refers to the organized knowledge structures corresponding between how the knowledge content is represented in the mind and how the knowledge representation can be modeled by the researcher (DeChurch and Mesmer-Magnus 2010). Techniques include:

- Pathfinder – This technique is used to produce appropriate psychological scaling based on the underlying structure between concepts. It provides algorithm that can transform raw paired comparison ratings into a network structure where these concepts are represented as nodes, while the relatedness of the concepts are represented as links.
- UCINET - This technique is developed by Borgatti et al. (2002) to support social network data analysis. It comes with a complete software package for data visualization.
- Multidimensional scaling – This technique uses geometric models to represent proximity data spatially. It is used to identify unknown underlying dimensions in organizing cognitive stimuli.
- Concept mapping/card sorting - as described.

c. *Representation of Emergence*

It refers to representation technique used "to reveal the structure of data or determine the relationships between elements in an individual's mind" (Mohammed et al. 2000, p. 129). Techniques include; concept mapping, path-finder, UCINET, interactively elicited cause mapping, text-based cause mapping and Euclidean distance.

Of all the techniques described in each steps for measuring SMM, Mohammed et al. (2000) recommended only four (4) techniques because they encompassed both elicitation and representation. These techniques are referred to as pathfinder, multidimensional scaling, interactive elicited mapping and text-based cause mapping.

Next section describes Cognitive Task Analysis as another approach of evaluation in SMM.

4.2 Cognitive Task Analysis Approach

Cognitive task analysis (CTA) focuses on the difficulties in cognitive structures such as knowledge-based and representational skills as well as processes such as attention, problem solving and decision making (Stanton et al. 2005). The aspects of cognitive structures and processes in the CTA can provide a description of the knowledge and thought processes that are required at the expertise level (Schraagen et al. 2008; Seamster et al. 1997). They can also lead to a process for designing, developing and evaluating a better human–computer interface intended to amplify and extend the human ability to make good decisions (Crandall et al. 2006).

Most studies in CTA are concerned with expertise (Klein and Militello 2001, p. 180). Cognitive study is designed to elicit the knowledge and wisdom acquired (Crandall et al. 2006, p. 134). For example, during CTA interviews, interviewers will appreciate the nature of expertise when responses and feedback received are probed in detail. Some related expertise studies that have been conducted using CTA include experienced air warfare coordinators unpacking their expertise and coaching skills for the development of shipboard-based on-the job training for the Navy (Pliske et al. 2000), certified cytotechnologists detecting questionable cells and making sense of the clinical picture for the process documentation of tissue biopsies and cell samples for pathology (McDermott and Crandall 2000) and army ranger squad or platoon leaders describing the required skills for clearing buildings in urban combat settings for the development of training software (Phillips et al. 1998).

Cooke (1994) found more than hundred (100) types of CTA methods and techniques. Due to the growing number of CTA methods, extensive CTA reviews by Stanton et al. (2005), Schraagen et al. (2000) as well as Wei and Salvendy (2004) offer a broad exploration of the difference among these methods and techniques in a number of ways. Stanton et al. (2005) present five (5) selected CTA methods based upon their popularity and the application used, while Schraagen et al. (2000) described a comprehensive review of reviews and classifications to guide researchers interested in exploring and applying the CTA techniques. On the other hand, Wei and Salvendy (2004) classify the CTAs into four (4) broad families, namely: 1) observation and interview 2) process tracing 3) conceptual techniques and 4) formal models. This CTA family classification is meant to guide researchers who are aiming for particular outputs, to select appropriate techniques.

5 Role of Shared Visualisation to Support Shared Mental Model

One way to externalize the individual mind is through the representation of visualization. Many studies have also demonstrated that visual representation that is shared among the users can lead to the development of SMM. Visualization is referred to as "a method of computing...offers a method for seeing the unseen, enriches the process and unexpected insights" (National Science Foundation's Visualization 1987). According to McGrath et al. (2012), visualization is a graphical representation of data to aid human cognition. These two (2) definitions explain that visualization that is shared can enrich the process as well as the unexpected insights performed by many users.

The application and effect of visualization to shared cognition have been studied by many researchers in the domain of cognition and design studies. For example, the effectiveness of visual representation for the purpose of externalizing and communicating the design process has been demonstrated by Goldschmidt (2007). In this study, two (2) experiments are conducted to clarify how the visual representations have created a SMM of a new bicycle accessory meant to carry a backpack. The result shows that in order for all team members to arrive at a shared task model, it is necessary for them to see the design entity eye to eye in order to progress.

Other studies which had demonstrated the important of visual representation includes; collaborative knowledge construction via visual graphical representation (Suthers 2005), reducing the effort of explicit communication via shared white boards in emergency department (Xiao et al. 2007), and understanding different kinds of video representation and analysis via the use of video story (McNeese 2004).

Arias et al. (2000) opined that SMM can be visualized through the use of external artifacts. External representation can be used to make the knowledge available to all members explicitly as well as able to transcend the cognitive limitation across individual minds. This externalization is important as it creates what is vaguely resides in one's mental efforts. In other word, artifact represents an externalization which can be communicated visually to the users.

Next section describes the significant role of using cognitive artifact to externalize visualisation.

5.1 Cognitive Artifacts

Artifact that is used as a tool for cognitive activities is called as the "cognitive artifacts". According to Visser (2006), cognitive artifact comes in two (2) forms:

- *Internal cognitive artifacts or mental cognitive artifacts (i.e. mental representation)* – Examples; such as rules of thumb, mnemonics, shopping lists, and other kinds of procedures.
- *External cognitive artifacts* – There are two(2) types; physical being such as buildings, cars or garments or any results from mental representations and symbolic such as software, route plan, drawings, mock-ups or any results that symbolize the mental representation.

Visser (2006) explains that due to the nature of the mental cognitive artifacts that are less vague in terms of ideas or images that designers have "in their heads", the use of external cognitive artifact to visually externalize the emergence of mental cognitive artifacts are therefore required. A research example of using external cognitive artifact is conducted by Nemeth et al. (2004) to study communication and information sharing among healthcare providers. In this study, external cognitive artifacts which are related to operating room or scheduling are used such as the availabilities sheet, master schedule, graph and board. This study finds that better computer-supported cognitive artifacts should benefit patient safety by making teamwork processes, planning, communications and resource management more resilient. Figure 1 shows the emergence process of external artifact to represent the visual externalization of internal cognitive artifact.

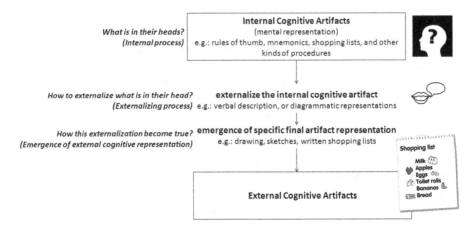

Fig. 1. The emergence process of an external cognitive artifact

5.2 The Technological Approach Using Collaborative Visualisation

Salas et al. (2008) suggested that SMM can be improved using technological development and implementation. One (1) way to support SMM is by using technological visualization approach called as collaborative visualization (CoVis). In CoVis, a shared use of any forms of visual representations is required to enable any cognitive activities collaboration.

CoVis is referred to as "the shared use of computer-supported, interactive, visual representations of data by more than one (1) person with the common goal of contribution to joint information processing activities" (Isenberg et al. 2011, p. 312). CoVis is an approach that emphasis the shared use of interactive visual representations, which could be in a form of joint viewing, interacting with, discussing, or interpreting the presentation.

According to Isenberg et al. (2011), one (1) of the important aspects in CoVis is the focus on cognition and results. It emphasizes that CoVis is not concern about the creation of a "product" i.e. the shared representation, but the focus should involve unique cognitive activities i.e. shared cognition. Besides, Isenberg et al. (2011) also emphasizes the need to support social interaction process around the data. It concerns on the ability for a team to build each other's insights, which in turn could reach a common understanding of the datasets. Examples of social interaction include arriving at a common understanding of the data as well as enhance knowledge construction by making use of interaction of data.

CoVis also includes distributed design environment which can be directed to the division or spread of resources such as design artifacts, design knowledge or design team. Distributed design can be operated in either synchronous or asynchronous mode. Synchronous mode in distributed design enables real-time communication and collaboration in a "same time-different space" environment; whereas, asynchronous mode enables the communication and collaboration in distributed design to be operated over a period of time through a "different time-different space" mode.

Table 2. Analogy of shared cognitive process in shared mental model and collaborative visualisation

Major process	Shared mental model	Collaborative visualisation
Input	Knowledge elicitation from the mind of individual/ group	Data acquisition from the users
Processing	Structure representation	Data presentation
Output	Emergence of presentation	Emergence of visualisation

In order to understand the analogy of cognitive process in SMM and CoVis, we present the major processes of input, processing and output as shown in Table 2. As seen in CoVis process, data within the mind of users is acquired, represented and emerged using a form of visual representation (Isenberg et al. 2011). Similarly, SMM process also show the data or knowledge from the mind of individual or group is elicitated, represented in a structure form and finally emerge to form SMM (DeChurch and Mesmer-Magnus 2010).

These processes however do not show how the role of artifact can be used for each process. As such, we come out with a cognitive data process of SMM using the role of artifact. Based on these two (2) processes, the role of artifact can be used to map with each of the major process in order to understand what cognitive data can be acquired as an input, how the cognitive data can be externalized as a process, and what form of visualization can display the emergence of that cognitive data. From the previous review, we have mentioned the focus on elements of knowledge and/or needs which can lead to the development of individual mental model derived from Badke-Schaub et al. (2007)'s framework. The cognitive data process of SMM using the role of artifact is as follows:

- *Cognitive Data Acquisition* - is the input process to identify internal cognitive artifact that represents the data from knowledge and needs of users.
- *Cognitive Data Process* - refers to how that internal cognitive artifacts being processed.
- *Cognitive Data Emergence* - is the output process which displays the emergence of that cognitive data in a form of visualization. It can be referred as the Visser (2006)'s symbolic form of external cognitive artifact.

Looking the importance of shared visualization for SMM, a systematic review study of shared visualizations focusing on SMM has been conducted by Nor'ain and Siti Salwah (2015). This paper is aimed to understand what strategy or techniques being applied in shared visualization to achieve SMM, which include how these strategies and techniques are being used in shared visualization.

Next section describes user interaction in shared visualization technology.

5.3   User Interaction in Shared Visualisation Technology

Brodlie et al. (2004) provide four (4) aspects that relate to how users interact when participating in visualization systems. The four aspects are:

- *Joining/leaving* – shared visualization systems should have a facility to allow users to join and leave at any time.
- *Floor control* – Shared visualization systems should offer different levels of access to a session for individual users such as allowing editing, or sharing or both editing and sharing authority.
- *Privacy* – Shared visualization systems should allow users to work privately and at the same time, still remain in the conference. This is to protect some information and at the sometime can share other information.
- *Global view* - Shared visualization systems should be able to allow users to view the network editor of other users in order to reassure that they understand what other user is doing.

6   Future Research Directions

This review discusses the fundamental understanding of SMM processing in visualization technologies. There are number of areas that can benefit from the SMM approach, which is briefly discuss in this section.

6.1   Shared Visualisation in Big Data and Visual Analytics

For big data processing, SMM can be explored on how it should work effectively and efficiently in both distributed and collaborative visualization. According to Brodlie et al. (2004), distributed visualization has some resource allocation problems such as location of processing close to data for minimizing data traffic. One example of an enabling technology is cloud computing and the link to web services could provide better enhancement for distributed visualization in the any visualization technology system, such as close coupling of simulations and visualizations in a real-time, interactive steering environment. A model of visual management system has been devised to support Lean Production in construction sites (Valente et al. 2018). Big data visualisation is needed to support intuitive tools as well as practices involved in very large and changing construction environments, teams and equipment that are often spread in large areas. This work shows that shared visualisation is needed to manage wide construction sites where there is a gap of visual languages, design, infrastructure, and mechanics of visual perception among the construction workers (Valente et al. 2016).

According to Alharthi (2016), a selected team can be visually transformed to enhance collaborative ideas for a specific purpose to enhance mental and overall capabilities as well as to maintain high performance of task and reduces likelihood of failure on a work mission. Mixed-initiative visual analytics system (MIVAS) is another work conducted by Makonin et al. (2016). The conceptual architecture of MIVAS consists of five (5) key components that support data wrangling, alternative discovery and comparison, parametric interaction, history tracking and exploration, system agency and adaptation to support human-human collaboration, multi-user collaboration as well as autonomous revisit system solution.

Work by Seipp et al. (2019) shows that combining information visualization in data mining techniques is an interactive decision-making process that can offer a visual guide to communicate uncertainty. This GeoVisual Analytics should assist the analysis of increasing availability and complexity of geographical information. A review by Chapeton et al. (2018) shows that GeoVisual analytics support visualisation sharing in hybrid collaborative scenarios, cross-device collaboration, time-critical and long-term analysis. Similar work in geographical domain is studied by Ruda (2015) about cartography visualisation to obtain precise results for spatial decision making.

6.2   Shared Visualisation in Multimedia Interface

Another area that requires great attention is involving the user interfaces design construction and development. For example, SMM processing can be increased using visualization and multimedia output capabilities through sophisticated multimodal interaction. According to Oviatt and Cohen (2000), multimodal input facility in a system could give more powerful interfaces for the user to access and manipulate information. Example of future work may include designing multimodal inputs such as speech and handwritten recognition from the user. These recognition techniques should be able to read, interpret and translate integrated data inputs in a form of visualization can provide better multimodal interaction facility in the visualization technology system. Redlich et al. (2017) found that creative virtual tools enhance SMM but still lack in perceived efficiency compared to physically present teamwork. It is suggested that to increase the quality, efficiency and satisfaction of virtual creative processes (such as in virtual agent, chatbot, live chat), further investigation on the usage of information communication technology (ICT), multimedia and virtual tools need to be conducted.

6.3   Shared Visualisation for Mobility

On the other hand, the support for SMM processing in the visualization technology system can be extended through the use of mobile-based application with particular interest to the shared visualization data. For example, the capability of the framework can be enhanced by allowing both desktop and mobile clients to simultaneously visualize the same data visualization in sharing a common view. A research contribution in this area has been done by Craig and Huang (2015). Interactive animated mobile information visualization in "Mobile Tree Browser" is developed to browse labelled hierarchies on mobile devices. The display is able to optimize readability and navigation on devices with limited space. This browser can be used to support multi-device co-located synchronous collaboration using animation to smooth the transition between views (Craig 2015; Craig et al. 2015).

6.4   Shared Visualisation for Disability, Awareness and Learning Analytics

In terms of multi-users participations, SMM can be further explored by incorporating different types of users such as those with certain kind of disabilities. Krishna et al. (2010) stated that social situational awareness is one of the important aspects for new technologies to enrich the social presence among remote disability users. For visually impaired users, portable and wearable interfaces can provide better access to non-verbal social cues through non-visualization medium. Specific research can be done to identify techniques that provide accessible SMM processing among different types of disabilities.

A shared route navigation that combines mobile devices with large display interfaces is designed to allow metro system navigation more convenient and at the same time, allowing a passenger to plan their route in the station using a large-display interface and follow the route using notifications on a smart-watch or similar wearable device (Craig and Liu 2019). This alternative view coordination methods that link smart-phone and large-display multi-device environments is a pervasive approach to facilitate different degrees of autonomous and collaborative working (Craig and Liu 2018). A Multi-Dimensional Visualization is an awareness visualisation system that reveals the level of perception of students towards learning of innovative skills in university (Muraina and Ibrahim 2016). This awareness system focuses on bringing the perception of students towards the learning of innovative skills into reality prior to the commencement of teaching.

Learning analytics is another specialized work in SMM to support education. Middleton (2020) examines the constructed visualization formats of small groups in a library learning commons. This study generates a hybrid and ubiquitous approach to integrating multiple visualization modalities within spatial settings in order to understand how visual and social affordances of group generated visualization formats. Liu and Nesbit (2020) study that visual dashboard facilitate group awareness, shared mental models and group cognition, and in turn fosters effective teaching and learning strategies.

As a conclusion, SMM provides a powerful predictive and explanatory for understanding the interaction in a common way. In order to interact with the world, people form unique share representations or shared mental models with which they can interact to perform better task and work performance.

Acknowledgments. This work is sponsored by the Ministry of Higher Education Malaysia under the Exploratory Research Funding Scheme EP20120612006 and from the Multimedia University Malaysia Research Funding IP20110707004 and IP20120511020.

References

Ruda, A.: Cartographic visualization of outputs for spatial decision-making in regional development. Geod. Cartogr. **41**(4), 174–184 (2015). https://doi.org/10.3846/20296991. 2015.1120431

Alharthi, S.Z.: Forming, measuring, optimizing, human capital with enhancement concerning dynamic capability: Update on theoretical framework (2016)

Badke-Schaub, P., Neumann, A., Lauche, K., Mohammed, S.: Mental models in design teams: a valid approach to performance in design collaboration? CoDesign **3**(1), 5–20 (2007). https://doi.org/10.1080/15710880601170768

Borgatti, S., Everett, M., Freeman, L.: Ucinet for Windows: Software for Social Network Analysis. Analytic Technologies, Harvard (2002)

Brodlie, K.W., Duce, D.A., Gallop, J.R., Walton, J.P.R.B., Wood, J.D.: Distributed and collaborative visualization. Paper Presented at the Computer Graphics Forum (2004)

Burtscher, M.J., Kolbe, M., Wacker, J., Manser, T.: Interactions of team mental models and monitoring behaviors predict team performance in simulated anesthesia inductions. J. Exp. Psychol.: Appl. **17**(3), 257–269 (2011). https://doi.org/10.1037/a0025148

Nemeth, C.P., Cook, R.I., O'Connor, M., Klock, P.A.: Using cognitive artifacts to understand distributed cognition. IEEE Trans. Syst. Man Cybern. - Part A: Syst. Hum. **34**(6), 726–735 (2004). https://doi.org/10.1109/tsmca.2004.836798

Cannon-Bowers, J.A., Salas, E.: Reflections on shared cognition. J. Organ. Behav. **22**(2), 195–202 (2001)

Cannon-Bowers, J.A., Salas, E., Converse, S.: Shared mental models in expert team decision making. Individual and group decision making: current issues. In: Jr. John Castellan, N. (ed.), Individual and Group Decision Making: Current Issues, pp. 221–246. Lawrence Erlbaum Associates, Inc., Hillsdale (1993)

Cooke, N.J.: Varieties of knowledge elicitation techniques. Int. J. Hum.-Comput. Stud. **41**(6), 801–849 (1994). https://doi.org/10.1006/ijhc.1994.1083

Craig, P., Liu, Y.: Coordinating user selections in collaborative smart-phone large-display multi-device environments. In: Luo, Y. (ed.) CDVE 2018. LNCS, vol. 11151, pp. 24–32. Springer, Cham (2018). https://doi.org/10.1007/978-3-030-00560-3_4

Craig, P., Liu, Y.: A vision for pervasive information visualisation to support passenger navigation in public metro networks. In: 2019 IEEE International Conference on Pervasive Computing and Communications Workshops (PerCom Workshops), pp. 202–207. IEEE (March 2019)

Crandall, B., Klein, G., Hoffman, R.R.: Working Minds: A Practitioner's Guide to Cognitive Task Analysis. MIT Press, Cambridge (2006)

DeChurch, L.A., Mesmer-Magnus, J.R.: The cognitive underpinnings of effective teamwork: a meta-analysis. J. Appl. Psychol. **95**(1), 32–53 (2010)

Salas, E., Cooke, N.J., Rosen, M.A.: On teams, teamwork, and team performance: discoveries and developments. Hum. Factors: J. Hum. Factors Ergon. Soc. **50**(3), 540–547 (2008). https://doi.org/10.1518/001872008x288457

Endsley, M.R., Bolstad, C.A., Jones, D.G., Riley, J.M.: Situation awareness oriented design: from user's cognitive requirements to creating effective supporting technologies. Proc. Hum. Factors Ergon. Soc. Ann. Meet. **47**(3), 268–272 (2003). https://doi.org/10.1177/154193120304700304

Entin, E.B., Entin, E.E.: Assessing team situation awareness in simulated military missions. Proc. Hum. Factors Ergon. Soc. Ann. Meet. **44**(1), 73–76 (2000). https://doi.org/10.1177/154193120004400120

Arias, E., Eden, H., Fischer, G., Gorman, A., Scharff, E.: Transcending the individual human mind—creating shared understanding through collaborative design. ACM Trans. Comput.-Hum. Interact. **7**(1), 84–113 (2000). https://doi.org/10.1145/344949.345015

Espinosa, J.A.: Shared mental model: accuracy and visual representation. Paper Presented at the Seventh Americas Conference on Information Systems, Boston, Massachusetts, USA (2001)

García-Chapeton, G.A., Ostermann, F.O., de By, R.A., Kraak, M.J.: Enabling collaborative GeoVisual analytics: systems, techniques, and research challenges. Trans. GIS **22**(3), 640–663 (2018)

Goldschmidt, G.: To see eye to eye: the role of visual representations in building shared mental models in design teams. CoDesign **3**(1), 43–50 (2007). https://doi.org/10.1080/15710880601170826

Grimstead, I.J., Walker, D.W., Avis, N.J.: Collaborative visualization: a review and taxonomy. Paper Presented at the Ninth IEEE International Symposium on Distributed Simulation and Real-Time Applications, 2005 (10–12 October 2005)

Haig, K.M., Sutton, S., Whittington, J.: SBAR: a shared mental model for improving communication between clinicians. Joint Comm. J. Qual. Patient Saf. **32**(3), 167–175 (2006). https://doi.org/10.1016/S1553-7250(06)32022-3

Isenberg, P., Elmqvist, N., Scholtz, J., Cernea, D., Ma, K.-L., Hagen, H.: Collaborative visualization: definition, challenges, and research agenda. Inf. Vis. **10**(4), 310–326 (2011). https://doi.org/10.1177/1473871611412817

Seipp, K., Gutiérrez, F., Ochoa, X., Verbert, K.: Towards a visual guide for communicating uncertainty in visual analytics. J. Comput. Lang. **50**, 1–18 (2019). https://doi.org/10.1016/j.jvlc.2018.11.004. ISSN:2590-1184

Klein, G., Militello, L.: 4. Some guidelines for conducting a cognitive task analysis. Advances in Human Performance and Cognitive Engineering Research, vol. 1, pp. 163–199. Emerald Group Publishing Limited (2001)

Krishna, S., Balasubramanian, V., Panchanathan, S.: Enriching social situational awareness in remote interactions: insights and inspirations from disability focused research. Paper Presented at the Proceedings of the 18th ACM International Conference on Multimedia, Firenze, Italy (2010)

Liu, A.L., Nesbit, J.C.: Dashboards for computer-supported collaborative learning. In: Virvou, M., Alepis, E., Tsihrintzis, G.A., Jain, L.C. (eds.) Machine Learning Paradigms. ISRL, vol. 158, pp. 157–182. Springer, Cham (2020). https://doi.org/10.1007/978-3-030-13743-4_9

Makonin, S., McVeigh, D., Stuerzlinger, W., Tran, K., Popowich, F.: Mixed-initiative for big data: the intersection of human + visual analytics + prediction. In: 2016 49th Hawaii International Conference on System Sciences (HICSS) (2016). https://doi.org/10.1109/hicss.2016.181

Marks, M.A., Sabella, M.J., Burke, C.S., Zaccaro, S.J.: The impact of cross-training on team effectiveness. J. Appl. Psychol. **87**(1), 3–13 (2002). https://doi.org/10.1037/0021-9010.87.1.3

Mathieu, J.E., Heffner, T.S., Goodwin, G.F., Salas, E., Cannon-Bowers, J.A.: The influence of shared mental models on team process and performance. J. Appl. Psychol. **85**(2), 273–283 (2000). https://doi.org/10.1037/0021-9010.85.2.273

McComb, S.A.: Shared mental models and their convergance. In: Letsky, M.P., Warner, N.W., Fiore, S.M. (eds.) Macrocognition in Teams: Theories and Methodologies. Ashgate Publishing Group, Abingdon (2008)

McDermott, P.L., Crandall, B.: Uncovering expertise: how cytotechnologists screen pap smears. In: The Proceeding of 5th Conference on Naturalistic Decision Making, Tammsvik, Sweden, May 26–28, 2000 (2000)

McGrath, W., Bowman, B., McCallum, D., Ramos, J.D.H., Elmqvist, N., Irani, P.: Branch-explore-merge: facilitating real-time revision control in collaborative visual exploration. Paper Presented at the Proceedings of the 2012 ACM International Conference on Interactive Tabletops and Surfaces. Cambridge, Massachusetts (2012)

McNeese, M.: How video informs cognitive systems engineering: making experience count. Cogn. Technol. Work **6**(3), 186–196 (2004). https://doi.org/10.1007/s10111-004-0160-4

Middleton, D.A.: Seeking cognitive convergence: small group collaborative visualization in the library learning commons. In: Ahram, T., Taiar, R., Colson, S., Choplin, A. (eds.) IHIET 2019. AISC, vol. 1018, pp. 143–149. Springer, Cham (2020). https://doi.org/10.1007/978-3-030-25629-6_23

Mohammed, S., Ferzandi, L., Hamilton, K.: Metaphor no more: a 15-year review of the team mental model construct. J. Manag. **36**(4), 876–910 (2010). https://doi.org/10.1177/0149206309356804

Mohammed, S., Klimoski, R., Rentsch, J.R.: The measurement of team mental models: we have no shared schema. Organ. Res. Methods **3**(2), 123–165 (2000). https://doi.org/10.1177/109442810032001

Muraina, I.D., Ibrahim, H.: Student's perception to learning of innovative skills through multidimensional visualization system: reliability and validity tests of some measurements. In: Knowledge Management International Conference (KMICe), 29–30 August, 2016, Chiang Mai, Thailand, pp. 162–167 (2016)

National Science Foundation's Visualization. Scientific Computing Workshop Report (1987)

Nor'ain, M.Y., Siti, S.S.: Social-based versus shared situation awareness-based approaches to the understanding of team cognitive research in HCI. In: The Proceedings of the 3rd International Conference on User Science and Engineering 2014 (i-USEr 2014), 2nd–5th September, 2014, Shah Alam, Malaysia, pp. 281–286 (2014)

Nor'ain, M.Y., Siti, S.S.: A systematic review of shared visualization to achieve common ground. J. Vis. Lang. Comput. **28**, 83–99 (2015). https://doi.org/10.1016/j.jvlc.2014.12.003. Open Access. Elsevier Science

Oviatt, S., Cohen, P.: Perceptual user interfaces: multimodal interfaces that process what comes naturally. Commun. ACM **43**(3), 45–53 (2000). https://doi.org/10.1145/330534.330538

Craig, P., Huang, X.: The mobile tree browser: a space filling information visualization for browsing labelled hierarchies on mobile devices. In: 2015 IEEE International Conference on Computer and Information Technology; Ubiquitous Computing and Communications; Dependable, Autonomic and Secure Computing; Pervasive Intelligence and Computing, Liverpool, pp. 2240–2247 (2015). https://doi.org/10.1109/cit/iucc/dasc/picom.2015.33

Craig, P., Huang, X., Chen, H., Wang, X., Zhang, S.: Pervasive information visualization: toward an information visualization design methodology for multi-device co-located synchronous collaboration. In: 2015 IEEE International Conference on Computer and Information Technology; Ubiquitous Computing and Communications; Dependable, Autonomic and Secure Computing; Pervasive Intelligence and Computing, Liverpool, pp. 2232–2239 (2015). https://doi.org/10.1109/cit/iucc/dasc/picom.2015.330g

Craig, P.: Interactive animated mobile information visualisation. In: SIGGRAPH Asia 2015 Mobile Graphics and Interactive Applications (SA 2015), 6p. ACM, New York (2015). https://doi.org/10.1145/2818427.2818458. Article no. 24

Payne, S.J.: Mental models in human-computer interaction. In: Andrew, S., Jacko, J.A. (eds.) The Human-Computer Interaction Handbook. Fundamentals, Evolving Technologies and Emerging Applications, Second edn., pp. 63–75. Lawrence Erlbaum Associates, Inc., Mahwah (2003)

Phillips, J., McDermott, P.L., Thordsen, M., McCloskey, M., Klein, G.: Cognitive requirements for small unit leaders in military operations in urban terrain. Research Report 1728: U.S. Army Research Institute for Behavioural and Social Science (1998)

Pliske, R.M., Green, S.L., Crandall, B.W., Zsambok, C.E.: The collaborative development of expertise (CDE): a training program for mentors. society for industrial and organizational psychology, New Orleans, LA (2000)

Redlich, B., Siemon, D., Lattemann, C., Robra-Bissantz, S.: Shared mental models in creative virtual teamwork. In: Proceedings of the 50th Hawaii International Conference on System Sciences (January 2017)

Resick, C.J., Dickson, M.W., Mitchelson, J.K., Allison, L.K., Clark, M.A.: Team composition, cognition, and effectiveness: examining mental model similarity and accuracy. Group Dyn.: Theory Res. Pract. 14(2), 174–191 (2010). https://doi.org/10.1037/a0018444

Schraagen, J.M.C., Militello, L., Ormerod, T.C., Lipshitz, R.: Macrocognition and naturalistic decision making (2008)

Schraagen, J.M., Chipman, S.F., Shute, V.J.: State-of-the-art review of cognitive task analysis techniques. In: Schraagen, J.M., Chipman, S.F., Shute, V.J. (eds.) cognitive Task Analysis, pp. 467–489. Erlbaum Associates, Mahwah (2000)

Seamster, D., Redding, R.E., Kaempf, D.: Cognitive Task Analysis, p. 386. Wright State University, Dayton (1997)

Staggers, N., Norcio, A.F.: Mental models: concepts for human-computer interaction research. Int. J. Man-Mach. Stud. 38(4), 587–605 (1993)

Stanton, N., Salmon, P., Walker, G.: Human Factors Methods: A Practical Guide for Engineering and Design. Ashgate Publishing Group, Abingdon (2005)

Suthers, D.D.: Collaborative knowledge construction through shared representations. Paper Presented at HICSS 2005, Proceedings of the 38th Annual Hawaii International Conference on the System Sciences (03–06 January 2005)

Thompson, L., Fine, G.A.: Socially shared cognition, affect, and behavior: a review and integration. Pers. Soc. Psychol. Rev. 3(4), 278–302 (1999)

Undre, S., Sevdalis, N., Healey, A.N., Darzi, S.A., Vincent, C.A.: Teamwork in the operating theatre: cohesion or confusion? J. Eval. Clin. Pract. 12(2), 182–189 (2006)

Valente, C.P., Pivatto, M.P., Formoso, C.T.: Visual management: preliminary results of a systematic literature review on core concepts and principles. In: Proceedings of 24th Annual Conference of the International Group for Lean Construction, Boston, MA, USA, sect. 1, pp. 123–132 (2016)

Valente, C.P., Brandalise, F.M., Formoso, C.T.: Model for devising visual management systems on construction sites. J. Constr. Eng. Manag. 145(2), 04018138 (2018)

Visser, W.: The Cognitive Artifacts of Designing. CRC Press, Boca Raton (2006)

Wei, J., Salvendy, G.: The cognitive task analysis methods for job and task design: review and reappraisal. Behav. Inf. Technol. 23(4), 273–299 (2004)

Xiao, Y., Schenkel, S., Faraj, S., Mackenzie, C.F., Moss, J.: What whiteboards in a trauma center operating suite can teach us about emergency department communication. Ann. Emerg. Med. 50(4), 387–395 (2007). https://doi.org/10.1016/j.annemergmed.2007.03.027

Zou, T., Lee, W.B.: A study of the similarity in mental models and team performance. In: Proceedings of the International Conference on Intellectual Capital, Knowledge Management & Organizational Learning, pp. 536–544 (2010)

Whether Information Source Should Be Provided in the Response of Voice Interaction System?

Yaping Zhang[1], Ronggang Zhou[1,2(✉)], Yanyan Sun[3], Liming Zou[3],
Huiwen Wang[1,2], and Min Zhao[3]

[1] School of Economics and Management, Beihang University,
Beijing 100191, China
zhrg@buaa.edu.cn
[2] Beijing Advanced Innovation Center for Big Data and Brain Computing,
Beihang University, Beijing 100191, China
[3] Baidu, Beijing 100085, China

Abstract. The information source of voice interaction system response may influence users' affective experience. This research is to get a better understanding of whether information source should be provided in the response of voice interaction system and what types of information source should be used to offer a better affective experience to users. In this study, we explored the effect of three different information source types (no information source, information source from professional organizations, information source from internet users) of voice interaction system's responses on users' affective experience from five different application scenarios (music query, news query, health query, travel query, and restaurant query). Three questions, including affection, acceptance and satisfaction, were used to measure users' affective experience. All quantitative data were collected based on the E-Prime experimental platform and 21 participants took part in this study. The results showed that, as a whole, participants preferred the responses with information sources from professional organizations, while there was no significant difference between information source from internet users and no information source. In different application scenarios, the types of information sources preferred by users were different. In music query, news query, and health query scenarios, it was recommended to consider information source from professional organizations; In travel query and restaurant query scenarios, it was suggested to consider information source from internet users.

Keywords: Response of voice interaction system · Information source · Users' affective experience

1 Introduction

Users' affective experience is their attitudinal experience and cognitive process towards objective things. Users' affection and cognition have effects on their behaviors [1, 2]. Norman put forward the concept of emotional design, pointing out that individual

© Springer Nature Switzerland AG 2020
D. Harris and W.-C. Li (Eds.): HCII 2020, LNAI 12186, pp. 257–269, 2020.
https://doi.org/10.1007/978-3-030-49044-7_21

emotional state depends on whether objective things meet their needs [3]. Emotional experience provided by products can build more connections between users according to Norman. The satisfaction level can lead to emotion change of users, and also the final and most intuitive manifestation of users' experience with the product. User's affective experience is also affected by various factors, such as the usability of the product [4, 5]. Improving the affective experience of users can also improve user's acceptance and satisfaction of the products [6]. If users' affective experience is not good, they will prefer to put up with the temporary troubles to adapt to other alternative products rather than give up a better experience [4].

Voice interaction has gradually become a common form of human-computer interaction. With the continuous realization of voice interaction system functions, users pay more attention to its affective experience [7]. People always want to use their way of speaking to communicate with the voice system, expecting it can easily understand their intention and give a satisfactory response. Thus, in the process of interaction with the voice system, users tend to concern about whether the response of the voice system can help them to solve the problem directly. However, users have made a lot of negative comments on the responses of current voice interaction products, such as "*Its response was not what I wanted, I doubted the reliability of its reply*", "*I just listen to it and don't take it too seriously. If the question is important, I will look for more professional responses from other sources*" or "*I use it more as an entertainment tool than an assistant*", etc.

Obviously, current voice interaction system cannot fully meet the needs of users, let alone the affective experience. In many cases, such as health inquiry, users would rather do online search on their own than take the answer provided by voice systems, even though the voice systems and the search engines share the same web resources and provide the same answers. Why? One possible reason is that voice system's responses are not expressed in a trustworthy way to users. During the process of interpersonal communication, different ways of word expression will generate different affective experience for listeners [8, 9]. How to improve users' affective experience based on current voice interaction system? A better approach is to change the expression of the voice system's response to make the idiotic system interact more naturally. It brings a new challenge to the existing natural language processing technology. But what kind of expression means can improve users' affective experience are unrevealed. Based on the experience of presentation of web information retrieval results, an effective way is to add an information source flag to the search result page to let users know the information source of each result [10–12], then users can filter the system's recommendation based on their personal information source preferences and make a final choice. This method effectively improves the user experience of web search [13]. But at present, the response of voice interaction system does not provide relevant information source.

Based on above, this study explored the influence of three different voice system's response means with different information source types (i.e. no information source, information source from professional organizations, information source from internet users) on user's affective experiences in five commonly-used scenarios (i.e. music query, news query, health query, travel query, and restaurant query) respectively. Through the experimental method, quantitative data were collected on the E-Prime platform. And based on the previous researches [14–16], a 3-item questionnaire, including affection, acceptance and satisfaction, was designed to measure users' affective experience with different voice system response. This research will contribute to the emotional experience design of the response of voice interaction system and share some Chinese users' experience of voice interaction system.

2   Method

2.1   Questionnaire Design

Many scholars have tried to measure users' affective experience through different methods, such as extracting users' affective variables from online reviews [17], collecting users' physiological responses to observe users' affective experiences [18], and acquiring users' self-reported emotional data through the measurement questionnaires [19–21]. Such as PAD scale, the commonly-used emotional measurement scale, uses three dimensions to distinguish and explain the specific human emotions [21], while the AttrakDiff scale was usually used to measure users' specific affective responses, not affective experiences [19, 20]. At present, there is no general scale for the measurement of users' affective experience with voice system response. Therefore, this study used direct questions to ask users how they feel about the voice response.

In this study's questionnaire, three factors, namely affection, acceptance, and satisfaction, were included. Among them, the dimension of affection was used to evaluate users' liking of the voice system responses [15, 16], the dimensions of acceptance and satisfaction were used to measure user's overall satisfaction with the content presentation of a specific voice system response. The three-item questionnaire was asked as "In this scenario, I think the voice system's response was pleasing. In this scenario, I think the voice system's response was acceptable. In this scenario, I think the voice system's response was satisfactory". Each item was rated on a seven-point scale ranging from 1 (strongly disagree) to 7 (totally agree).

2.2   Design and Synthesis of Voice System's Responses

According to the purpose of this study, we wrote a preliminary experiment script of the responses of voice interaction system. Through a comprehensive investigation on the brand, sales volume and usability of smart speakers in the Chinese domestic market, Baidu's products were selected as the basis of experimental script design. After three rounds of review and modification, the final experimental script was completed.

In music query scenario, when a user inputs a query, "*Xiaodu Xiaodu, play the best pop vocal album of the year for me*", the voice system answers, "*Best pop vocal album for you*" (no information source), or "*Best pop vocal album from the Grammys for you*" (information from professional organizations), or "*Best pop vocal albums from internet users' recommendation for you*" (information from internet users).

In news query scenario, when a user inputs a query, "*Xiaodu Xiaodu, play the latest social news*", the voice system answers, "*Beijing will implement a comprehensive garbage classification policy next year*" (no information source), or "*According to state media, Beijing will implement a comprehensive garbage classification policy next year*" (information from professional organizations), or "*According to internet users revealed, Beijing will implement a comprehensive garbage classification policy next year*" (information from internet users).

In health query scenario, when a user inputs a query, "*Xiaodu Xiaodu, how to remove dampness*", the voice system answers, "*Dampness is a theoretical concept of traditional Chinese medicine, and it can be taken with proprietary Chinese medicines, such as xiangsha liujunzi pill*" (no information source), or "*According to professional medical institutions, dampness is a theoretical concept of traditional Chinese medicine, and it can be taken with proprietary Chinese medicines, such as xiangsha liujunzi pill*" (information from professional organizations), or "*According to the recommendation by internet users, dampness is a theoretical concept of traditional Chinese medicine, and it can be taken with proprietary Chinese medicines, such as xiangsha liujunzi pill*" (information from internet users).

In travel query scenario, when a user inputs a query, "*Xiaodu Xiaodu, what famous tourist attractions are there in Yunnan*", the voice system answers, "*Famous tourist attractions in Yunnan include: shangri-la, lijiang old town, and stone forest scenic spot*" (no information source), or "*According to the recommendation by professional scenic spot rating agencies, famous tourist attractions in Yunnan include: shangri-la, lijiang old town, and stone forest scenic spot*" (information from professional organizations), or "*According to the recommendation by internet users, famous tourist attractions in Yunnan include: shangri-la, lijiang old town, and stone forest scenic spot*" (information from internet users).

In restaurant query scenario, when a user inputs a query, "*Xiaodu Xiaodu, what good restaurants are there in Beijing*", the voice system answers, "*Delicious restaurants in Beijing include: Jubaoyuan, Dadong roast duck restaurant, and Donglaishun restaurant*" (no information source), or "*According to recommendation by professional food review agencies, delicious restaurants in Beijing include: Jubaoyuan, Dadong roast duck restaurant, and Donglaishun restaurant*" (information from professional organizations), or "*According to recommendation by internet users, delicious restaurants in Beijing include: Jubaoyuan, Dadong roast duck restaurant, and Donglaishun restaurant*" (information from internet users).

Based on the experimental script, audio experimental materials were synthesized in the Text to Speech column of Baidu AI open platform. Standard male voice was used to compose the audio of users' questions, standard female voice was used to compose the audio of voice interaction system response, and then the user's question audio and corresponding response audio of the voice interactive system were spliced into a complete conversational interaction audio by an audio editing software. A complete conversation audio was used as an experimental material. 15 experimental materials were synthesized in this study.

2.3 Building of Experimental Procedures

E-Prime software is an experimental platform for researchers in psychology design and analyze their own experiments. E-Prime promises to become the standard for building experiments in psychology, it is possible to construct a Web-based resource that uses

E-Prime as the delivery engine for a wide variety of instructional materials [22]. Based on the E-Prime platform, we built a 7-point scoring experimental procedure for data collection. In this procedure, participants were asked to listen to a conversation (experimental audio material) first and then grade each voice system response based on the affective experience questionnaire. Before the start of the experiment procedure, participants would be reminded to ignore their preference of physical parameters such as tone and timbre of the voice in the experiment, because all the experimental audio material's physical parameters were same. And the participants should pay attention to the content of the voice interaction system's response.

The experimental procedure consisted of two parts. The first part was the exercises, and the second was the formal experiment. The exercises part could be repeated. Every participant needed to do the exercises first and started the formal experiment after they were familiar with the experiment.

2.4 Design of Personal Information Questionnaire

The Personal Information Questionnaire was designed to collect the participants' personal information, including their gender, age, educational background, employment status, marital status and the usage of intelligent voice interactive devices.

2.5 Data Collection

This study was approved by the ethics review committee of Beihang University in China. Each participant was required to sign a consent form before the experiment, and was informed of the experiment purpose and the data categories to be collected. Personal-generated data would be used for experimental research, and participants had the right to revoke and the obligation to keep secret. The researchers promised to keep each participant's private information strictly confidential, and cash reward (RMB 60 yuan) was paid to each participant at the end of the experiment. A total of 21 participants were recruited. All the participants had experiences in using smart speakers or smart voice assistants on mobile phones, and they were interested in voice interaction systems.

In experimental procedure, the five application scenarios were given in a random order, and the three experimental audio materials with different information source types in every application scenario all given in a random order as well. The test data were collected on the E-Prime experimental platform.

3 Results

3.1 Demographic Statistics

All of 21 participants' experimental data were valid in this study. Of these participants, 10 people were male and 11 people were female. With respect to age group, 38.1% were aged 18–25 years, 57.1% were aged 26–30 years, and 4.8% were aged 31–40 years. In terms of educational level, 14.2% of the participants had bachelor's degrees or below, 42.9% had master's degree or were postgraduates, and 42.9% had doctor's degree or were PhD students. With respect to employment status, 47.6% of the participants were students, and 52.4% were employed.

3.2 Reliability and Validity Test of the Affective Experience Questionnaire

The reliability and validity of affective experience questionnaire were tested with the data collected from the scoring procedure. 315 sample points were collected and SPSS22.0 software was used for reliability analysis and exploratory factor analysis of the data. The results showed that the Kaiser-Meyer-Olkin measure of sampling adequacy and the Bartlett's test of sphericity all indicated that this scale was appropriate for factor analysis (KMO = 0.743, Chi-square = 866.33, df = 3, p = 0.000). Table 1 presents the validity test results of the scale. Only one principal component was extracted which explained the 88.97% of the variance. And the factor loads of the three variables were all greater than 0.80, indicating that the validity of the scale was good. The reliability analysis result showed that the questionnaire was reliable (Cronbach's α = 0.94). Therefore, the questionnaire met the requirements of reliability and validity.

Table 1. Reliability and validity test of the affective experience questionnaire.

Items	Variables	Factor load	Variance interpretation rate	Cronbach's α
1. In this scenario, I think the voice system's response was pleasing	Affection	0.92	88.97%	0.94
2. In this scenario, I think the voice system's response was acceptable	Acceptance	0.96		
3. In this scenario, I think the voice system's response was satisfactory	Satisfaction	0.95		

3.3 Data Analysis of 7-Point Scoring Procedure

We used one-way analysis of variance (ANOVA) to test the differences in the average scores for participants' affective experience with the three information sources (i.e. no information source, information sources from professional organizations, information sources from internet users) provided in voice system's response. Response type was used as factor, and the three variables of the affective experience questionnaire were used as dependent variables.

The data analysis results of all five experimental scenarios are shown in Fig. 1. In terms of the affection, the results showed that there was a significant difference between the average score of the voice system's responses in which information sources from professional organizations were provided and the average score of the responses in which information sources from internet users were provided. Participants preferred the responses with information sources from professional organizations. Regarding the acceptance and satisfaction, there were significant differences between the average scores of the responses with no information source and the responses with information sources from professional organizations. The differences of the average scores between responses with information sources from professional organizations and responses with information source from internet users were significant, too. The average scores of acceptance and satisfaction were higher when responses with information sources from professional organizations were provided. Therefore, on the whole, participants preferred responses with information sources from professional organizations the most, followed by responses with no information sources and information sources from internet users. There was no significant preference difference between the average scores of the responses with no information source and the responses with information sources from internet users.

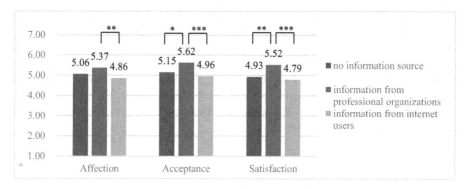

Fig. 1. Average scores of the three variables in all five scenarios (N = 21). ***p < 0.001, **p < 0.01, *p < 0.05.

Then, response type was used as factor, the three variables of affective experience questionnaire were used as dependent variables. The ANOVA analysis was conducted in the five application scenarios, including music query, news query, health query, travel query, and restaurant query, respectively to test the differences among the average scores of the three response types.

In the music query scenario (see Fig. 2), there were no significant differences among the average scores of the three different responses types. In terms of scoring trends, participants preferred responses with information sources from professional organizations the most, followed by responses with no information sources and information sources from internet users.

In the news query scenario (see Fig. 3), in terms of affection variable, there was a significant difference between the average score of the responses with information sources from professional organizations and the average score of the responses with information sources from internet users. Participants preferred the responses with information sources from professional organizations the most. Regarding the acceptance

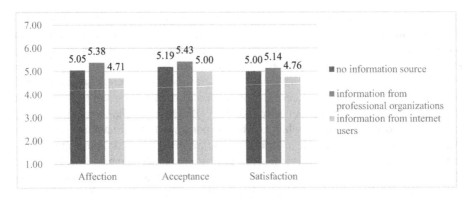

Fig. 2. Average scores of the three variables in music query scenario (N = 21).

and satisfaction, the significance of multiple comparisons between scores of responses with no information source and scores of responses with information sources from professional organizations met the requirement of $p < 0.01$. The significance of multiple comparisons between scores of responses with information sources from professional organizations and scores of responses with information sources from internet users met the requirement of $p < 0.001$. Participants preferred the responses with information sources from professional organizations the most. According to the scoring trends, participants preferred the responses with no information sources than the responses with information sources from internet users.

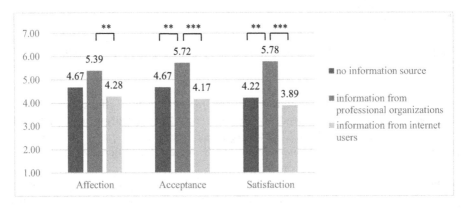

Fig. 3. Average scores of the three variables in news query scenario (N = 21). ***$p < 0.001$, **$p < 0.01$.

In the health query scenario (see Fig. 4), for about affection, acceptance and satisfaction, the multiple comparison differences between the responses with information sources from professional organizations and information sources from internet users were significant, and participants preferred the responses with information sources from professional organizations. In terms of acceptance and satisfaction variables, the differences between the average scores of the responses with no information source and information sources from internet users were significant, and participants preferred the responses with no information sources. It indicated that the participants enjoyed the responses with information sources from internet users the least. There were no significant differences between the average scores of the responses with no information source and information sources from professional organizations.

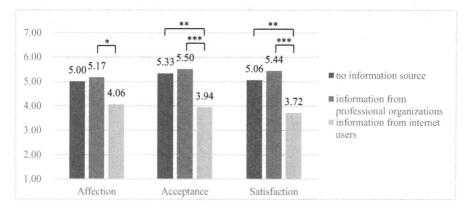

Fig. 4. Average scores of the three variables in health query scenario (N = 21). ***p < 0.001, **p < 0.01, *p < 0.05.

In the travel query scenario (see Fig. 5), there were no significant differences among the average scores of the three different responses types. In terms of scoring trends, participants preferred responses with information sources from internet users the most, followed by responses with information sources from professional organizations and responses with no information sources.

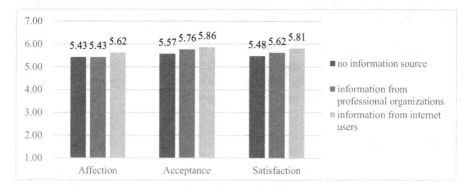

Fig. 5. Average scores of the three variables in travel query scenario (N = 21).

In the restaurant query scenario (see Fig. 6), for about affection variable, there were no significant differences among the average scores of the three different responses types. In terms of acceptance variable, the difference between the average score of the response with no information source and the response with information sources from internet users was significant, and participants preferred the responses with information sources from internet users. Regarding the satisfaction variable, there was a significant difference between the average score of the responses with information sources from professional organizations and the average score of the responses with no information sources. The difference between the average score of the responses with information sources from internet users and the average score of the responses with no information sources was significant, too. And the participants preferred the responses with no information sources the least.

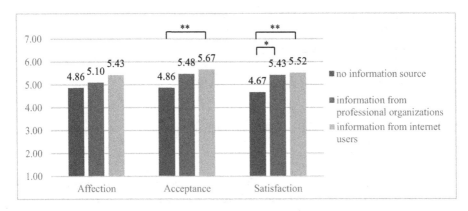

Fig. 6. Average scores of the three variables in restaurant query scenario (N = 21). **p < 0.01, *p < 0.05.

In conclusion, in music query, news query and health query scenarios, participants preferred the responses with information sources from professional organizations most, followed by the responses with no information source and then the responses with information sources from internet users. In travel query and restaurant query scenarios, participants preferred the responses with information sources from internet users the most, followed by the responses with information sources from professional organizations and then the responses with no information source.

4   Discussion

This study examined the scores of participants' affective experience with the voice system responses, in which different types of information sources were provided in five different application scenarios. The three variables of the affective experience questionnaire were used to explored the participants' affective experience with different response types.

In human-computer interaction, the user's perception of credibility can enhance their experience with the product [23–25]. And it's the same for voice systems. It can be seen from the experimental results that users have different preferences for the responses with different information source types in the five experimental scenarios. In relaxing and entertaining scenes, such as tourism query and restaurant query, users prefer the responses with information sources from internet users. In scenarios with specialized requirements, such as news query and health query, users prefer the responses with information sources from professional organizations. The music query scenario was originally a scene of entertainment, but the user's needs in this study were designed professionally (*"Xiaodu Xiaodu, play the best pop vocal album of the year for me"*), so in this case, the participants preferred the responses with information sources from professional organizations. Further, if the voice system cannot provide a satisfactory information source to the response, it is better to provide a response with no information source to users, otherwise the effort will only be backfire.

These achievements can be supported by the results of the qualitative interviews in the early stage. During our qualitative interview, we found that our participants had a need to know the information sources of the voice system responses, but not all the responses need to provide information sources accordingly in different scenarios. For examples, P6 commented as *"I hope the answer of professional question can be supported by more professional and accurate evidences. Providing the information sources is a form of support"*. But how to provide information sources for voice system responses? P3 told us *"If it is a restaurant query scenario, I prefer the responses with information sources from internet users, rather than the responses with information sources from professional organizations. If it's a travel query scenario, I prefer the recommendation from net friends. If it is a medical query, I am more inclined to the responses with information sources from professional organizations"*. P10 addressed *"I have some requirements of professional for the responses to medical questions, and the voice system response can be given according to a certain doctor, which is more convincing than internet users' recommendation"*.

5 Conclusions and Prospects

In this paper, we found that, in application scenarios with professional requirements, users prefer the voice system responses with the information sources from professional organizations. Nonetheless, in the relaxing and entertaining application scenarios, users prefer the voice system responses with the information sources from internet users. In different application scenarios, different types of voice system responses, which the different information sources were provided, have different effects on users' affective experiences. Our work brings to light the role of the information source in the effective experiences with the voice interaction system responses in five different experimental scenarios. This work provides implications for the emotional design of future voice system responses to improve users' experience of voice interaction system.

This study explored how information sources play its role in users' affective experience of voice system responses only in five different experimental scenarios respectively. Further research can also verify the results in the real application scenarios and explore the voice system responses with other information source types on users' experiences in other application scenarios.

References

1. Ji, Z., Li, B., Zhu, J., Chen, W.: Mechanism of social media users' fatigue behavior from the dual-perspective of emotional experience and perceived control. Inf. Stud.: Theory Appl. **42**(4), 129–135 (2019)
2. Kostov, V., Fukuda, S.: Emotion in user interface, voice interaction system. In: 2000 IEEE International Conference on Systems, Man and Cybernetics, pp. 798–803 (2000)
3. Norman, D.A.: Emotional Design: Why We Love (or Hate) Everyday Things. Basic Books, New York (2004)
4. Desmet, P.M.A., Hekkert, P.: Framework of product experience. Int. J. Design **1**(1), 13–23 (2007)
5. Li, X., Xiao, Z., Cao, B.: Effects of usability problems on user emotions in human–computer interaction. In: Long, S., Dhillon, Balbir S. (eds.) MMESE 2017. LNEE, vol. 456, pp. 543–552. Springer, Singapore (2018). https://doi.org/10.1007/978-981-10-6232-2_63
6. Prastawa, H., Ciptomulyono, U., Laksono-Singgih, M., Hartono, M.: The effect of cognitive and affective aspects on usability. Theor. Issues Ergon. Sci. **20**(4), 507–531 (2019)
7. Yang, X., Aurisicchio, M., Baxter, W.: Understanding affective experiences with conversational agents. In: CHI Conference on Human Factors in Computing Systems, pp. 1–12 (2019)
8. Newberg, A., Waldman, M.R.: Words Can Change Your Brain: 12 Conversation Strategies to Build Trust, Resolve Conflict, and Increase Intima. Hudson Street Press, The Penguin Group, New York (2012)
9. Scherer, K.R., Fontaine, J.R.J.: The semantic structure of emotion words across languages is consistent with componential appraisal models of emotion. Cogn. Emot. **33**(4), 673–682 (2019)
10. Aladhadh, S., Zhang, X., Sanderson, M.: Location impact on source and linguistic features for information credibility of social media. Online Inf. Rev. **43**(1), 89–112 (2019)
11. Ushigome, R., et al.: Establishing trusted and timely information source using social media services. In: 16th IEEE Annual Consumer Communications and Networking Conference (2019)
12. Westerman, D., Spence, P.R., Van Der Heide, B.: Social media as information source: recency of updates and credibility of information. J. Comput.-Mediat. Commun. **19**(2), 171–183 (2014)
13. Hussain, S., Ahmed, W., Jafar, R.M.S., Rabnawaz, A., Jianzhou, Y.: eWOM source credibility, perceived risk and food product customer's information adoption. Comput. Hum. Behav. **66**, 96–102 (2017)
14. Adiga, N., Prasanna, S.R.M.: Acoustic features modelling for statistical parametric speech synthesis: a review. IETE Tech. Rev. **36**(2), 130–149 (2019)
15. Lee, E.J., Nass, C., Brave, S.: Can computer-generated speech have gender? An experimental test of gender stereotype. In: CHI 2000 Extended Abstracts on Human Factors in Computing Systems, pp. 289–290 (2000)

16. Richard, L.: Street: evaluation of noncontent speech accommodation. Lang. Commun. **2**(1), 13–31 (1982)
17. Kim, W., Ko, T., Rhiu, I., Yun, M.H.: Mining affective experience for a kansei design study on a recliner. Appl. Ergon. **74**, 145–153 (2019)
18. Chen, S., Epps, J.: Automatic classification of eye activity for cognitive load measurement with emotion interference. Comput. Methods Programs Biomed. **110**(2), 111–124 (2013)
19. Baumgartner, J., Sonderegger, A., Sauer, J.: No need to read: developing a pictorial single-item scale for measuring perceived usability. Int. J. Hum. Comput. Stud. **122**, 78–89 (2019)
20. Hart, J., Sutcliffe, A.: Is it all about the apps or the device? User experience and technology acceptance among iPad users. Int. J. Hum. Comput. Stud. **130**, 93–112 (2019)
21. Mao, Y., Fan, Z., Zhao, J., Zhang, Q., He, W.: An emotional contagion based simulation for emergency evacuation peer behavior decision. Simul. Model. Pract. Theory **96**, 101936 (2019)
22. MacWhinney, B., St. James, J., Schunn, C., Li, P.: Schneider W: STEP—a system for teaching experimental psychology using E-Prime. Behav. Res. Methods Instr. Comput. **33**, 287 (2001)
23. Kang, J.-W., Namkung, Y.: The information quality and source credibility matter in customers' evaluation toward food O2O commerce. Int. J. Hospit. Manage. **78**, 189–198 (2019)
24. Nayak, S., et al.: Integrating user behavior with engineering design of point-of-care diagnostic devices: theoretical framework and empirical findings. Lab Chip **19**(13), 2241–2255 (2019)
25. Shin, D.-H., Lee, S., Hwang, Y.: How do credibility and utility play in the user experience of health informatics services? Comput. Hum. Behav. **67**, 292–302 (2017)

The Effects of Face Inversion and the Number of Feature Differences on Eye-Movement Patterns

Min-Fang Zhao[1]([⊠]) and Hubert D. Zimmer[2]

[1] Huizhou University, Huizhou 516007, China
minfang_zhao@foxmail.com
[2] Saarland University, 66123 Saarbruecken, Germany

Abstract. Recognition of the inverted faces (compared to upright faces) becomes much more difficult, this phenomenon is known as "facial inversion effect". The role of face inversion on the eye-movement patterns is still under debate. In the current research, we aimed to investigate the effects of face inversion and the number of feature differences on eye-movement patterns during judgement of similarity between two faces. The participants were asked to judge the pairs of faces similarity by a 4-point Likert scale. Eye-movements were recorded during similarity ratings. The faces were presented either upright or inverted. The pairs of the faces were either the same or different. The different pairs of faces were manipulated via the number of feature differences (i.e., one feature or two features). As expected, we found that the inversion of the faces resulted in a higher level of estimated similarity in overall. However, when only the hairs were different between two faces, the inversion of the faces resulted in a lower level of estimated similarity. Surprisingly, the inversion of the faces resulted in a significant decrease of transitions between the pairs of inverted faces. The less eye gaze transitions during similarity rating of inverted pairs suggested that the feature-based processing strategy of inverted faces was less likely to be used in our rating task that presumably exerts no memory load. In conclusion, our findings suggested that both face inversion and the number of feature differences influence the eye-movement patterns during the similarity rating of two simultaneously presented faces.

Keywords: Eye-movements · Face perception · Face inversion effect · Similarity rating · Eye gaze transitions

1 Introduction

The ability to read human face is critical for our living adaption and smooth social communication. It has been shown that the faces "all look the same" when they are presented upside-down. This perceptual asymmetry is generally known as "facial inversion effect"—recognition of the inverted faces (compared to upright faces) becomes much more difficult, that is, poorer accuracy and longer reaction times when faces are upside down [1–3]. This phenomenon has been investigated in a very large number of studies over the past decades. However, it is still unclear what causes the

D. Harris and W.-C. Li (Eds.): HCII 2020, LNAI 12186, pp. 270–280, 2020.
https://doi.org/10.1007/978-3-030-49044-7_22

facial inversion effect. Most of the researchers agree that the holistic-based strategy of processing faces is disrupted when faces are presented upside-down [1, 2, 4].

Numerous researchers attempted to investigate the mechanisms underlying facial inversion effects via Eye-tracking techniques. The eye movements pattern of people looking at a normal (upright) face is very specific. Internal features such (i.e., Eyes, nose and mouth) are usually the focuses when people looking at a face [5–8]. However, the diversity of the eye-movement results does not yield a sound conclusion about whether face inversion influences the gaze patterns. Some studies demonstrated that different inspection strategies were used to process upright and inverted faces [9–11]. In a study employing a categorization task [9], results demonstrated that upright faces receive more fixations than inverted ones when the participants were required to judge whether a face is normal (upright) or not (inverted or scrambled). Similarly, Van et al. found that face inversion leads to fewer fixations on the inverted faces and eye area, which might play an important role on face-specific effect [12]. In contrast, some studies found that participants fixate more gazes on the inverted faces than upright faces [10, 11]. In addition, some studies showed that same inspection strategy is used to process upright and inverted faces. For example, in the study of Williams and Henderson [8], they asked the participants to view and memorize a set of upright and inverted faces one by one, and eye-tracking data was recorded. They found that face inversion did not affect the distributions of fixations; internal features received more fixations when inspecting both upright and inverted faces. Therefore, the authors concluded that the relative difficulty in recognizing inverted faces may not due to specific inspection strategies but probably more complex perceptual mechanisms [8].

It should be noted that the tasks in most of these above-cited studies either explicitly required the participants to memorize or recognize the faces which exerts memory load. Under these conditions, it is possible to process a face by encoding specific pictorial properties of a face image that could be easily remembered, rather than by specific face perception skills [13–15]. Thus, not only face inversion, but also the demands of the task or task difference may lead to different eye-movement strategies [16]. Unlike the tasks that employed in these above-cited studies, face matching task exerts less memory load.

Face matching is a theoretical and practical important topic on face processing. In our daily life, we might need to decide whether two simultaneous presentations of a face photo belong to the same person or different people. Securities at national border have to check whether the face photo in passport is matched to the face of its purported owner. Many factors might result in failure of face matching, especially for unfamiliar faces, for example, changes in lighting, viewpoint, facial expression, hairstyle, and so on. When time pressure is high, perceptually similarity causes high categorization demands, face recognition become much more difficult [17]. In a study on unfamiliar upright face matching task [18], participants were asked to make match/mismatch decisions to pairs of faces shown for 200, 500, 1000, or 2000 ms, or for an unlimited duration. The results showed that the participants switched more quickly between faces and fixated fewer facial features if time pressure was high [18]. Such a matching task is suitable for testing differences in the perceptual processing of upright and inverted faces. Popivanov and Mateeff [19] reported that face inversion leads to a higher level of estimated similarity and more eye fixations on the external features (forehead, cheeks and chin) at the cost of less fixations on the internal features (forehead, cheeks and chin) in a matching task of the similarity rating of pairs of faces. However, it remains unresolved whether face inversion and the number of feature differences also have influences on eye gaze switching patterns in a matching task.

In the current research, we aimed to investigate examine whether the face inversion and the increased number of feature differences affect similarity rating and eye gaze transitions between pairs of faces. Our task comparing two simultaneously presented faces potentially exerts no memory load. Thus, our present research could focus on facial processing or face perception. Upright face processing is mainly based on holistic processing, in contrast, if face pairs are inverted, holistic processing breaks down and a feature by feature processing should happen [20], the participants should make much more eye gaze transitions between the pairs of inverted faces than between the pairs of upright ones. More feature differences between the pairs of faces should be rated less similar and force participants to make more eye gaze transitions between the pairs.

2   Methods

2.1   Participants

A total of 20 students (10 females; Mean age = 25.3 years, SD = 2.98 years) from Saarland University volunteered to participate in this experiment. All of them had normal or corrected-to-normal vision and received course credit for participation. The experiment was approved by the local ethics committee of Saarland University. Informed consent was required. All participants were paid at a rate of €8/h.

2.2   Stimuli and Apparatus

The face stimuli were generated by PhotoFit software FACES 4.0 (IQ Biometrix, Redwood Shores, CA) at a size of approximately 350 × 450 pixels, subtended 9° × 12° of visual angle at a viewing distance of 60 cm. This program enables the generation of composites of facial features (i.e., hair, head shape, ears, eyes, eyebrows, nose, mouth, and facial markings) that can be freely replaced. 18 faces were generated as the prototype of faces. We created our stimuli set based on 18 prototype faces, six corresponded changed faces were created based on each prototype face. For each of six changed faces, one of the face features (i.e., eye, nose, mouth, hair, ear or chin) is replaced with a different feature almost in the same size (also same colors of eyes) and in the same position (cf. Fig. 1). The inverted faces were generated by rotating the upright faces upside down. In total, 126 upright and 126 inverted face stimuli were generated. The viewing distance was 60 cm. Pairs of faces were presented on white background. The face pairs were evenly split to depict female and male face stimuli. The pairwise combinations of face stimuli in all conditions of the design were all distinct.

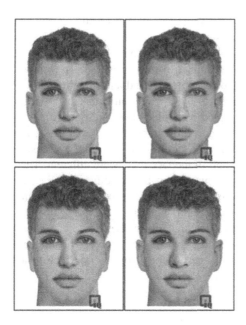

Fig. 1. Examples of face stimuli. The top panel shows an example of one feature (e.g., chin) difference between a pair of faces. The panel below depicts an example of two features (e.g., nose and ears) differences between a pair of faces. The symbol appearing on the bottom of each face picture is a trademark of the Faces TM software (IQ Biometrix, Inc., http://www.iqbiometrix. com/products_faces_40.html).

The eye movements were recorded by a Tobii TX300 eye-tracker (Tobii Technologies, 2011) at a sampling rate of 300 Hz. It is built in a 23″ TFT display with resolution of 1920 × 1080 pixels. The display was used for presentation of the face stimuli. Two images of the face pair were defined as two screen regions of interest (ROIs). The ROIs included the face and the margin around it. The two face pictures of each pair were presented vertically centered and horizontally 10 cm apart. Throughout the experiment, both eyes were calibrated and validated. Calibration was repeated if the criterion was not met. A drift correction was performed before each trial. Trials were excluded from further analysis when the eye tracker was unable to detect one or both eyes exceeded 30% of data points in one trail. For each trial and for each condition, we extracted the total number of transitions made between the two faces of each pair. A transition occurred when successive fixations were not within the same ROIs (the same face); that is when an eye movement is made from one face to another face picture.

2.3   Procedure

Each study trial began with a fixation cross (+) displayed in the center of the screen for 1000 ms. Then, a pair of faces was displayed centrally against a gray background for 6000 ms. After the faces disappeared, the subject was asked to judge their similarity by a 4-point Likert scale (1-dissimilar; 2-rather dissimilar; 3-rather similar; 4-similar). The similarity rating was self-terminated, i.e. the rating question remained on the screen until the response was given. The participants were requested to take a break for 3 min every 72 trials. The total number of trials was 288. They were organized in 4 blocks according to an ABBA order. The inverted and upright conditions were carried out in separate blocks. The order of blocks was randomised across the subjects.

The experiment adopted a within-subject design with the factors of orientation (upright, inverted) and feature difference (same, eyes, nose, mouth, hairs, ears, chin, ear-chin, hair-ear, nose-mouth, mouth-hair, eye-chin, eye-nose). Half of the face pairs were the same, another half of the face pairs were different. The different pairs of faces were manipulated via the number of feature difference (i.e., one feature or two features). In "one feature difference" condition, there is one feature difference (i.e., eyes, nose, mouth, hairs, ears or chin) between two faces. In "two features differences" condition, there are two features differences (i.e., ear-chin, hair-ear, nose-mouth, mouth-hair, eye-chin, eye-nose) between two faces. The number of transitions between two faces was computed, i.e. switches from to one face to another face during the presentation of a face pair.

3 Results

3.1 Behavioral Data

A two-way ANOVA with orientation and feature difference as within-subject factors on rating scores revealed a significant main effect of orientation (F (1, 19) = 24.28, $p < .001$, $\eta_p^2 = .56$), with a higher rating score for inverted face pairs than upright face pairs ($MD = .30$, $SE = .06$, $p < .001$), suggesting that the inverted faces were rated significantly more similar than the upright faces. The main effect of feature difference was also significant (F (4.2, 78.9) = 25.85, $p < .001$, $\eta_p^2 = .58$). Importantly, the interaction between orientation and feature difference was also significant (F (12, 228) = 10.34, $p < .001$, $\eta_p^2 = .35$).

To resolve this interaction, t tests revealed that only in the "same" and "nose" difference condition, similar rating was not influenced by the orientation of face pairs, that is, similar rating score was not significantly different between upright and inverted face pairs (see Fig. 2). Only in the "hairs" difference condition, face pairs in the upright condition were rated more similar than those in the inverted condition ($p < .05$). In contrast, in all other feature difference conditions (i.e., eyes, mouth, ears, chin, ear-chin, hair-ear, nose-mouth, mouth-hair, eye-chin, eye-nose), face pairs in the inverted condition were rated more similar than those in the upright condition ($p < .05$).

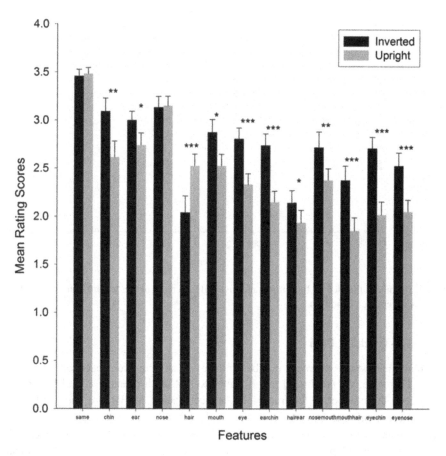

Fig. 2. Similar rating scores for upright and inverted face pairs in different feature difference (i.e., same, eyes, nose, mouth, hairs, ears, chin, ear-chin, hair-ear, nose-mouth, mouth-hair, eye-chin, eye-nose) conditions.

We further examined the influence of the number of feature difference on similar ratings, t tests revealed that in both upright (t (19) = 13.18, p < .001) and inverted (t (19) = 8.83, p < .001) condition, pairs of faces that have "one feature difference" were rated more similar than those have "two features differences" (see Fig. 3).

A two-way ANOVA with orientation and the number of feature difference as within-subject factors on reaction times revealed that the main effect of orientation (F (1, 19) = 1.76, p = .20, η_p^2 = .09) and the number of feature differences (F (2, 38) = 1.37, p = .27, η_p^2 = .07), and the interaction (F (2, 38) = .57, p = .55, η_p^2 = .03) all did not reach significance.

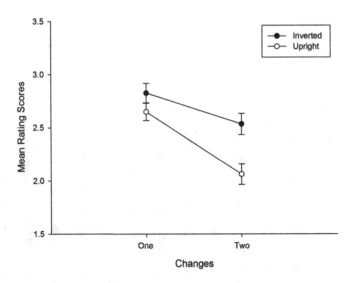

Fig. 3. Similar rating scores for upright and inverted face pairs in one feature difference and two features differences conditions.

3.2   Eye-Tracking Data

Our main interest was the mean number of the eye gaze transitions between the pairs of faces. Therefore, we conducted a two-way ANOVA with orientation and feature difference on the number of the transitions between face pairs. Results showed that the main effect of orientation was significant (F (1, 19) = 7.94, $p < .05$, $\eta_p^2 = .3091$), with more eye gaze transitions between upright face pairs than between inverted face pairs ($MD = .56$, $SE = .20$, $p < .05$). The main effect of feature difference was also significant (F (2, 228) = 2.68, $p < .01$, $\eta_p^2 = .12$). The interaction between orientation and feature difference did not reach significance.

Similar to behavior data, we further checked the influence of the number of feature difference on the number of eye gaze transitions, a two way ANOVA with the number of feature differences (one, two) and orientation (upright, inverted) revealed that the main effect of the number of feature differences was also significant (F (1, 19) = 5.39, $p < .05$, $\eta_p^2 = .22$). Further t tests revealed that the mean transitions for upright pairs of faces was not significantly different from those for inverted faces in "one feature difference" condition (t (19) = 1.51, $p = .15$). In contrast, the number of the transitions for upright pairs of faces was significantly higher than those for inverted faces in "two features differences" condition (t (19) = 3.38, $p < .05$) (see Fig. 4).

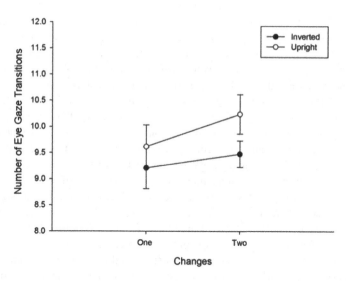

Fig. 4. The number of eye gaze transitions between the pairs of upright faces and inverted faces under one feature and two features differences conditions.

4 Discussion

In the current study, we aimed to investigate the effects of face inversion and the number of facial feature differences on the eye-movement transitions pattern during the similarity rating of two simultaneously presented faces. Consistent with the previous research [19], our results showed that the inverted faces were rated significantly more similar than the upright faces. This robust inversion effect on similarity ratings corresponds to the findings that upright (normal) faces are recognized more easily than inverted faces [1–3]. In this case, the difficulty of processing inverted faces may lead to a higher similarity rating score for the pairs of inverted faces than those of upright faces.

However, when only the hairs were different between two faces, the pairs of upright faces were rated significantly more similar than the pairs of inverted faces. This is in line with the previous finding that a significant increase of fixations on the external features (e.g., hair) at the cost of fixating the internal features when faces are presented upside-down [19]. In other word, a specific strategy may be used during similarity rating of inverted faces. In our study, the participants may make more fixations on the hair (the most salient external feature) when faces are inverted (compared to upright), therefore, they can more easily find out the hair difference between the pairs of inverted faces compared to upright faces. Moreover, research on the differential effects of external or internal features in recognition and matching tasks has shown that performance with external features is generally less affected by inversion [21–23]. Moreover, there are recent findings that face matching while attending internal features is strongly modulated by orientation and by viewpoint, but hardly for attending the external features of the same faces [22]. These results indicate that external features were less likely influenced by face inversion, while in contrast, internal features are more likely affected by face inversion in recognition and matching tasks. Based on this, external features are more easily processed than internal features when faces are upside-down. Combined with the finding of more fixation on external features for inverted faces [19], thus, external features such as hair should be more salient in inverted faces. These previous findings together support our findings that when only hairs are different between two faces, two upright faces look much more similar than two inverted face.

In both upright and inverted face conditions, those pairs of faces that have one feature difference were evaluated significantly more similar than those that have two features differences. The number of eye gaze transitions between the pairs of upright faces was significantly higher than those between the pairs of inverted faces. The number of eye gaze transitions in "two features differences" condition was significantly higher than those in "one features difference" condition for both upright and inverted faces pairs. As expected, the increased number of feature differences between the pairs of faces do make two faces look less similar and force the participants to make more eye gaze transitions between two faces. This suggested that the participants do notice the differences between the pairs of faces even only one more feature was different.

Surprisingly, the inversion of the faces resulted in a significant decrease of eye-movement transitions between the pairs of inverted faces (compared to upright faces) in "two features differences" condition. It was suggested that the difficulty of processing inverted faces forced the subjects to be less willing to search for differences between the pairs of faces. The less eye gaze transitions during similarity rating of inverted pairs suggested that the feature-based processing strategy of inverted faces was less likely to be used in our rating task that presumably exerts no memory load. This finding was contrast with the notion that face inversion effect is caused by a more feature-based strategy of processing inverted faces, indeed, the faces inversion effect could be caused by less eye gaze transitions during processing inverted faces, at least in our current research. Some previous studies in which require the participants either explicit memorize, or recognition of faces, probably, fixations on internal features both for upright inverted and faces, may provide the sufficient information for a memorizing task [6]. However, our study requires less memory involvement, different task demands may probably give rise to different eye-movement patterns of processing faces. Therefore, the task differences should be taken into account when discussing the contradictions in literature about the role of the face inversion on the eye movement patterns.

By investigating the effects of feature changes and inversion on face similarity rating and eye gaze transitions for face pairs, our current study sheds light on how people process the upright and inverted face. As well as being of theoretical significance, our findings also have practical implications. A face is a complex stimulus and rich in social information, which provides a person's identity, emotional state, age, gender and so forth. Understanding face perception is therefore of high significance for the public. For example, in security area and passport control at national borders, matching a photograph to the face of its purported owner is a typical affordance. It has been showed that face matching is an error-prone process [24, 25].

Our findings suggested that both face inversion and the number of feature differences influence the eye-movement patterns during the similarity rating of two simultaneously presented faces. The inversion of the faces resulted in a higher level of estimated similarity in overall. In contrast, when only the hairs were different between two faces, the inversion of the faces resulted in a lower level of estimated similarity. That is, hair difference could be more easily noticed when faces are presented upside-down. It suggests that different eye-movement strategies are used when inspecting upright and inverted faces in our current comparison task. Upright face pairs receive more eye gaze transitions than inverted faces during similarity rating. We suggest that it may be more meaningful to explore when and how the face inversion influences the eye-movements instead of whether it happens. Face inversion affects similarity processing of face pairs during a matching task, which reflects on the difference on the number of eye gaze transitions.

Acknowledgments. This project was supported by National Natural Science Foundation of China (31700955), Guangdong Planning Office of Philosophy and Social Science Project (GD17CXL03) and Guangdong Innovative Young Talents in Universities and Colleges Project (2018WQNCX175).

References

1. Haxby, J.V., Hoffman, E.E.A., Gobbini, M.I.: The distributed human neural system for face perception. Trends Cogn. Sci. **4**(6), 223–233 (2000)
2. Tanaka, J.W., Simonyi, D.: The "parts and wholes" of face recognition: a review of the literature. Q. J. Exp. Psychol. **69**(10), 1876–1889 (2016)
3. Yin, R.K.: Looking at upide-down faces. J. Exp. Psychol. **81**(1), 141–145 (1969)
4. Rossion, B., Gauthier, I.: How does the brain process upright and inverted faces? Behav. Cogn. Neurosci. Rev. **1**(1), 63–75 (2002)
5. Althoff, R.R., Cohen, N.J.: Eye-movement-based memory effect: a reprocessing effect in face perception. J. Exp. Psychol. Learn. Mem. Cogn. **25**(4), 997–1010 (1999)
6. Henderson, J.M., Williams, C.C., Falk, R.J.: Eye movements are functional during face learning. Mem. Cogn. **33**(1), 98–106 (2005)
7. Stacey, P.C., Walker, S., Underwood, J.D.M.: Face processing and familiarity: evidence from eye-movement data. Br. J. Psychol. **96**(Pt 4), 407–422 (2005)
8. Williams, C.C., Henderson, J.M.: The face inversion effect is not a consequence of aberrant eye movements. Mem. Cogn. **35**, 1977–1985 (2007)
9. Lê, S., Raufaste, E., Démonet, J.F.: Processing of normal, inverted, and scrambled faces in a patient with prosopagnosia: Behavioural and eye tracking data. Cogn. Brain. Res. **17**(1), 26–35 (2003)
10. Barton, J.J.S., et al.: Information processing during face recognition: the effects of familiarity, inversion, and morphing on scanning fixations. Perception **35**(8), 1089–1105 (2006)
11. Gallay, M., et al.: Qualitative differences in the exploration of upright and upside-down faces in four-month-old infants: an eye-movement study. Child Dev. **77**(4), 984–996 (2006)
12. Der Geest, J.N.V., et al.: Gaze behavior of children with pervasive developmental disorder toward human faces: a fixation time study. J. Child Psychol. Psychiatry **43**(5), 669–678 (2002)
13. Bruce, V.: Changing faces: visual and non-visual coding processes in face recognition. Br. J. Psychol. **73**(1), 105–116 (1982)

14. Duchaine, B., Nakayama, K.: Developmental prosopagnosia and the Benton Facial Recognition Test. Neurology **62**(7), 1219–1220 (2004)
15. Longmore, C.A., Liu, C.H., Young, A.W.: Learning faces from photographs. J. Exp. Psychol. Hum. Percept. Perform. **34**(1), 77–100 (2008)
16. Armann, R., Bülthoff, I.: Gaze behavior in face comparison: the roles of sex, task, and symmetry. Atten. Percept. Psychophys. **71**(5), 1107–1126 (2009)
17. Zhao, M.-F., et al.: Exploring the cognitive processes causing the age-related categorization deficit in the recognition of facial expressions. Exp. Aging Res. **42**(4), 348–364 (2016)
18. Özbek, M., Bindemann, M.: Exploring the time course of face matching: temporal constraints impair unfamiliar face identification under temporally unconstrained viewing. Vis. Res. **51**(19), 2145–2155 (2011)
19. Popivanov, I.D., Mateeff, S.: Eye-movements during similarity judgement of upright and inverted faces. Comptes Rendus De L Acad. Bulgare Des Sci. **64**(3), 425–430 (2011)
20. Tanaka, J.W., Farah, M.J.: Parts and wholes in face recognition. Q. J. Exp. Psychol. **46**(2), 225–245 (1993)
21. Nachson, I., Shechory, M.: Effect of inversion on the recognition of external and internal facial features. Acta Physiol. (Oxf) **109**(3), 227–238 (2002)
22. Meinhardt-Injac, B., Meinhardt, G., Schwaninger, A.: Does matching of internal and external facial features depend on orientation and viewpoint? Acta Physiol. (Oxf) **132**(3), 267–278 (2009)
23. Rakover, S.S., Teucher, B.: Facial inversion effects: parts and whole relationship. Percept. Psychophys. **59**(5), 752–761 (1997)
24. Bindemann, M., Sandford, A.: Me, myself, and I: different recognition rates for three photo-IDs of the same person. Perception **40**(5), 625–627 (2011)
25. Kemp, R.I., Towell, N., Pike, G.: When seeing should not be believing: photographs, credit cards and fraud. Appl. Cogn. Psychol. **11**(3), 211–222 (1997)

A Visual-Based Approach for Manual Operation Evaluation

Yiyao Zhao[ID], Zhen Wang$^{(\boxtimes)}$, Yanyu Lu, and Shan Fu

School of Electronic Information and Electrical Engineering,
Shanghai Jiao Tong University, Shanghai, People's Republic of China
b2wz@sjtu.edu.cn

Abstract. In order to improve the human-machine interface design and monitor the operational performance, hand behavior detection and analysis are very important. However, there is a lack of effective indicators of hand operation behavior. This paper introduces a vision-based approach for manual operation evaluation. K-curvature is employed to detect the operation key points on hand and the palm center is located combining image region moments and palm maximum circle fitting algorithm. The inter-frame velocity cosine is used as a representation of the motion sequence and we extracted the features of approximate entropy (ApEn) and maximum duration ratio (MDR) from the sequences to distinguish different operation statuses. A digit clicking experiment with different difficulty is designed to study operation behavior characteristics and we found the features of ApEn and MDR had a significant difference between position-determined operation and position-unknown operation. When the operation process is aimless and inconsistent, the ApEn value tends to be larger and the MDR value tends to be smaller. Our approach could effectively describe hand operation and assist manual operation performance evaluation and usability improvement.

Keywords: Usability · Operation keypoints · Performance evaluation · Hand motion analysis

1 Introduction

Operational performance is closely related to the design of the control interface. The human hand plays a vital role in the most manual operation process, for which reason the performance could be assessed by the hand motion. The operator's hand tends to react according to his operating habits or experience degree, so it could also help to analyze the cognition of the participant. The result of hand performance would turn to reflect the rationality and assist Human-centered Design (HCD) improvement. Thus, it is necessary to effectively evaluate the operating performance of human hands.

Generally, performance evaluation methods are divided into subjective evaluation and objective evaluation. The latter is not affected by the subjective consciousness and has the characteristics of real-time evaluation, thus it receives more attention. Scholars have researched the operation completion process and the physiological indicators from aspects of EEG signal [1], EMG signal [2], eye movement signal [3], etc. Human hand

© Springer Nature Switzerland AG 2020
D. Harris and W.-C. Li (Eds.): HCII 2020, LNAI 12186, pp. 281–292, 2020.
https://doi.org/10.1007/978-3-030-49044-7_23

behavior is also essential during the interaction with the computer but could not be monitored efficiently by these signals. To enable the direct extraction of operational characteristics, vision-based human hand analysis is needed.

Although the research on the operator cognitive process has made great progress, the analysis of hand behavior did not acquire enough achievements. Operation key point detection plays an important role in action analysis, performance evaluation, gesture recognition and human-computer interaction [4–7]. Generally, fingertips are taken as the key points to be located. Zhang et al. [8] use the positional characteristics and simply adopted the highest point of hand as the fingertip. Some researchers tried to find the furthest point from palm centroid [9–11]. In some papers, the geometry of the hand region is used for convex polygon fitting and the fitted vertices are treated as candidate points of fingertips [12]. Other researchers use the hand contour curvature to search fingertips [13–15]. With the improvement of depth cameras, the depth information is widely used. Liang et al. [16] introduced three depth-based features based on the geodesic distance and got satisfactory results on some challenging gestures such as bending and side-by-side. Some studies use CNN and attention-based methods to detect fingertips and achieve good results [17–19]. This kind of approach needs a lot of marked data and is too complex for our application.

Some studies have noticed that fingertips cannot accurately represent operating points, so attempts have been made to use the positional relationship between fingertips and screen to determine the position of key points. Agarwal et al. [20] fit the part of the fingertip into an ellipsoid and uses its center as the point of intersection. [21] introduced a Complementary Fingertip Model (CFM) to match the finger and estimate the operation point. In [22] researchers treat the intersection of the projection direction at the fingertip and the operation plane as the operating point. These papers realized the difference between operation point and fingertip but their application environment is distinct from ours.

Human hand motion (HHM) analysis is an essential research topic. It could apply in training assessment, action recognition and other fields [23–25]. Traditionally the hand motion is modeled as Hidden Markov model (HMM) [26], Gaussian Mixture Model (GMM) [27] or Dynamic Time Wrapping (DTW) [28]. Recently, a few studies tried to solve the problem using a deep neural network such as RNN [29] and got favorable results. Although these models could distinguish hand movements based on hand movement trajectories, they require much data for training, and the results are difficult to use in behavioral cognition. Zia et al. [30] compared the Approximate Entropy (ApEn) and Cross-Approximate Entropy (XApEn) method with the results of the features of Sequential Motion Texture (SMT), Discrete Cosine Transform (DCT) and Discrete Fourier Transform (DFT), where acceleration is acquired by two accelerometers. In order to use hand motion analysis to study the influence of human cognition, [31] explored the educational impact of hand motion analysis using the indexes of duration, working volume, number of movements and hand path length. However, its application scenario and workload is far from our experiment.

In this paper, we monitor the operation process in egocentric interactions. Operation key points are detected based on fingertips located by the k-curvature method and the palm center is located combining the image region moments and palm maximum circle fitting algorithm. The velocity of the palm center and the key point is extracted from the hand image captured. The features of maximum duration ratio (MDR) and the Approximate Entropy (ApEn) is introduced to discover the differences between position-determined operation and position-unknown operation.

The remainder of the paper is organized as follows. Section 2 introduces the vision-based method. In Sect. 3, we present our experiment setup and give the results and discussions in Sects. 4 and 5. Conclusion and future work are given in Sect. 6.

2 Method

2.1 Hand Detection

Although the depth-based segmentation method could cover the effects of different lighting and color temperature, it would cause failure when contacting with other objects. By analyzing the application environment, we found that skin color distribution has obvious aggregation characteristics. Therefore, we utilize a color-based approach to conduct hand area detection and segmentation.

YCbCr color space has a similar advantage with HIS and other color space which could separate the influence of the luminance component. The components of YCbYr could be obtained by converting the RGB values using Eq. (1)

$$\begin{cases} Y = 0.299 * R + 0.587 * G + 0.114 * B \\ Cr = -0.147 * R - 0.289 * G + 0.436 * B \\ Cb = 0.615 * R - 0.515 * G - 0.100 * B \end{cases} \tag{1}$$

The Y component represents the brightness of the pixel. Therefore, ignoring the Y component could make the color space less affected by brightness, in which case the skin color could cluster well. In our method, we refer to a YCbCr skin color model with a narrow ribbon [10] as shown in (2). The ribbon is an experience value and could detect the skin color well in most instances.

$$\begin{cases} 77 \leq Cb \leq 127 \\ 133 \leq Cr \leq 173 \end{cases} \tag{2}$$

To detect the hand more accurately, we perform color estimation on each subject's skin tone. We interactively select an area from the hand and calculate the mean μ_{Cb}, μ_{Cr} and standard deviation σ_{Cb}, σ_{Cr} of Cb and Cr components within it. The skin color is modeled as (3). The variable a is set artificially and equals to 10 in our experiment. After skin-color extraction, median filtering is performed to remove the noise and search for the maximum connected region as the hand target.

$$\begin{cases} \mu_{Cb} - a * \sigma_{Cb} \leq Cb \leq \mu_{Cb} + a * \sigma_{Cb} \\ \mu_{Cr} - a * \sigma_{Cr} \leq Cr \leq \mu_{Cr} + a * \sigma_{Cr} \end{cases} \tag{3}$$

2.2 Keypoint Detection

The locations of operation key points could help analyze the movement mode of the operator and further evaluate the operation performance. This paper aims at the behavior analysis during human-computer interaction so the click gesture is focused. The operating key point is at the contact point between the finger and the screen.

We first locate the fingertip because the target key point on the belly is obscured. The k-cosine curvature algorithm is employed to detect the fingertip. The algorithm is shown in Fig. 1. In the figure, p_i is the i-th point at the 8-connected pixel contour retrieved by the border tracing algorithm proposed by Suzuki et al. [32]. p_{i-k} and p_{i+k} are the k-th points before and after p_i in the contour list. Two vectors are formed by the three points:

$$\begin{cases} \boldsymbol{\alpha}_{ik} = \overrightarrow{p_i p_{i+k}} = (x_{i+k} - x_i, y_{i+k} - y_i) \\ \boldsymbol{\beta}_{ik} = \overrightarrow{p_i p_{i-k}} = (x_{i-k} - x_i, y_{i-k} - y_i) \end{cases} \tag{4}$$

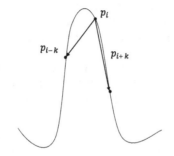

Fig. 1. The k-cosine curvature diagram.

Therefore, the cosine curvature at the point p_i could be calculated as:

$$c_{ik} = \cos\langle \boldsymbol{\alpha}_{ik}, \boldsymbol{\beta}_{ik} \rangle = \frac{\boldsymbol{\alpha}_{ik} \cdot \boldsymbol{\beta}_{ik}}{\|\boldsymbol{\alpha}_{ik}\| \|\boldsymbol{\beta}_{ik}\|} \tag{5}$$

The cosine curvature c_{ik} depicts the degree of unevenness of the contour curve at the point p_i. Candidate points $\{u_i\}_m$ are selected as fingertips by k-cosine curvature algorithm but generally should be filtered to avoid interference caused by finger depressions. In this paper, we add two limitations to finish the selection process. Firstly, we fit the convex hull of the hand area and get the vertices of the convex hull $\{v_j\}_l$. Euclidean distance from the candidate point to the vertices $\{d_{ij} = \|\overrightarrow{u_i v_j}\| \ |i = 1, 2, \ldots, m, j = 1, 2, \ldots, l\}$ is calculated. We only keep the points whose d_{ij} to any vertex is within the threshold χ. Secondly, we utilize the palm maximum circle fitting result and limit the Euclidean distance to the palm center greater than αR_{in}, where R_{in} is the radius of hand maximum inscribed circle and α is an empirical constant which is set as 1.2.

Next, we use the position of fingertips to find the operation points. The curvature of the contour points is arranged in reverse order to find γ points around the fingertips with the limitation of Euclidean distance to the tips. γ is a constant and equals 14 in this paper. The least-square circle fitting approach is conducted to find the center as the coordinate right above the finger belly. For the same subject, the difference between the operating keypoint and the detected point could be approximated as a constant. Therefore, we could simply adopt the key point above the finger belly.

To locate the overall movement of the hand, the palm center coordinate is also needed. We apply two algorithms to estimate the center of the palm. The image region moments are used to calculate the centroid (x_{c1}, y_{c1}). The formula is:

$$\begin{cases} x_{c1} = \dfrac{M_{10}}{M_{00}} = \dfrac{\sum_{(i,j)\in S} i I(i,j)_k}{\sum_{(i,j)\in S} I(i,j)_k} \\ y_{c1} = \dfrac{M_{01}}{M_{00}} = \dfrac{\sum_{(i,j)\in S} j I(i,j)_k}{\sum_{(i,j)\in S} I(i,j)_k} \end{cases} \tag{6}$$

where M_{00}, M_{10}, M_{01} are the moments of the binary image of the hand obtained by the hand detection algorithm introduced before. S represents the set of hand region coordinates and $I(i,j)_k$ is the pixel intensity at point (i, j) in the k-th channel.

The second estimation method takes the shape of the hand into account. Generally, the palm part could be approximated as a circle whose center is the hand centroid. Therefore, we use the hand maximum inscribed circle algorithm to find the hand centroid (x_{c2}, y_{c2}).

The final hand center coordinate (x_c, y_c) is acquired by averaging the results of the two estimation methods. The palm center detection results are shown in Fig. 2. The combination strategy could minimize the interference of palm shape irregularities on palm location.

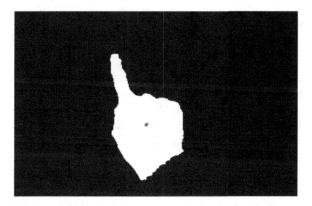

Fig. 2. The calculation of palm center. The yellow point is the hand moment center. The green point is the hand maximum inscribed circle center. The red point is the final coordinate obtained as the mean of the two methods. (Color figure online)

2.3   Operation Pattern Analysis

The 3D coordinates of the palm center and operating point of each frame could be obtained after hand detection and keypoint location. Their velocities could be calculated by the coordinates and timestamps. To indicate the behavior continuity, we designed a variable using the cosine of the key points' velocities between consecutive frames. The variables are named based on the point names as *palm_v_cos* and *keypoint_v_cos*. The variables are in value range [−1, 1] and the angle decreases as the cosine value increases.

We introduce the features of the approximate entropy and the maximum duration ratio to characterize different time series and evaluate different states of people during operation. Approximate entropy (ApEn) is a non-linear dynamic parameter to quantify the regularity and unpredictability of time series fluctuations. It uses a non-negative number to represent the complexity of a time series, reflecting the possibility of new information. [30] used this method to classify different surgical skill statements, but there are few studies applying it in other fields.

We design a feature named maximum duration ratio (MDR) according to the operation process. It is based on the observation of the search behavior. The subject tends to move back and forth when there is a search process. MDR is defined as the ratio of the longest hold time of a pre-defined state to the total duration of the operation. The definition formula of MDR is:

$$\tau_{MDR} = \max_i \frac{\sum_{\psi \in \{\psi(i, limitation)\}} \Delta t}{\sum_\psi \Delta t} \tag{7}$$

where i is the index of the segment that satisfies the state limitation and ψ is the monitored state of the key point. Δt represents the time period during the state to the next one and the $\{\psi(i, limitation)\}$ is the set of the i-th consequent segment that meets the requirement.

3   Experiment

3.1   Task

The digital click experiment is to tap the digits on the touch screen in order. The whole experiment contains two sets of tasks with different initial number *NumStart* of 1 and 26. In each experiment set, there are three difficulty levels determined by the number of digits. The layouts of 3*3, 4*4 and 5*5 in the game correspond to distinct difficulties. Since single digits are relatively obvious for visual search, we add digit *0* in front of them to make all digits look the same. To eliminate the impact of the decreasing numbers of digits after clicking, the digits will change to another set of non-crossover data in our experiments. The distribution of digits in each game is random so as to reduce the proficiency effect of experimental repetition. The experimental interface is shown in Fig. 3.

Fig. 3. The digit click experimental environment with a layout of 5*5. Each set begins with different initial number *NumStart*. The digit arrangement is generated randomly.

In the experiment, the participants need to perform both sets of tasks. During each part of process, the subject could decide to start by himself after complete preparation and click displayed digits in the right order from the number *NumStart*.

3.2   Apparatus

This digital experiment is conducted using a laptop connected with an external touch screen on a table. The vision-based hand behavior analysis approach is implemented with an Intel RealSense™ depth camera D435. The resolution of the collected image information, both in the RGB channel and in depth channel, is 1280*720. The PC responsible for the recording of the camera data and the computer for the digital display are time-synchronized before starting the experiments. All calculations are conducted on a PC with a 4.0 GHz Core i7 processor and 8 GB of RAM.

3.3   Participants

We collected the experiment data from 7 participants from Shanghai Jiao Tong University whose age ranges from 22 to 27. All participants tried the digital click process only once before the experiment to ensure a sufficient understanding of the task and avoid the impact of proficiency.

3.4   Procedure

The RealSense™ camera is fixed at the right rare of the subject and records the whole operation process. The vision data are saved to a computer in a *rosbag* format that stores compressed information such as RGB images, depth data and timestamps. During the experiment, we keep quiet to eliminate the interference to participants. The laptop is responsible for controlling the progress of the experiment and recording the digits and timestamps of the participants' clicks. All timestamps are accurate to milliseconds for data fusion.

The images extracted from the rosbag file are processed the hand segmentation based on the YCrCb color space. The 3D coordinates of the palm center and key points are calculated by the approach introduced in Sect. 2. Data fusion is conducted according to the timestamps of behavior data and the click data. The behavior data are cut into sequences by the click data and marked with the participant name, experiment index and target digit to further analyze the behavior characteristics.

4   Result

4.1   Hand Detection and the Key Point Location

After hand segmentation based on the estimated skin color interval in the YCbCr color space, the mean of the hand moment center and the maximum inscribed circle center are calculated as the palm coordinate. We combine the k-cosine curvature algorithm and circumscribed circle fitting method to detect the operation keypoint. Satisfactory detection results are obtained as shown in Fig. 4. After obtaining the pixel coordinates of the target points, the camera's internal and external parameters are utilized to convert them into 3D world coordinates.

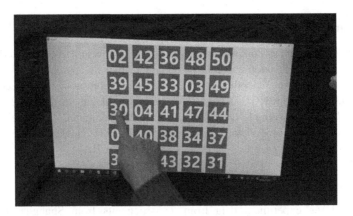

Fig. 4. The detected palm center and key points of the hand.

4.2   Operation Pattern Analysis

Different digit position awareness plays a vital role in the behavior of the subject. Here we divide operation sequences into position-determined and position-unknown ones according to the time delta ΔT between two clicks. In this paper, there are 654 sequences of digital click behavior in total from 39 experiments of 7 subjects. We consider that the click processes with ΔT of fewer than 0.4 s are position-determined operations while those more than 1.8 s position-unknown operations. There are 106 pieces of data in the position-determined operation sequences and 113 in the position-unknown sequences after classification.

The ApEn and MDR features for the time series *palm_v_cos* and *keypoint_v_cos* are extracted to distinguish the two types of operation. The student t-test is applied to determine whether they are significantly different from each other. The result of the t-test is shown in Table 1. From the table we could see that the values of probability P are much smaller than α, so the null hypothesis is rejected. Therefore, the position-determined and position-unknown operation could be well separated.

Table 1. The t-test results of ApEn and MDR, $\alpha = 0.05$

Sequence	ApEn	MDR
palm_v_cos	$2.4364*10^{-31}$	$6.7450*10^{-21}$
keypoint_v_cos	$2.3624*10^{-30}$	$8.4575*10^{-12}$

5   Discussion

The distribution of each sequence feature is compared to analyze operational behavior. The approximate entropy (ApEn) of the two kinds of sequences are compared in Fig. 5. We could find that there is some similarity between *palm_v_cos* and *key-point_v_cos*. It may be caused by the finger moving together with the palm for most of the time. Since the feature of ApEn characterizes the irregularity of the sequence and the mean value of the entropy of the transformed sequences from the position-unknown points is significantly larger than that of position-determined ones, we could see that the degree of confusion of the former is much more than that of the latter. Considering the definition of sequence transformation, we think that the direction of hand movement during the search process is more random.

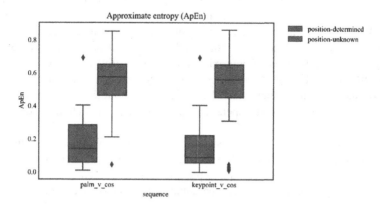

Fig. 5. The boxplot of approximate entropy (ApEn) features extracted from the *palm_v_cos* sequence and *keypoint_v_cos* sequence. Its value characterizes the irregularity of the sequence.

The comparison of the maximum duration ratio (MDR) is shown in Fig. 6. This feature represents the consistency and stability of the sequence. In this paper, we set the limitation of the angle cosine as $0.7 < v\_cos \leq 1$ and regard those actions with cosine in this interval having consistent movement direction. From the contrast, we could see that the mean of position-determined MDR is obviously larger than that of position-unknown. That means that the movement direction of the former is more consistent and continuous. The phenomenon may be due to the unconscious hand reciprocation during the search process.

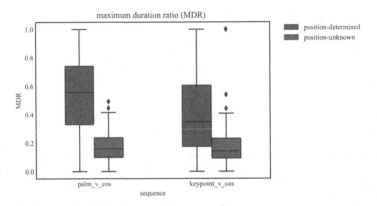

Fig. 6. The boxplot of maximum duration ratio (MDR) features extracted from the *palm_v_cos* and *keypoint_v_cos* sequence. Its value represents the consistency and stability of the movement direction between the previous frame and the current one.

6   Conclusion

In this paper, we introduce a vision-based approach to monitor the state of hand operations. The coordinates of palm center and operation key points are detected from images. We apply this approach in a digit click experiment and find the *palm_v_cos* and *keypoint_v_cos* sequences that could reflect the hand hovering behavior in the search process. The features of approximate entropy (ApEn) and maximum duration ratio (MDR) could classify different operation categories and well mirror the discontinuity of the participant behavior. The behavior monitoring approach could be employed to assist manual operation performance evaluation and usability improvement.

However, our approach has some shortcomings. First of all, the hand segmentation method might fail when there are objects with similar skin tones around. To overcome this problem, we consider combining depth and color information for more accurate and precise segmentation. Secondly, we could only monitor one hand which is not suitable when two hands are operating at the same time.

As introduced previously, the vision-based hand operation performance evaluation approach has broad application fields. In the future, we will continue to improve the information extraction process and combine a variety of information to study the relationship between the operator's cognition and behavior from the hand movement trajectory.

References

1. Fox, N.A., et al.: Assessing human mirror activity with EEG mu rhythm: a meta-analysis. Psychol. Bull. **142**(3), 291–313 (2016)
2. Alba-Flores, R., Hickman, S., Mirzakani, A.S.: Performance analysis of two ANN based classifiers for EMG signals to identify hand motions. In: SoutheastCon 2016. IEEE, pp. 1–5 (2016)
3. Martinez, F., Pissaloux, E., Carbone, A.: Towards activity recognition from eye-movements using contextual temporal learning. Integr. Comput.-Aided Eng. **24**(1), 1–16 (2017)
4. Proença, H.: Performance evaluation of keypoint detection and matching techniques on grayscale data. Sig. Image Video Process. **9**(5), 1009–1019 (2015). https://doi.org/10.1007/s11760-013-0535-1
5. Ravikiran, J., Mahesh, K., Mahishi, S., Dheeraj, R., Sudheender, S., Nitin, V.P.: Finger detection for sign language recognition. In: International Multi Conference of Engineers and Computer Scientists, pp. 18–20 (2009)
6. Yuan, S., et al.: Depth-based 3D hand pose estimation: from current achievements to future goals. In: Computer Vision and Pattern Recognition, pp. 2636–2645. IEEE (2018)
7. Kang, S.K., Nam, M.Y., Rhee, P.K.: Color based hand and finger detection technology for user interaction. In: 2008 International Conference on Convergence and Hybrid Information Technology, pp. 229–236. IEEE (2008)
8. Zhang, Z., Shan, Y.: Visual screen: transforming an ordinary screen into a touch screen. In: IAPR Workshop on Machine Vision Applications, pp. 215–218 (2000)
9. Lee, L.H., Braud, T., Bijarbooneh, F.H., Hui, P.: TiPoint: detecting fingertip for mid-air interaction on computational resource constrained smartglasses. In: The 23rd International Symposium on Wearable Computers, pp. 118–122 (2019)
10. Wu, G., Kang, W.: Robust fingertip detection in a complex environment. IEEE Trans. Multimed. **18**(6), 978–987 (2016)
11. Ren, Z., Yuan, J., Meng, J., Zhang, Z.: Robust part-based hand gesture recognition using kinect sensor. IEEE Trans. Multimed. **15**(5), 1110–1120 (2013)
12. Li, C., Zhang, R., Liu, Z., Hang, C., Li, Z.: Algorithm of fingertip detection and its improvement based on kinect. In: 2017 International Conference on Industrial Informatics-Computing Technology, Intelligent Technology, Industrial Information Integration (ICII-CII), pp. 63–66. IEEE (2017)
13. Cheng, J., Wang, Q., Song, R., Wu, X.: Fingertip-based interactive projector–camera system. Sig. Process. **110**, 54–66 (2015)
14. Wang, J., Qian, J., Ying, R., Jin, K., Wang, W., Liu, P.: Hand motion recognition based on a 3D fingertip detection fusion method. In: 2017 International Conference on Computational Science and Computational Intelligence (CSCI), pp. 510–515. IEEE (2017)
15. Ma, X., Peng, J.: Kinect sensor-based long-distance hand gesture recognition and fingertip detection with depth information. J. Sens. **2018**, 1–9 (2018)
16. Liang, H., Yuan, J., Thalmann, D.: 3D fingertip and palm tracking in depth image sequences. In: The 20th ACM International Conference on Multimedia, pp. 785–788. ACM (2012)
17. Bambach, S., Lee, S., Crandall, D.J., Yu, C.: Lending a hand: detecting hands and recognizing activities in complex egocentric interactions. In: International Conference on Computer Vision, pp. 1949–1957. IEEE (2015)
18. Suau, X., Alcoverro, M., Lopez-Mendez, A., Ruiz-Hidalgo, J., Casas, J.R.: Real-time fingertip localization conditioned on hand gesture classification. Image Vis. Comput. **32**(8), 522–532 (2014)

19. Guo, H., Wang, G., Chen, X.: Two-stream convolutional neural network for accurate RGB-D fingertip detection using depth and edge information. In: 2016 IEEE International Conference on Image Processing (ICIP), pp. 2608–2612. IEEE (2016)

20. Agarwal, A., Izadi, S., Chandraker, M., Blake, A.: High precision multi-touch sensing on surfaces using overhead cameras. In: Second Annual IEEE International Workshop on Horizontal Interactive Human-Computer Systems (TABLETOP 2007), pp. 197–200. IEEE (2007)

21. Son, Y.J., Choi, O., Lim, H., Ahn, S.C.: Depth-based fingertip detection for human-projector interaction on tabletop surfaces. In: 2016 IEEE International Conference on Consumer Electronics-Asia (ICCE-Asia), pp. 1–4. IEEE (2016)

22. Choi, O., Son, Y.J., Lim, H., Ahn, S.C.: Co-recognition of multiple fingertips for tabletop human-projector interaction. IEEE Trans. Multimed. **21**(6), 1487–1498 (2018)

23. Vemulapalli, R., Arrate, F., Chellappa, R.: Human action recognition by representing 3D skeletons as points in a lie group. In: Computer Vision and Pattern Recognition (CVPR), pp. 588–595. IEEE (2014)

24. Das, N., Ohn-Bar, E., Trivedi, M.M.: On performance evaluation of driver hand detection algorithms: challenges, dataset, and metrics. In: 2015 IEEE 18th International Conference on Intelligent Transportation Systems, pp. 2953–2958. IEEE (2015)

25. Xue, Y., Ju, Z., Xiang, K., Chen, J., Liu, H.: Multimodal human hand motion sensing and analysis-a review. IEEE Trans. Cogn. Dev. Syst. **11**(2), 162–175 (2018)

26. Wang, S., Hou, Y., Li, Z., Dong, J., Tang, C.: Combining ConvNets with hand-crafted features for action recognition based on an HMM-SVM classifier. Multimed. Tools Appl. **77** (15), 18983–18998 (2018)

27. Lin, L., Cong, Y., Tang, Y.: Hand gesture recognition using RGB-D cues. In: 2012 IEEE International Conference on Information and Automation, pp. 311–316. IEEE (2012)

28. Reyes, M., Dominguez, G., Escalera, S.: Featureweighting in dynamic timewarping for gesture recognition in depth data. In: 2011 IEEE International Conference on Computer Vision Workshops (ICCV Workshops), pp. 1182–1188. IEEE (2011)

29. Shahtalebi, S., Atashzar, S.F., Patel, R.V., Mohammadi, A.: HMFP-DBRNN: real-time hand motion filtering and prediction via deep bidirectional RNN. IEEE Robot. Autom. Lett. **4**(2), 1061–1068 (2019)

30. Zia, A., Sharma, Y., Bettadapura, V., Sarin, E.L., Essa, I.: Video and accelerometer-based motion analysis for automated surgical skills assessment. Int. J. Comput. Assist. Radiol. Surg. **13**(3), 443–455 (2018)

31. Zago, M., et al.: Educational impact of hand motion analysis in the evaluation of fast examination skills. Eur. J. Trauma Emerg. Surg. **45**, 1–8 (2019)

32. Suzuki, S.: Topological structural analysis of digitized binary images by border following. Comput. Vis. Graph. Image Process. **30**(1), 32–46 (1985)

Hand Movements Influence Time Perception of Visual Stimuli in Sub or Supra Seconds Duration

Weiqi Zheng[✉], Han Zhao, Yichen Zhang, Jiaxin Ma, and Ziyuan Ren

School of Psychology, Beijing Sport University, Beijing 100084, China
zhengweiqi@bsu.edu.cn

Abstract. Time perception is critical in human-computer interaction to guarantee precise action performance. The duration of visual stimuli can sometimes be perceived longer or shorter than its actual duration. Hand movements are frequently seen when we operate computers. The current study aimed to address whether hand movements influence time perception of visual stimuli in sub or supra seconds duration. The study adopted the method of constant stimuli belonging to the psychophysical method. The experiment was divided into a hand movement part and a static part. In the movement part, participants put their left hands in the carton and moved in an anticlockwise circle at a constant speed. In the static part, they rested their left hands in the carton. Participants were asked to make a judgment with their right fingers about whether the standard or probe stimuli lasted longer. Results showed a compressed effect for the perception of sub-second stimuli and an expanded effect for the supra-second stimuli. Furthermore, for sub-second, hand movements enhanced the compressed effect as compared to a stationary state, while for supra-second, there was no significant difference between movement state and stationary state. It also indicated that the processing mechanisms of the two timing scales are distinct and could be modulated by hand movements.

Keywords: Hand movements · Time perception · Sub- and supra-second · Visual stimuli · Psychophysical method

1 Introduction

A critical aspect of human-computer interaction is that individuals must be able to precisely determine when to execute actions to achieve a certain goal. For example, in virtual games, players often need to decide when to move or fight with enemies so that they could click a mouse or press keys timely; in human-robot interaction, users need to accurately estimate when to send an instruction to robot according to some feedback information. Therefore, it is essential to explore time perception during a dynamic human-computer interaction and to investigate the possible influential factors.

The duration of visual stimuli can sometimes be perceived longer or shorter than its actual duration. Previous studies showed that many visual features can distort subjective duration, including stimulus size, brightness, number, complexity, and spatial

D. Harris and W.-C. Li (Eds.): HCII 2020, LNAI 12186, pp. 293–301, 2020.
https://doi.org/10.1007/978-3-030-49044-7_24

frequency [1, 2]. Furthermore, static images with implied motion or body movement can affect subjective time [3]. Besides the features of visual stimuli per se, recent evidence suggests that our perception of time also depends on whether we are moving or not, which turns out an important external information interacting with visual sensation, especially when dealing with information on computer/machine interface in a state of motion [4]. Moreover, moving state can lead to compressed or expanded effect of duration perception, and influence stability or accuracy of time prediction [5].

Taken together, some properties of body movements would influence subjective time perception, including motion speed, motion duration, motion direction, action stage and some individual factors during motion. We summarized the relevant experiments and results in Table 1.

Table 1. Properties of motion that influence time perception and the related experiments

Motion properties	Study	Main results
Motion speed	Yokosaka et al. [6]	Fast hand movements reduced the apparent time interval between visual events
Motion duration	Yon et al. [7]	Judgments of tone duration were attracted toward the duration of executed movements, namely, tones were perceived to be longer when executing a movement of longer duration
Motion direction	Tomassini and Morrone [8]	Visual time depends on the movement direction, being expanded for hand movements pointing away from the body and compressed in the other direction
Action stage	Preparation stage: Hagura et al. [9]	A reduction of perceived frequency for flickering stimuli and an enhanced detection of rapidly presented letters during action preparation, suggesting increased temporal resolution of visual perception during action preparation
	Execution stage: Tomassini et al. [10]	The results indicate that time intervals are compressed around the time of hand movements
Sense of agency	Imaizumi and Asai [5]	Minimal delay of the video feedback resulted in longer perceived duration than the actual duration and stronger agency, while substantial feedback delay resulted in shorter perceived duration and weaker agency
Long-term exercise experience	Chen and Cesari [11]	The level of precision can be finely tuned through long-term sport training: Elite athletes, independently from their sport domains, generate better time estimates than nonathletes by showing higher accuracy and lower variability, particularly for sub-second time

The neural mechanism of body movements' influence on time perception might stem from the tight relationship between time and space representation in our brain [12]. Growing number of studies revealed that motoric brain structures may form the core component of a neural network supporting a wide range of timed behaviors. Supplementary motor area (SMA) is not only a part of motoric brain structures, but also plays a key role in time processing as part of the striato-cortical pathway verified by animal studies, human neuropsychology and neuroimaging studies [13].

In human-computer interaction, our time perception of visual stimuli often involves durations lasting below or above a second, namely the sub-second or supra-second timing scale. A large body of neuroimaging studies suggested that distinct neural systems are recruited for the measuring of sub-second and supra-second intervals [14–17]. For example, Pouthas et al. [18] found that the activation of some brain regions increases with increasing time, which demonstrated that sub-second and supra-second might be based on different internal clocks. Furthermore, previous studies pointed out that sub-second durations are processed in the motor system, whereas supra-second durations are processed in the parietal cortex by utilizing the ability of attention and working memory to keep track of time [19]. However, no conclusive theories have been presented to explain how body movements influence time perception of different timing scales.

Hand movements are frequently seen when we operate computers. When we play games or immerse ourselves in VR, we often use hands to manipulate or interact with some equipment. The present study aimed to verify how hand movements influence time perception of visual stimuli; and to probe whether there is a difference in duration perception between sub-second and supra-second stimuli and how hand movements modulate them.

2 Method

2.1 Participants

A total of 11 students (3 males) with a mean age of 20.55 years old (range 18–27) participated in the experiment. All participants were right-handed, with normal or corrected-to-normal vision, had no known abnormalities of their motor systems. They gave written informed consent prior to the experiment and were paid for their participation.

2.2 Stimuli and Design

Visual stimuli in the experiment were generated by a computer and presented on a computer screen. The viewing distance to the monitor was about 50 cm. Participants sat at a table and stared horizontally at the center of the screen. Their left hands placed in an opaque carton on the left of the keyboard on the table, and they used right fingers to press keys to respond.

The experiment was divided into a hand movement part and a static part, and the order of them was balanced among participants. The study adopted the method of constant stimuli belonging to the psychophysical method. In the movement part, participants put their left hands in the opaque carton and moved in an anticlockwise circle at a constant speed. In the static part, they rested their left hands in the carton.

The standard visual stimulus was a blue circle which lasted 700 ms at the sub-second scale, and 1400 ms at the supra-second scale. The two different standard visual stimuli scales were presented randomly in both experiment parts, in case of the practice effect. Each standard stimulus was corresponding to five probe durations (Table 2). Each probe stimulus presented ten times in both parts randomly.

Table 2. Standard and probe stimuli durations in the test phase.

	Sub-second (ms)	Supra-second (ms)
Standard stimuli	700	1400
Probe stimuli	600	1300
	650	1350
	700	1400
	750	1450
	800	1500

Before the test phase, there was a training phase which also contained movement and static parts. The duration of standard visual stimulus was 800 ms at the sub-second scale and 1500 ms at the supra-second scale in the training phase (Table 3). Each probe stimulus presented once in both parts. Participants would get their accuracy feedbacks in training phase, while in the test phase, there was no feedback.

Table 3. Standard and probe stimuli durations in the training phase.

	Sub-second (ms)	Supra-second (ms)
Standard stimuli	800	1500
Probe stimuli	700	1400
	750	1450
	800	1500
	850	1550
	900	1600

2.3 Procedure

In each trial, a fixation appeared firstly at the center of the screen in a uniform gray field, and a blue circle as a standard stimulus appeared for a while, followed by a masking stimulus; next, an identical circle as the previous one appeared for some time. Participants need to make a judgment with their right fingers on which circle lasted longer by pressing "1" or "2" keys on numeric keypad. A timeline chart of the experiment is shown in Fig. 1, it illustrated the procedure of the movement part. And the procedure of static part is similar except that participants' left hands were static for the whole experiment.

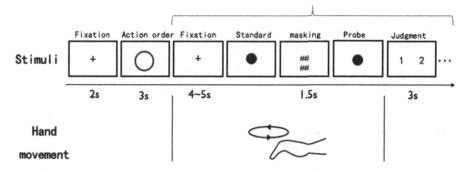

Fig. 1. Schematic illustration of the experimental procedure of hand movement part.

2.4 Data Analysis

The main dependent variables were the proportion participants judged the probe stimulus longer than the standard stimulus for each scale, as well as the point of subjective equality (PSE: the 50% point of the cumulative logistic distribution function fitted to the psychometric function by the maximum likelihood method to estimate the 50% correct point) for each standard stimulus. Logistic regression was used to relate the percentage of 'longer' judgment responses to overall stimulus duration in each condition for each participant. We also calculated the Weber ratio to analyze the temporal precision or sensitivity to the visual stimuli. This ratio is obtained by dividing the difference limen (half of the difference between the duration giving rise to 75 and 25% longer responses of probe stimuli) by the PSE.

3 Results

3.1 Proportion of Longer Probe Durations

For the proportion of probe stimuli being responded as longer ones, a two-way repeated-measures ANOVA was used to analyze the factors of movement states (motional vs. static) and timing scales (sub-second vs. supra-second). The index reflects a tendency of whether the duration perception was expanded or compressed (The main results were depicted in Fig. 2). Results showed that the main effect of movement state was significant ($F = 11.1$, $p = 0.01$, $\eta_p^2 = 0.582$), the main effect of timing scales was significant ($F = 181$, $p < 0.001$, $\eta_p^2 = 0.958$), and the interaction effect of the two factors was also significant ($F = 12.8$, $p = 0.007$, $\eta_p^2 = 0.615$). The post-hoc tests showed that the proportion of longer times of probe stimuli of sub-second was smaller than supra-second, which implied a compressed effect for the perception of sub-second stimuli than the supra-second stimuli.

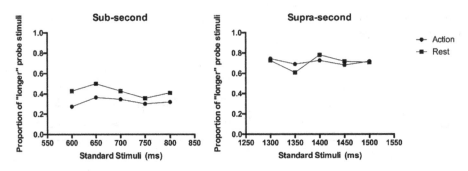

Fig. 2. Proportion of longer responses of probe duration plotted for each standard stimulus in movement and static condition (left: sub-second scale; right: supra-second scale).

The simple effect demonstrated that how hand movement could modulate the distortion of two timing scales, for sub-second, hand movements enhanced the compressed effect as compared to a stationary state ($t = -4.89$, $p = 0.005$), while for supra-second, there was no significant difference between movement state and stationary state ($t = 0.17$, $p = 0.998$).

3.2   Point of Subjective Equality (PSE)

The responses for each individual were modeled by fitting cumulative Gaussians and were used to calculate the PSE value of each timing scale. The PSE results can further examine the compressed effect for sub-second and the expanded effect for supra-second time perception (sub-second: PSE = 820 ms > 700 ms; supra-second: PSE = 1000 ms < 1400 ms). In addition, a two-way repeated-measures ANOVA was used to analyze the factors of movement states (motional vs. static) and timing scales (sub-second vs. supra-second) on PSE. Results showed that the main effect of movement state did not reach significant ($F = 0.18$, $p = 0.68$, $\eta_p^2 = 0.018$), the main effect of timing scales was significant ($F = 1175.45$, $p < 0.001$, $\eta_p^2 = 0.992$). However, because the standard stimuli durations were actually different between sub-second and supra-second, the statistical analysis had no actual meaning about the main effect of timing scale. And the interaction effect of the two factors did not reach significant ($F = 0.09$, $p = 0.77$, $\eta_p^2 = 0.009$).

3.3   Weber Ratio

As for the Weber ratio results, the two-way repeated-measures ANOVA showed that the main effect of movement state did not reach significant ($F = 0.001$, $p = 0.99$, $\eta_p^2 = 0$). The main effect of timing scales was significant ($F = 37.95$, $p < 0.001$, $\eta_p^2 = 0.791$), which indicated that participants were more sensitive to sub-second duration. And the interaction effect of the two factors did not reach significant, which showed that there was no obvious modulation of hand movements on temporal perception precision in different timing scales ($F = 0.523$, $p = 0.486$, $\eta_p^2 = 0.05$) (The main results were depicted in Fig. 3).

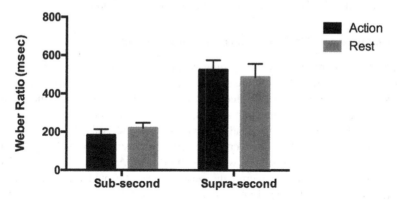

Fig. 3. Weber ratio plotted for sub- and supra-second scales in movement and static condition.

4 General Discussion

The present study examined whether hand movements influence perceived presentation duration of visual stimuli in sub- and supra-seconds. We used the blue circle as the visual stimuli presented on screen and asked participants to judge whether the standard circle or probe circle lasted longer, when their left hands moved in an anticlockwise circle at a constant speed or just rested on table.

As a general result, we found that individuals perceived sub-second visual stimuli shorter than the actual duration, which could be viewed as the compressed effect. While for supra-second visual stimuli, individuals perceived them longer than actual duration, namely the expanded effect. And the temporal precision or sensitivity was much higher when perceiving sub-second visual stimuli. In addition, hand movements enhanced the compressed effect of sub-second visual stimuli.

4.1 Different Timing Scales

The bulk of the analyses we performed on our data set yielded results consistent with the previous studies that for all participants, independent with the movement condition, two different "clocks" for evaluating time below and above a second, respectively, were applied in time perception of visual stimuli [20]. We supported the idea from two scalar properties: (1) the results from proportion of longer probe durations and PSE revealed different duration distortions of two timing scales (a compressed effect for sub-second duration and an expanded effect for supra-second duration); (2) the results from Weber ratio suggested different perceiving precision of two timing scales (the precision of sub-second was higher than supra-second).

However, there is no uniform conclusion about which time point is the cut-off point of different duration processing mechanisms. For example, Michon proposed that below 1/2 s, temporal information processing has the attribute of perception processing [21]. While Pöppel emphasized that below 2–3 s, temporal processing should be viewed as time perception, and above 2–3 s, temporal processing should be viewed as time estimation [22]. Therefore, more research is needed to address the question.

4.2   Modulation of Hand Movement on Time Perception

Another main finding of the current study was that hand movements could modulate the time perception of visual stimuli, especially for sub-second scale. Compared with static condition, the compressed effect was enhanced by hand movement condition. According to previous neuroimaging studies, different brain areas were associated with the ability to discriminate sub-second and supra-second intervals [19]. The estimation of the sub-second range tends to recruit the primary sensorimotor and supplementary motor cortices and the cerebellum, which is more like "Automatic timing". Whereas supra-second timing tends to recruit the posterior parietal, which is associated with "cognitive timing" [17]. Therefore, we can speculate that hand movements involve more activation of primary sensorimotor areas and might lead to the modulation of sub-second estimation.

4.3   The Limitation of the Study

A major limitation in our study is that only one temporal duration in each range of time (sub- and supra-second time) has been considered. In future, we can try to apply more different time durations of the two timing scales, just like in Chen and Cesari [11]'s study.

5   Conclusion

In summary, although the distortion of sub-second and supra-second is different when individuals perceive a duration of visual stimuli, hand movements can modulate the distortion and enhanced the compressed effect of sub-second visual stimuli. The results further indicate that the processing mechanisms of the two timing scales are distinct. The findings have implications for time perception mechanisms during dynamic human-computer interaction involving body movements.

Acknowledgements. This research was supported by the Fundamental Research Funds for the Central Universities of China (Grant No. 2019QD012).

References

1. Yamamoto, K., Miura, K.: Time dilation caused by static images with implied motion. Exp. Brain Res. **223**(2), 311–319 (2012). https://doi.org/10.1007/s00221-012-3259-5
2. Jia, L., Shi, Z., Zang, X., Müller, H.J.: Watching a real moving object expands tactile duration: The role of task-irrelevant action context for subjective time. Atten. Percept. Psychophys. **77**(8), 2768–2780 (2015). https://doi.org/10.3758/s13414-015-0975-5

3. Nather, F.C., Bueno, J.L.O.: Static images with different induced intensities of human body movements affect subjective time. Percept. Motor Skills **113**(1), 157–170 (2011)
4. Wiener, M., Zhou, W., Bader, F., Joiner, W.M.: Movement improves the quality of temporal perception and decision making. ENeuro **6**(4), 1–17 (2019)
5. Imaizumi, S., Asai, T.: My action lasts longer: potential link between subjective time and agency during voluntary action. Conscious. Cogn. **51**(April), 243–257 (2017)
6. Yokosaka, T., Kuroki, S., Nishida, S., Watanabe, J.: Apparent time interval of visual stimuli is compressed during fast hand movement. PLoS ONE **10**(4), 1–11 (2015)
7. Yon, D., Edey, R., Ivry, R.B., Press, C.: Time on your hands: perceived duration of sensory events is biased toward concurrent actions. J. Exp. Psychol. Gen. **146**(2), 182–193 (2017)
8. Tomassini, A., Morrone, M.C.: Perceived visual time depends on motor preparation and direction of hand movements. Sci. Rep. **6**, 1–12 (2016)
9. Hagura, N., Kanai, R., Orgs, G., Haggard, P.: Ready steady slow: action preparation slows the subjective passage of time. Proc. Roy. Soc. B Biol. Sci. **279**(1746), 4399–4406 (2012)
10. Tomassini, A., Gori, M., Baud-Bovy, G., Sandini, G., Morrone, M.C.: Motor commands induce time compression for tactile stimuli. J. Neurosci. **34**(27), 9164–9172 (2014)
11. Chen, Y., Cesari, P.: Elite athletes refine their internal clocks. Mot. Control **19**(1), 90–101 (2015)
12. Merchant, H., Yarrow, K.: How the motor system both encodes and influences our sense of time. Curr. Opin. Behav. Sci. **8**, 22–27 (2016)
13. Macar, F., Coull, J., Vidal, F.: The supplementary motor area in motor and perceptual time processing: fMRI studies. Cogn. Process. **7**(2), 89–94 (2006). https://doi.org/10.1007/s10339-005-0025-7
14. Wiener, M., Turkeltaub, P., Coslett, H.B.: The image of time: a voxel-wise meta-analysis. Neuroimage **49**(2), 1728–1740 (2010)
15. Morillon, B., Kell, C.A., Giraud, A.L.: Three stages and four neural systems in time estimation. J. Neurosci. **29**(47), 14803–14811 (2009)
16. Jahanshahi, M., Jones, C., Dirnberger, G., Frith, C.: The substantia nigra pars compacta and temporal processing. J. Neurosci. **26**(47), 12266–12273 (2006)
17. Lewis, P.A., Miall, R.C.: Remembering the time: a continuous clock. Trends Cogn. Sci. **10**(9), 401–406 (2006)
18. Pouthas, V., et al.: Neural network involved in time perception: an fMRI study comparing long and short interval estimation. Hum. Brain Mapp. **25**(4), 433–441 (2005)
19. Hayashi, M.J., Kantele, M., Walsh, V., Carlson, S., Kanai, R.: Dissociable neuroanatomical correlates of subsecond and suprasecond time perception. J. Cogn. Neurosci. **26**(8), 1685–1693 (2014)
20. Buhusi, C.V., Meck, W.H.: What makes us tick? Functional and neural mechanisms of interval timing. Nat. Rev. Neurosci. **6**(10), 755–765 (2005)
21. Michon, J.A.: The compleat time experiencer. In: Michon, J.A., Jackson, J.L. (eds.) Time: Mind and Behavior, pp. 20–52. Springer, Berlin (1985). https://doi.org/10.1007/978-3-642-70491-8_2
22. Pöppel, E.: Lost in time: a historical frame, elementary processing units and the 3-second window. Acta Neurobiol. Expr. **64**(3), 295–301 (2004)

Managing Human Energy with Music? An Explorative Study of Users' Energy-Related Listening Behaviours

Mourad Zoubir[(⊠)] and Thomas Franke

Universität Zu Lübeck, Ratzeburger Allee 160, 23562 Lübeck, Germany
{zoubir, franke}@imis.uni-luebeck.de

Abstract. Music has been shown to increase activation and could be used to offset resource-demand-discrepancies in knowledge workers, thereby increasing mental well-being and job performance. The present research sought to explore how students use music to manage human energy states to identify use cases for a human energy state management system, which would use musical stimuli to achieve users' energy goals. An online survey ($N = 224$) assessed typical practices related to music-based management of human energy states and desired state transitions (i.e. users' energy goals). In addition, spontaneous suggestions of possible features were examined and key variables to characterize user diversity were assessed. Users sought to avoid low and strived for high (but not maximum) energy levels. Genre was the most used selection characteristic for this function, but users varied in their need for novelty. Assistance systems should work within personal preferences to dynamically adjust energy levels and avoid peaks. Future work should examine benefits of interaction depth for energy management strategies.

Keywords: Human energy · Arousal · Activation · Music · Academic learning · Knowledge work

1 Introduction

To counteract negative psychological effects of work, mHealth applications have been shown to successfully answer employees stress and mental-health concerns [26], suggesting that smartphone applications are a low-cost and feasible method to support workers mental well-being. Prominent examples include meditation or reflection apps, which offer *post-hoc interventions*, i.e. treating the symptoms of stress [26]. However, *prophylactic methods* (i.e. prevention) which prevent demand-resource discrepancies (i.e. experiencing a lack of means to achieve ends; a key contributor to burnout [6]) in the first place should be favoured over post-hoc treatment, to minimise discomfort and constraints on quality of life, as well as improve job performance and satisfaction. In the best case, such a prophylaxis should not require large changes in ongoing activities or habits (which would result in additional workload or distractions) and should therefore utilize existing assets.

One very promising method is background music. For many knowledge workers, music offers a personalised, flexible supplement to daily tasks, which can be used in

© Springer Nature Switzerland AG 2020
D. Harris and W.-C. Li (Eds.): HCII 2020, LNAI 12186, pp. 302–313, 2020.
https://doi.org/10.1007/978-3-030-49044-7_25

conjunction with work, even when sharing workspace (e.g. via headphones); this is not the case with e.g. lighting, olfactory stimuli or room design). Other individualized options, such as physical activity, would be an interruption of ongoing activity. So, how can we leverage music to increase resources, and with which theoretical framework?

One emerging concept which can be applied for this purpose is that of human energy. For example, Quinn et al. [20] point to definitions of human energy comprising two elements: physical energy (i.e. the potential energy, in the form of adenosine triphosphate (ATP), and kinetic energy, in the form of human action) and energetic activation (i.e. the subjective component of a biobehavioural system of activation, e.g. feeling too tired to jog, even though one is well nourished and rested). In their work, an integrated model of human energy which builds upon corresponding, established frameworks was introduced. This includes the theories of Conservation of Resources [9], Interaction Ritual Chain [4], Attention Restoration [15], Ego Depletion [1], Tense/Energetic Activation [27] and Self-Determination [23].

At the centre of this model, Quinn et al. focus on the job demand-resource discrepancy, arguing that a surplus of resources can lead to energetic activation; a deficit leads to tense activation. As described by the corresponding theory [27], tense activation is a reaction to negative emotions, focusing our attention on negative (and potentially dangerous) situations, which need to be handled. On the other hand, energetic activation accompanies positive emotions and can lead to broader thoughts which increase a person's ability to recognise available resources. Some work has established methods of measuring energetic activation, including intuitive pictograms, which grade activation into "energy levels" which can be measured by the state-of-charge of a battery metaphor [28].

But what exactly are resources? Resources are globally defined as anything that helps an individual to achieve a goal (e.g. ATP) [19], and the integrated model differentiates between total resources (e.g. entirety of ATP available in a day), remaining resources (e.g. ATP left after midday) and resources-in-use (e.g. amount of ATP used for holding a conversation). Of particular interest here is the activation of remaining resources into resource-in-use to address demands. The efficiency of this transfer can be improved with practice – a conclusion taken from the ego-depletion theory [1] -, this assumption of resource allocation as a variable implies that supportive interventions may be used to improve individual's ability to wilfully increase energetic activation.

One possible intervention is the use of music. While, to the best of our knowledge, no research has addressed the use of music in the context of the above human energy model, there is a large body of work on cognitive functioning and music, e.g. on spatial task performance [22] or pupil's school task performance [8]. Of particular interest is research surrounding the mood-and-arousal hypothesis [11, 12], which stipulates that music listening only indirectly affects cognitive functioning and is mediated by individuals' reactions of arousal and emotion to music, which may support the idea that music enables the activation of remaining resources. This is further supported by other research, which found that positive effects of music at work carry is also improved by practice [16].

Previous research often examined time-constrained tasks, e.g. in validated experimental tests [8, 11, 12, 22], vigilance tasks, e.g. in a driving context [29], or the effects of background music during work, e.g. as a possible distraction [10]. However, there remains much potential for further research on influencing human energy with music in knowledge work.

The objective of the present research was to pave the way for a human energy management system by identifying possible use cases for induced energy transitions in particular knowledge workers' tasks. In this first explorative pilot study, we explicitly assessed university students, who often experience job demand-resource discrepancies during examination study periods. Specifically, we aimed to explore how students use music to manage human energy states and energy state transitions in study situations. In addition, users' spontaneous suggestions of possible features of a music-based human energy management assistant were examined and key variables to characterize user diversity (such as music interaction) were assessed.

2   Method

2.1   Participants

Using the mailing lists of the University of Lübeck, we recruited a sample of students ($N = 226$) to complete an online survey, implemented via LimeSurvey 3.8.1 [24]. As compensation, participants took part in a raffle of three 20€ cash prizes. Participants were informed that the survey conformed to the requirements of the EU's GDPR and all data collected was anonymized.

Two participants were excluded, as both had no variance in their responses (i.e. only selected the outer most answer category) and only entered question marks into answer fields. Therefore, a total of $N = 224$ were included in the further analysis.

The average age of the sample was $M = 23.50$ ($Mdn = 23$, $SD = 3.67$), with 62 reporting their gender as male (one subject chose not to disclose). To better understand their study background, we assessed participants total number of semesters studied ($M = 6.48$, $Mdn = 6$, $SD = 4.52$).

2.2   Procedure

Before beginning the questionnaire, participants gave informed consent. Hereafter, "human energy" as a concept for the survey was defined as referring to the subjective component of energy (e.g. feeling tired even when rested) and therefore differentiated from physical energy (e.g. amount of available ATP or glucose). This is in line with proposed definitions [20] and was deemed necessary, as "human energy" is often used colloquially.

As our aim was to achieve insights into a somewhat homogenous cognitive learning task, we asked participants to think of a typical session, in which they were studying for an exam, for example within or shortly before an examination period. The questionnaire was dispersed two weeks before such an examination period of the university, thereby increasing the likelihood of sampling students who were currently studying, conceivably improving the validity of input received.

Contained within the online study were two groups of items: those referring to listening habits and personal preferences, or those assessing the current usage of digital tools.

Listening Habits. Three groups of items assessed music listening, especially during study. Unless stated otherwise, participants signalled their agreement to statements on a 6-point Likert scale ("completely disagree" to "completely agree").

Study Listening Habits. Two statements assessed a) the usage of energy management strategies in general and b) with music in particular. Two further statements sampled the assumed quality which music influenced: did participants a) feel that music activated them or b) changed their mood; these reflect the qualities addressed by the arousal-and-mood-hypothesis. Participants who stated they completely disagreed to using music to affect energy levels ($n = 20$), were excluded from the subsequent questions, as they may not have had the necessary experience to answer expertly.

Furthermore, six statements assessed the individual selection criteria for use of music to affect subject energy. These were indicated either by previous literature or the authors own personal experience, and included Tempo (speed of a track, see e.g. [12]), Genre (style classification see e.g. [5]), Vocals (presence or absence of a singing voice, see e.g. [17]), Familiarity (if a song is known or not, see e.g. [5, 17]), Similarity (similarity or dissimilarity of tracks in a sequence) or Supposed Value (purported ability to increase performance, i.e. the Vivaldi effect, see e.g. [18]).

Energy-Level Use Cases. Participants were sequentially asked to envision two scenarios: a typical point during a study session in which they were dissatisfied with their performance and a typical point in a study session, after they had successfully used music to influence their performance. In each, students selected their most typical performance emotions and energy level in each. A validated battery visual scale [28] was used for energy levels, which comprised five "states of charge" (SoC) as a metaphor for human energy (Fig. 1).

Fig. 1. Visual scale of human energy using a battery metaphor. Participants were instructed to select the state of charge which would best describe their typical energy level in specific study situations. Reprinted with kind permission from Weigelt et al. (in press).

Music Interaction. In order to assess how deeply participants interacted with music (i.e. if they were more likely to seek profound or superficial connections to music), we adapted the affinity for technology interaction (ATI) scale [7] for this context. The ATI has a strong focus on the use of object interactions as a personal resource and is therefore applicable to the theoretical framework employed in this study.

By assessing music interaction, we seek to identify needs of user diversity, such as differences in applicable use cases. The five items covered engagement, immersion, the search for novel or appealing music, and the search for understanding one's own music taste. The Music Interaction Scale was internally consistent (Cronbach's α = .872), with each item's α-if-removed lower than this value (range: .832–.863).

Digital Tool Features. Here participants were asked in open questions to report their most-used music program as well as which features thereof supported their attempts to influence subjective energy. Furthermore, participants were asked to imagine – without constraints – any features which could possibly assist them in influencing subjective energy.

3   Results

A total of 224 students' responses were analysed, while the items pertaining to music characteristics, energy level use cases and music interaction were only analysed for those who did not completely disagreed to using music to influence energy (n = 204). The following section examines listening habits in a quantitative manner, and then continues with qualitative analysis of responses regarding digital tools. Statistical analysis was carried out with R [21] and the Likert package, as well as with Jamovi Version 1.1 [13].

3.1   Listening Habits

Study Listening Habits. Generally, 91.7% of participants agreed (either slightly, largely or completely), that they influenced their energy level at all. 76.8% of the sample said they used music for this purpose. Specifically, during studying sessions, 88.3% said they used music to change their mood, and 67.8% to activate themselves; 59.8% reported doing both. This indicates that the wide majority of students sampled used music to affect their mood or activation, indicating that there may indeed be potential for a human energy management system.

Regarding the characteristics of music used to influence energy levels (Fig. 2), we found that genre was the most common selection criteria (89% agreement). However, genre did not significantly correlate with familiarity ($r_s(202)$ = −.10, p = .137), which may indicate that genres are selected for this specific task, rather than based on general preference. On the other hand, supposed value (i.e. knowledge that a music's properties have been reported to increase performance), was the least common selection criteria (11% agreement).

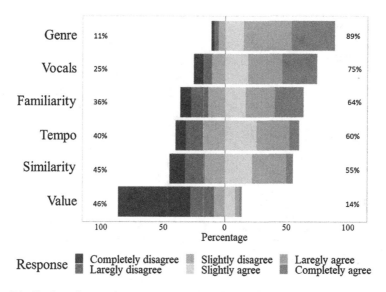

Fig. 2. Distribution of agreement to statement "In order to influence my subjective energy level, I actively select songs or playlists based on their...".

Energy-Level Use Cases. For situations where students were dissatisfied with their study performance, the reported typical energy level was most often low, but not "depleted" ($M = 2.02$, $Mdn = 2.00$, $SD = 0.81$). Conversely, for situations where performance was successfully influenced by music, the typical energy level was high, but not "full" ($M = 3.86$, $Mdn = 4.00$, $SD = 0.74$). For a full comparison of energy levels, see Fig. 3.

The used situations did not imply that they were contained within the same study session, in order to ascertain use cases for digital energy management systems. Responses suggest that the vast majority of use cases seek to increase energy levels (89.8%). A minority represent an equilibrium (7.8%), perhaps suggesting a wish to uphold an energy level, or satisfaction achieved independent of energy level. Finally, 2.4% of use cases comprise a reduction in energy, which may indicate a wish to reduce tense activation.

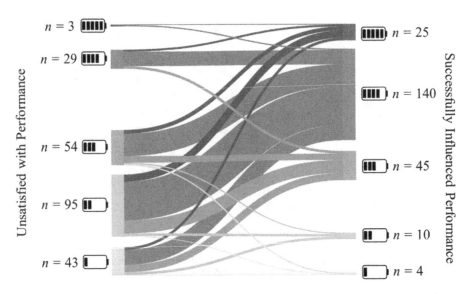

Fig. 3. Reported energy levels when students were dissatisfied with study performance, contrasted with energy levels after music had been used to influence performance. ($n = 204$). *Note:* Categories shifted out of alignment to allow for ease of reading: ascending bar = increase of energy level; descending bar = decrease of energy level; horizontal bar = no change.

Music Interaction Scale. The scale value ($M = 4.20$, $Mdn = 4.20$, $SD = 1.05$) was correlated with absolute change in energy levels ($M = 1.88$, $Mdn = 2.00$, $SD = 0.97$), as well as with the aggregated reported use of music to activate and to change mood ($M = 4.38$, $Mdn = 4.50$, $SD = 1.15$).

After Bonferroni corrections, the results of a Spearman correlation indicated that there was a significant positive association between the Music Interaction Scale and absolute change in energy levels ($r_s(202) = .16$, $p = .019$), which suggests a weak correlation, according to Cohen [3]. In other words, those who reported deeper interactions with music, were more likely to report a greater influence on energy levels through music.

A further Spearman correlation indicated that there was a significant positive association between Music Interaction Scale and reported use of music to activate and to affect mood ($r_s(202) = .21$, $p = .003$), which suggests a weak correlation. This may indicate that those who reported deeper interactions with music were more likely to report using music to influence mental processes.

3.2 Digital Tool Features

In response to our open questions, we collected 214 comments on currently used features and 208 comments on envisioned features for future applications. After familiarising ourselves with the data, we identified in vivo codes and categorized them according to similarity to existing technical feature. Rough themes and patterns are presented here.

Currently Used Features. Of the 252 derived codes, the most commonly named feature was playlists (132 codes). Out of these, 14 specifically mentioned the ability to create one's own playlist. 71 codes referred directly to ready-made playlists, which could be further subsetted into mood (18), concentration (17) or genre playlists (14). Another 23 codes explicitly mentioned song radios or discovery playlists, which generate playlists from specific songs or listening history. Beyond playlists, 48 codes specified some form of automation feature, most notably similarity suggestions (29). Taken together, this pattern suggests a spectrum of features spanning manual song curation to automation of playlists.

Besides these, song shuffle (i.e. the randomisation of play order) was frequently coded (17). Other categories with less than ten codes included fluid transitions, a wide range of available music, the ability to turn off ads or to repeat songs (i.e. endless loops).

Envisioned Features. While we established 263 codes altogether, some participants reported an inability to imagine a feature (30), already being satisfied with existing features (8) or mentioned a feature which was regarded by the authors as already widely implemented in music programs, e.g. genre playlists (32). Three themes to support energy-level management were generated.

(Automatic) Song Selection Based on State Specific Criteria. 70 codes were attributed to this category (possibly as a result of priming by previous questions). Noteworthy suggestions here included the automatic detection of mood or energy level, e.g. by way of physiological measures (e.g. heart rate variability, skin conductance or EEG), behaviour (e.g. subject being studied, typing speed or body movement), or time since begin of session or last break (i.e. implying a continual loss of energy over time).

Once an intervention is indicated, many entries suggest the adjustment of songs based on tempo, genre or scientific validity, which could then be combined with personal preferences or listening history. Alternatively, the use of machine learning techniques to identify which music had helped the user in the past was proposed. Finally, one entry suggested that users should be able to input their desired energy level transition, i.e. the end state they wished to achieve.

While some comments in this category suggested manual input, the vast majority implied automated functions, suggesting that minimal additional workload was a preferred goal.

Automated Volume Control. Parallel to the previous suggestion, 14 codes suggested the regulation of volume based on current energy level. Most commonly mentioned examples were quietening during periods of high energy and increased strength during low energy periods. This category may indicate an association of flow experiences with high energy, and thereby, a wish not to be distracted.

Break support. Besides the direct influence of energy, at least 7 codes envisioned the use of music as a non-verbal indicator for study breaks, where arguably energy could be increased. Suggested indicators included relaxing music or one's favourite "happy" song. The wish for breaks may be an indicator for previous experiences with energy generating activities, such as psychological detachment.

Miscellaneous. Uncommon codes included positive feedback for successful influence on energy, distraction reduction (e.g. hands-free use of application or long playlists) or the removal of vocals (i.e. the ad hoc creation of "karaoke" versions).

4 Discussion

4.1 Implications

Within the two-scenario comparison, a vast majority (89.8%) of use cases sought an increase of energy level, but only 16.1% reported a maximum or full SoC; most common was an SoC of 4 (58.0%). As the formulation of the item asked for SoC "after successfully influencing energy levels", this implies satisfactory conditions can be achieved before reaching the peak. This finding is in line with the arousal-mood-hypothesis, which postulates that moderately arousing music generally increase cognitive function [12]. Taken together with the finding that some use cases seek to maintain or actively reduce energy levels, we propose that energy management should seek to dynamically react to users current level in order to maintain an optimum, rather than perpetually aspire an increase, thereby facilitating long-term work periods which do not "burn out" users' resources. This is in line with Quinn's integrated model of human energy [20]. Furthermore, systems should seek to differentiate between energetic and tense activation, as the latter would hinder productivity.

Another finding of this study was the spectrum of manual control and automation, i.e. users endorsing either the personal curation of the known, or discovery of the new in playlists. Similarly, shuffle (i.e. randomization leading to new song placement) was contrasted with repetition (i.e. remaining on the same song or playlist). A possible explanation for this is an expression of a novelty dimension – users are diverse in their wish for old or new auditory stimuli. While previous psychological research has established novelty-seeking as a trait [2] – and this may be an expression of such – future work should determine the interaction between disposition and energy-level transitions, including preferences across states. We propose that energy management systems should include regulators of degrees of novelty and familiarity to encompass users' needs.

Finally, we found genre was the most common selection criteria for energy-influencing music (89% agreement) and did not correlate with familiarity, which has been shown in previous literature [5]. While Dillman et al. also showed that tempo had an effect on physiological arousal (and interacted with genre), here it received an almost even split among users (60% agreement, 40% disagreement). We therefore propose that a system should focus on the manipulation of energy-effecting variables within a genre, to maximise user acceptance.

4.2 Limitations and Future Work

Most noteworthy limitation here is the self-selected sample, fostered by the recruitment via a mailing list. While great care was given to formulate the email invitation in a neutral manner and to also address potential participants who may have negative experiences with music and energy (e.g. "Does music help or hinder energy levels while studying?"), it cannot be guaranteed that those with positive experiences were not oversampled. As this study serves as a basis for a future application, this bias may be irrelevant for a corresponding target audience. However, for the long-term goal of a generalized application, future work should include broader recruitment methods such as surveys of entire courses or teams. Finally, replication in workplaces should be conducted to incorporate discrepancies of other cognitive tasks and environments (e.g. longer sustained working hours, differences in structure and/or monotony).

In a feedback question at the end of the survey, four participants commented that they had difficulty remembering their energy levels retrospectively. While research on arousal (as a corresponding construct to energy) has shown positive influences on the memory of associated stimuli [25], conceivably this effect does not extend to memory of the arousal itself. Another hypothesis could be that some participants only experienced neutral levels of arousal (i.e. low or middling energy levels), which could be more difficult to ascertain or differentiate. Finally, difficulty in remembering energy levels may be caused by the form in which it was measured; e.g. a discrepancy between the battery metaphor and some participants' mental models of human energy caused disturbances in recall. Future work should consider comparing diverse tools, e.g. other pictograms and visualisations, and comparing their validity and ability to measure various levels of energisation. For example, one possibility currently being investigated by the authors of Weigelt et al. (in press), includes the use of colour to reflect SoC (personal communication, 24.02.2020).

Our work with the Music Interaction Scale showed that participants who interacted more strongly with music were more likely to have larger changes in energy level. Previous work suggests attentive music listening enables cognitive functions in general [14] and other forms of deep interactions (i.e. beyond music) may be a key component to enabling energy managing strategies, but further work should test this hypothesis specifically.

5 Conclusion

We assessed student's studying listening habits and found that the majority of students reported using music to influence their human energy levels. By comparing a situation where they were dissatisfied with their performance to a situation where they had successfully influenced their energy level, we identified that a common use cases for an energy management tool would be to assist the transition from a low (but not depleted) to a high (but not maximised) energy level, but may also include other situations, such as the reduction of energy levels in the case of tense activation. Furthermore, we found that students who reported deeper interactions with music also had greater differences in energy levels between these scenarios. Generally, participants most frequently used genre to influence energy, and envisioned automatic human energy detection features for future management applications.

We propose that such applications should strive to dynamically adapt to users' states to prevent over-exertion. Furthermore, they should account for novelty (or familiarity) seeking dispositions and work within preferred genres, in order to increase user acceptance. Future work should develop tools for human energy measurement, replicate findings in other cognitive work settings and determine the role of interaction depth for human energy generation strategies.

References

1. Baumeister, R.F., Bratslavsky, E., Muraven, M., Tice, D.M.: Ego depletion: is the active self a limited resource? J. Pers. Soc. Psychol. **74**, 1252–1265 (1998)
2. Cloninger, C.R., Svrakic, D.M., Przybeck, T.R.: Psychobiological model of temperament and character. Arch. Gen. Psychiatry **50**(12), 975–990 (1993)
3. Cohen, J.: Statistical Power Analysis for the Behavioural Sciences. Routledge, London (2013)
4. Collins, R.: Interaction Ritual Chains. Princeton University Press, Princeton (2004)
5. Dillman Carpentier, F.R., Potter, R.F.: Effects of music on physiological arousal: explorations into tempo and genre. J. New Music Res. **3**, 339–363 (2007)
6. Demerouti, E., Bakker, A.B., Nachreiner, F., Schaufeli, W.B.: The job demands-resources model of burnout. J. Appl. Psychol. **86**(3), 499 (2001)
7. Franke, T., Attig, C., Wessel, D.: A personal resource for technology interaction: development and validation of the affinity for technology interaction (ATI) scale. Int. J. Hum.-Comput. Interact. **35**(6), 456–467 (2010)
8. Hallam, S., Price, J., Katsarou, G.: The effects of background music on primary school pupils' task performance. Educ. Stud. **28**(2), 111–122 (2010)
9. Hobfoll, S.E.: Conservation of resources: a new attempt at conceptualizing stress. Am. Psychol. **44**, 513–524 (1989)
10. Huang, R.H., Shih, Y.N.: Effects of background music on concentration of workers. Work **38**(4), 383–387 (2011)
11. Husain, G., Thompson, W.F., Schellenberg, E.G.: Arousal, mood, and the mozart effect. Psychol. Sci. **12**(3), 248–251 (2001)
12. Husain, G., Thompson, W.F., Schellenberg, E.G.: Effects of musical tempo and mode on arousal, mood, and spatial abilities. Music Percept. Interdisc. J. **20**(2), 151–171 (2002)
13. jamovi [computer software] (2019). https://www.jamovi.org
14. Janata, P., Tillmann, B., Bharucha, J.J.: Listening to polyphonic music recruits domain-general attention and working memory circuits. Cogn. Affect. Behav. Neurosci. **2**(2), 121–140 (2002). https://doi.org/10.3758/CABN.2.2.121
15. Kaplan, R., Kaplan, S.: The Experience of Nature: A Psychological Perspective. Cambridge University Press, New York (1989)
16. Lesiuk, T.: The effect of music listening on work performance. Psychol. Music **33**(2), 173–191 (2005)
17. Loui, P., Bachorik, J.P., Li, H.C., Schlaug, G.: Effects of voice on emotional arousal. Front. Psychol. **4**, 675 (2013)
18. Mammarella, N., Fairfield, B., Cornoldi, C.: Does music enhance cognitive performance in healthy older adults? The vivaldi effect. Aging Clin. Exp. Res. **19**, 394–399 (2007). https://doi.org/10.1007/BF03324720
19. Pinder, C.C.: Work Motivation in Organizational Behavior. Prentice Hall, Upper Saddle River (1998)
20. Quinn, R.W., Spreitzer, G.M., Lam, C.F.: Building a sustainable model of human energy in organizations: exploring the critical role of resources. Acad. Manag. Ann. **6**(1), 337–339 (2012)

21. R [Computer software]. Retrieved from https://cran.r-project.org (2018)
22. Rauscher, F.H., Shaw, G.L., Ky, C.N.: Music and spatial task performance. Nature **365** (6447), 611 (1993)
23. Ryan, R.M., Deci, E.L.: Self-determination theory and the facilitation of intrinsic motivation, social development, and well-being. Am. Psychol. **55**, 68–78 (2000)
24. Schmitz, C.: LimeSurvey: an open source survey tool [computer software]. LimeSurvey Project Hamburg, Germany (2012)
25. Storbeck, J., Clore, G.L.: Affective arousal as information: how affective arousal influences judgments, learning, and memory. Soc. Pers. Psychol. Compass **2**(5), 1824–1843 (2008)
26. Stratton, E., Lampit, A., Choi, I., Calvo, R.A., Harvey, S.B., Glozier, N.: Effectiveness of eHealth interventions for reducing mental health conditions in employees: a systematic review and meta-analysis. PLoS One **12**(12) (2017)
27. Watson, D., Wiese, D., Vaidya, J., Tellegen, A.: The two general activation systems of affect: Structural findings, evolutionary considerations, and psychobiological evidence. J. Pers. Soc. Psychol. **76**, 820–838 (1999)
28. Lambusch, F., Weigelt, O., Fellmann, M., Seistrup, K..: Application of a pictorial scale of human energy in ecological momentary assessment research. In: Harris, D., Li, W.-C. (eds.) HCII 2020. LNAI, vol. 12186, pp. 171–189. Springer, Cham (2020)
29. Ünal, A.B., de Ward, D., Epstude, K., Steg, L.: Driving with music: Effects on arousal and performance. Transp. Res. Part F Traff. Psychol. Behav. **21**, 52–65 (2013)

Author Index

Printed in the United States
By Bookmasters